Pioneers in Arts, Humanities, Science, Engineering, Practice

Volume 16

Series editor

Hans Günter Brauch, Mosbach, Germany

More information about this series at http://www.springer.com/series/15230
http://www.afes-press-books.de/html/PAHSEP.htm
http://afes-press-books.de/html/PAHSEP_Stehr.htm

Marian T. Adolf
Editor

Nico Stehr: Pioneer in the Theory of Society and Knowledge

Springer

Editor
Marian T. Adolf
Department of Culture and Communication
Zeppelin University
Friedrichshafen, Baden-Württemberg
Germany

Acknowledgement: The cover photograph and most other photos in this volume were taken from the personal photo collection of the author who also granted the permission on their publication in this volume. A book website with additional information on Nico Stehr and his major book covers is at: http://afes-press-books.de/html/PSP_Stehr.htm.

ISSN 2509-5579 ISSN 2509-5587 (electronic)
Pioneers in Arts, Humanities, Science, Engineering, Practice
ISBN 978-3-030-08343-4 ISBN 978-3-319-76995-0 (eBook)
https://doi.org/10.1007/978-3-319-76995-0

Copyediting: PD Dr. Hans Günter Brauch, AFES-PRESS e.V., Mosbach, Germany
Language editing: Hella Beister, Marian T. Adolf

Printed on acid-free paper

This Springer imprint is published by the registered company Springer International Publishing AG part of Springer Nature
The registered company address is: Gewerbestrasse 11, 6330 Cham, Switzerland

For Barbara, Sebastian, Angelika, Leopold and Leander

Preface

What do democratic politics, economic production, modern markets and climate politics have in common? They all rely heavily on knowledge, its inherent properties and its practical application—so much so that knowledge can be identified as the one central factor that shapes modern society more than any other. But this is not the only feature that these phenomena share: all of them have been the focus of one of the most original and prolific social scientists of his generation, Nico Stehr. This volume brings together, for the first time, some of his most important texts published over the course of almost five decades and puts them in dialogue with research that draws on, and is inspired by, Nico Stehr's themes and theories. These texts, produced by notable contemporaries, all of them peers and collaborators of Nico Stehr's, reflect and apply central concepts of his work and demonstrate the wealth and plurality of the topic of knowledge as a central term of contemporary social theory and research.

The content of this book is organized according to three frames. First, the aim is to present some of the most important themes of Nico Stehr's work in chronological order. Moving from his earlier work on the sociology of knowledge tradition to the study of the production of scientific knowledge to the role of knowledge for and in society to the more concrete applications of his theory, it becomes possible to follow the progression in Nico Stehr's theoretical and conceptual work as he uncovers more and more facets of the notoriously evasive concept of knowledge.

The second aim that guides the way the texts of this volume are organized is to demonstrate the rich potential of an approach to questions of social and cultural research that applies the concept of knowledge to a variety of major contemporary issues such as employment, social equity and power, the global challenge of climate change or the question of the cultural embeddedness of modern consumer markets. There seems to be a particular, although not initially obvious rationale behind the progression of Stehr's studies: throughout his career, he has applied his general theory of knowledge to various social fields one by one, thus developing his concepts by tackling central challenges of modern society as he studies its general layout and, consequently, uncovering fundamental principles of the societal role of knowledge. The selection of contributions for the present volume thus strives to

Nico Stehr (second from the left) at Zeppelin University's Lake Campus on the occasion of his 70th birthday symposium in 2012 (with Sebastian Belser, former ZU president Stephan A. Jansen and Marian Adolf, left to right). *Source* Zeppelin University/Bertram Rusch. This photo was used with permission of the Zeppelin University, Friedrichshafen, Germany

assemble texts that demonstrate the wide applicability of this approach to the broader project of a theory of modern society.

The third idea we pursued in putting together this book was to add commentary and reflection by eminent scholars rather than let Nico Stehr's work stand by itself. We were fortunate to have secured the support of some of Nico's most trusted colleagues and co-authors from a variety of scholarly disciplines such as philosophy, sociology, and meteorology, scholars that have not only worked closely with him over the years but are themselves among the most important protagonists of their respective fields. Their contributions, reflecting on and employing some of Nico Stehr's central ideas, are published here for the first time. They demonstrate the theoretical influence as well as the practical applicability and analytical value of Nico Stehr's wide range of thinking.

In March 2012, Zeppelin University invited numerous colleagues and companions he worked with throughout his career to a symposium celebrating Nico Stehr's 70th birthday. Most of the contributions in this volume originate from this reunion. All of these texts are original works and refer in many ways to the productivity of Nico Stehr's thinking: from his early writings on the German tradition of the sociology of knowledge to his pioneering work on the theory of society, famously termed Knowledge Society, as well as his long-standing engagement with the social aspects of climate change or the impact of social change on democracy, markets and knowledge.

In addition, the volume also includes an interview with Nico Stehr that covers his academic career, a personal explication of his most important ideas and concepts and some rare private comments on his motives and experiences. Although it is difficult to gain an overview of the extraordinary extent of publications Nico Stehr has produced—and continues to produce—throughout his career, we have assembled a bibliography of the cornerstones of his pioneering quest to theorize modern society so far.

This book is the product of many people. Without the contributions of Gernot Böhme, Reiner Grundmann, Susan McDaniel, Alexander Ruser, Hans von Storch and Hermann Strasser, there would not have been much to edit. Thanks also to all the other participants of the symposium, especially Volker Meja and Steve Fuller. Andreas Hafenscher, Johanna Schabert, Falk Stratenwerth-Neunzig, Paul Malone, Linda Phillips and Hella Beister helped during the editorial process of these pages.

We also wish to thank the series editor Hans Günter Brauch for his close collaboration and support. We are grateful to the publishers of Nico Stehr's original texts for the permission to reproduce them here, and to all colleagues past and present, at Zeppelin University and elsewhere, who have been instrumental in bringing together this project. Finally, the editor wishes to thank Nico Stehr for many productive years of inspiration, collaboration and friendship.

München, Germany/Eisenstadt, Austria Marian T. Adolf
Summer of 2017

Contents

Part I
On Nico Stehr

Nico Stehr (in Duisburg, Germany, 2000). *Source* Personal archive

Chapter 1
A Short Biography of Nico Stehr

Marian T. Adolf

An economist and sociologist by training and a cosmopolitan at heart, Nico Stehr spent a large part of his career in the United States and Canada before returning to Europe around the turn of the millennium. His academic vita now spans nearly five decades and a vast geography of research institutions and professional appointments.

Nico Stehr was born in Berlin on March 19, 1942, and grew up in Bremerhaven in northern Germany. After attending secondary school, he earned a degree in economics at the University of Cologne where he studied with René König. König, who had survived Nazi rule under difficult circumstances in Zurich, not only acquainted Stehr with advanced empirical social research methods but may also have been instrumental in interesting him in the biographical aspects of forced emigrants. In 1967, Stehr relocated to the United States where he was awarded his Ph.D. in 1970 by the University of Oregon. After moving to the University of Alberta's Department of Sociology in Edmonton, he worked as an assistant, associate, and then full professor in 1979. As a member of one of Canada's leading Sociology departments he made a name for himself by launching the *Canadian Journal of Sociology*, for which he served as editor for many decades. During the 1980s, he held a number of visiting appointments in Europe, including Zurich and Munich, and finally returned to Germany in the late 1990s. During the academic year 2002–2003 he was Paul F. Lazarsfeld-Professor at the University of Vienna. In 2004, instead of retiring, Nico Stehr accepted an appointment at Zeppelin University, a private start-up university which had launched only a year earlier. There he holds the Karl Mannheim Chair of Cultural Studies and continues to play a pivotal role in providing visibility and funding as well as attracting international scholars from all over the world to the new academic venture at the shore of Lake Constance. Until today, Nico remains as productive as ever and only recently founded the European Center for Sustainability Research (ECS) at Zeppelin University.

M. T. Adolf (ed.), *Nico Stehr: Pioneer in the Theory of Society and Knowledge*, Pioneers in Arts, Humanities, Science, Engineering, Practice 16,
https://doi.org/10.1007/978-3-319-76995-0_1

His prolific work as well as his singular ability to connect and unite scholars from various academic backgrounds, countries and generations have introduced and circulated ideas, concepts and theories throughout the scientific community. Together with other European expats in the North American academic system, Nico was instrumental in familiarizing American social science with proponents of classic European social theory such as Karl Mannheim and Werner Sombart. He also had a keen interest in the lives and careers of European academic émigrés such as Paul F. Lazarsfeld and kept in close contact with some of the masterminds of 20th century social science, for example Robert K. Merton. A particularly memorable occasion was his invitation for Niklas Luhmann whom he familiarized with social phenomena of the New World such as ice hockey and the Hutterite settlements. While he introduced the North American community to important texts of German classics, he also put a lot of effort into the translation and publication of important works into German. Among the latter are classic texts by Lazarsfeld but also the rediscovery and translation of Paul Neurath's long forgotten study of his time in the Dachau and Buchenwald concentration camps in 1941 ("The Society of Terror", 2005).

Nico Stehr continues to study society as a dynamic and ever-changing entity and ventures into new and innovative fields and phenomena. Remarkably, Nico's output may not have yet reached its peak as he keeps publishing theoretical texts and topical studies of current societal developments while also engaging in often highly politicized debates about contemporary societal challenges. His regular comments on the social consequences of climate change stand out as his contribution and reverence to the social role of knowledge. To this day, Nico Stehr has produced well over 20 monographs that have been published in 9 languages, more than 120 articles in peer-reviewed journals, as many book chapters, and nearly 40 articles in his own edited volumes. And by the time you read this, the numbers are likely to have increased (see the bibliography in this volume).

References

Neurath, Paul M. 2004. *Die Gesellschaft des Terrors. Innenansichten der Konzentrationslager Dachau und Buchenwald*. Herausgegeben von Christian Fleck und Nico Stehr. Frankfurt am Main: Suhrkamp Verlag.

Neurath, Paul M. 2005. *The Society of Terror. Inside the Dachau and Buchenwald Concentration Camps*. Edited by Christian Fleck and Nico Stehr. Boulder, London: Paradigm Publishers.

Nico Stehr (third from the right) with fellow members of Zeppelin University's initial Scientific Advisory Board (left to right: Bill Ashcroft, Guy Peters, Friedrich Schneider, ZU founding president Stephan A. Jansen, Hans Ulrich Gumbrecht, Nico Stehr, Steve Fuller and Wolfgang Eßbach, 2011). *Source* ZU/Rainer Böhme. Rights granted by ZU/Rainer Böhme on 8 January 2018

Chapter 2
Nico Stehr, Sociology, Knowledge and the Sociology of Knowledge: An Introduction

Alexander Ruser

2.1 Introduction

Most readers find sociological terminology opaque, dry and hard to understand. Important insights in, and descriptions of, the functioning of modern societies tend to be formulated in a highly specific and hardly accessible scientific vocabulary. The breakdown of social bonds in the wake of economic and social modernization may have been a widespread experience in the late 19th and early 20th century. However, Emile Durkheim's famous term to describe this very experience, *anomie*, resonated (and still resonates) only within the small community of social scientists. Many other equally important concepts share this fate. Hardly anyone outside the specialized circles of professional sociologists has the faintest idea of the importance or the meaning of Talcott Parsons "AGIL scheme" or Max Weber's concept of "ideal types".

Sociological concepts and terms do not frequently become part of everyday language, and if they do they are often stripped of any substantial meaning in the process. Hence it is most noteworthy when sociological concepts manage to transcend the boundaries of the academic world, contributing to a new and deeper understanding of social conditions for an enlarged group of people who now use this very concept.

More than 30 years ago, one such concept was formulated. Yet when *The Knowledge Society* was published by Gernot Böhme and Nico Stehr in 1986, hardly anyone could have anticipated how influential the concept would become. And influential it became. In January 2017, Google listed over 40 million hits for the term "knowledge society," and Google Scholar identified 4,7 million papers, citations or cross-references for this search item. The term has not only drawn scholarly or public attention but has been at the center of political and economic thinking,

M. T. Adolf (ed.), *Nico Stehr: Pioneer in the Theory of Society and Knowledge*, Pioneers in Arts, Humanities, Science, Engineering, Practice 16,
https://doi.org/10.1007/978-3-319-76995-0_2

as well. The UNESCO adopted the goal of "building knowledge societies,"[1] while the World Bank reflects about how to 'build' knowledge societies (World Bank 2002). Going even further, a 179-page report entitled *Understanding Knowledge Societies* and issued by the United Nations Department of Economic and Social Affairs in 2005 introduced the IKS, the "Index of Knowledge Societies". More important than the (rather questionable) attempt to measure and rank "knowledge societies" is the rationale behind this undertaking. The introductory section lists ten answers to the question "Why is it important to understand knowledge societies?". While they differ with regard to which aspect of modern knowledge society they address, they all share two convictions that could have been copied from one of Nico Stehr's papers and books:

(1) Knowledge is the main transformative power in modern societies
(2) Knowledge is an (individual) capacity to act (cf. United Nations 2005: 4–5)

The "knowledge society" may well be the single most prominent concept associated with the work of Nico Stehr. However, it would be a flawed account of his thinking to reduce his scientific oeuvre to just this one term. It would be an inadmissible simplification for two reasons: The first reason is closely related to the 'discovery' of the knowledge society itself. In order to understand how Nico Stehr and his collaborators came to coin the term we will have to take a look back and carefully work out which scholarly tradition informed his thinking. Or perhaps we should rather look down to examine on whose shoulders Nico Stehr stood when he saw the coming of a "world made of knowledge" (Stehr 2001). In a second step we have to look ahead, that is, we have to include Nico Stehr's 'use' of the concept and his attempts to highlight and prove the particular importance of knowledge for understanding the contemporary problems of and future challenges for society. While Nico Stehr's work cannot be reduced to the concept of the knowledge society, the term itself serves as a suitable focal point for understanding the significance of 'knowledge' as an issue and the sociology of knowledge as a stimulus for his contribution to a comprehensive understanding of modern societies.

Yet, it is not an easy task to give a short introduction to such a comprehensive work. The problem is not so much the sheer number of books, scientific papers and talks but the variety of topics and social problems that are addressed. At first glance, his work seems to encompass a rather daunting list of issues: scientific assessments of the work of some of the most important theorists and empirical researchers such as Karl Mannheim (Kettler et al. 1980, 1990), Robert K. Merton (Meja/Stehr 1998) or Paul F. Lazarsfeld (Lazarsfeld/Stehr 1976; Fleck/Stehr 2007); more empirically oriented writings covering classical topics of sociological inquiry such as social inequality (Stehr 1996, 1999), the economy (Stehr 2002), the future and the viability of democracy (Stehr/Grundmann 2003; Stehr 2015a), science and technology

[1] See http://en.unesco.org/themes/building-knowledge-societies.

(Stehr 2005), consumption and morality (Stehr 2008a, b; Stehr/Adolf 2010) and, perhaps most importantly, anthropogenic climate change (Stehr/von Storch 2010).

This short introduction aims to open an approach to Nico Stehr's comprehensive work by tracing some of the major lines of influence and by highlighting selected issue areas. To focus on a limited number of theories and theorists as well as issue areas, however, necessarily involves a simplification. Instead of giving an exhaustive overview, this introduction seeks to achieve a more humble goal: to provide a roadmap and guideline for any reader who wants to get more thoroughly acquainted with the sociological work of Nico Stehr.

2.2 Sociological Traditions and the 'Discovery' of the Knowledge Society

Let us begin with a look back, picking out some of the 'giants' who lent a shoulder to stand on. The operative word here again is, of course, 'some'. Any attempt to present an exhaustive catalogue of scholars, theories and concepts that have found their way into the thinking of Nico Stehr in this rather short chapter is inevitably doomed to failure. Instead of trying to list the names of all the thinkers who have influenced scientific publications that span more than four decades, the following paragraphs will limit themselves to highlighting five scholars whose work is key to understanding the rationale and the driving force of Nico Stehr's thinking. These scholars are Norbert Elias, Karl Mannheim, Peter Drucker, Daniel Bell and Robert K. Merton.

Robert K. Merton published his *On the Shoulders of Giants* in 1965. In it he deals with the significance of an aphorism that is widely attributed to Sir Isaac Newton (other sources name 12th-century philosopher Bernard of Chartres): "If I have seen farther, it is by standing on the shoulders of giants". It is only appropriate to begin with a scholar Nico Stehr, I think, would happily admit served as a platform and a stepping stone for his own work. Moreover, as Eugene Garfield once observed with respect to Robert Merton: "[I]t is not without irony that today Merton himself is widely considered one of the giants of science [...]. However, this intellectual giant would be among the first to moderate this tribute" (Garfield 1977: 176). As will be discussed below, the same holds true for Nico Stehr. Rather than praising himself for having 'discovered' the concept of the knowledge society, he routinely points towards his many predecessors.

2.2.1 *Merton, Elias and the Threat of Losing History*

In his paper "Robert K. Merton's Structural Analysis: The Design of Modern Sociology", published together with Volker Meja in 1998, Nico Stehr expressed and explained the importance of Merton's thinking for his own sociological inquiry. Merton's sociology is described as neither an expression of 'modern' American sociology nor an example of a 'traditional' theoretical approach (Meja/Stehr 1998: 34). It is this very ambivalence, the ability to bridge the gap between traditional and modern sociology, which qualified Merton to not only "be seen [...] as an expression and as a motivating force of modern sociology" but also as a driving force for Nico Stehr's own writing.

In Merton's structural analysis, Nico Stehr found the means for coping with an internal tension of contemporary sociology so aptly described by German-born sociologist Norbert Elias. In Elias' view any progress in sociology was accompanied by and had to be attributed to the "narrowing of the sociologist's focus of attention and interest to the immediate present" (Elias 1987: 150, quoted in Meja/Stehr 1998: 34), which in turn sparked an "almost explosive profusion of empirical sociological investigation" (ibid.). This focus on the "immediate present" and the emphasis on advancing methods of empirical research and applying the findings of sociological research to concrete short-term problems is certainly not without its merits. However, Elias' fear of "loosing history", which was echoed in Merton's writings, seems to be a particularly important concern.

But why should sociologists bother to analyze historical developments when they were just becoming proficient at analyzing (and sometimes even solving) short-term problems of the immediate present? To answer this question, one has to turn to yet another sociologist who arguably stands out among the other giants: Karl Mannheim.

2.2.2 Reclaiming History and Conceptualizing Knowledge: The Influence of Karl Mannheim

In order to get a first hint of the importance of Karl Mannheim for Nico Stehr's scholarly work it is absolutely sufficient to go to room number '1.11' at the lake campus of Zeppelin University. A small doorplate identifies the room as the office of the "Karl Mannheim Chair of Cultural Studies" and Nico Stehr as its holder. This is no coincidence: When Nico Stehr joined Zeppelin University in 2004 only a year after it had opened, he deliberately chose Karl Mannheim to become the eponym for his newly established chair.

The sociological writing of Karl Mannheim is crucial for understanding why Nico Stehr consistently takes a historical perspective to understand contemporary societal problems. Moreover, and perhaps even more importantly, his reading of Karl Mannheim helps explain why he puts knowledge at the center of his theorizing.

In his seminal work *Ideology and Utopia*, Karl Mannheim outlines a distinctly social-constructivist understanding of the social world surrounding humans and of the history they "live in":

> Inasmuch as man is a creature living primarily in history and society, the 'existence' that surrounds him is never 'existence as such,' but is always a concrete historical form of social existence. For the sociologist, 'existence' is [...] a functional social order, which does not exist only in the imagination of certain individuals but according to which people really act. (Mannheim 1936: 341)

Neither social order nor historical developments which 'lead' to specific social configurations are simply 'given' but have to be 'known'. By following Mannheim in this principal understanding of social history, Nico Stehr quite naturally attributed greater importance to a (historicizing) sociology of knowledge.

In a joint paper with David Kettler and Volker Meja on *Karl Mannheim and Conservatism*, Nico Stehr not only stressed Mannheim's rejection of a value-free treatment and evaluation of ideas and social knowledge but also continued to emphasize that "[t]he question of assessing the validity of the social knowledge cannot, on this view, be separated from the work of historical interpretation itself" (Kettler et al. 1984: 75).

2.2.3 Understanding and Making Social Worlds—Knowledge as a Capacity to Act

We can now better understand what Nico Stehr means when he characterizes knowledge societies as "worlds *made* of knowledge" (Stehr 2001; emphasis added). Again in accordance with Karl Mannheim, knowledge is understood as necessarily containing an "activist element".

In fact, Karl Mannheim's description of 'knowledge' has foreshadowed and influenced Nico Stehr's own definition of "knowledge as a generalized *capacity to act* and as a model *for* reality" (Adolf/Stehr 2017: 18; emphasis in the original). Such a description of knowledge clearly rejects narrower definitions that reserve the term for products of theoretical, scientific thinking. A similar thought can, again, be found in Karl Mannheim's *Ideology and Utopia*:

> Idealistic philosophy was not upset by the discovery that the type of knowledge represented by pure theory was only a small segment of human knowledge, that in addition there can be knowledge where men, while thinking, are also acting, and finally, that in certain fields knowledge arises only when and in so far as it itself is action, i.e. when action is permeated by the intention of the mind, in the sense that the concepts and the total apparatus of thought are dominated by and reflect this activist orientation. (Mannheim 1936: 265)

It is the activist element, the ability to set something in motion that explains the singular importance of knowledge for our understanding of modern societies:

> What the sociology of knowledge seeks to reveal is merely that, after knowledge has been freed from the elements of propaganda and evaluation, it still contains an activist element which, for the most part, has not become explicit, and which cannot be eliminated, but which, at best, can and should be raised into the sphere of the controllable. (Mannheim 1936: 266)

Understanding the central role of the sociology of knowledge in Nico Stehr's work is key to fully grasping how he referred to and built upon his predecessors. It is precisely because he aimed at revealing the activist element of knowledge that he was able to follow up on and advance theories that envisioned the rise of the knowledge worker (Drucker 1969) or saw *The Coming of the Post-Industrial Society* (Bell 1973).

The more comprehensive understanding of knowledge as a capacity to act, as an activist element and means to set something in motion allows for an expansion of previous theories that emphasized the transformative power of knowledge for modern economies. Drucker and Bell realized that knowledge passed value chains would have the potential to fundamentally transform individual workplaces, organizational structures, and whole economic sectors. While Merton, Elias and Mannheim lend some key conceptual instruments, Bell and Drucker provide a starting point for Nico Stehr's empirical examination of central developments within modern societies.

2.3 Of Markets and Meteorologists: Challenges for Worlds Made of Knowledge

The concept of the knowledge society provides much more than a proper description of some central features of modern societies. The particular appeal of the concept and its success beyond purely academic debates cannot be explained without pointing towards its explanatory power.

Apart from being deeply rooted in sociological tradition, Stehr's work is characterized by its strong orientation to real-world problems. Following the example of Drucker and Bell, Stehr has aimed at contributing to a better understanding of pressing issues such as changing markets and consumer behavior (Stehr et al. 2006; Stehr/Adolf 2010), the viability of democratic societies (Stehr 2008b, 2015a) and anthropogenic climate change (Stehr/von Storch 2010).

The one topic that has demanded the most attention and has thus become the main issue of his writing is human-made climate change. Like no other issue anthropogenic climate change is emblematic for his contribution of a social science perspective to a *social, political* and *economic* problem usually described in natural-science terms (Stehr 2015b, 2016). Climate change is not immediately accessible but reveals itself in complex and, for lay people, opaque scientific data. Sophisticated experiments have to be conducted and advanced models are needed to detect and describe changes in local and regional weather patterns and the earth's atmospheric composition and to estimate the short- and long-term consequences of these developments. With most political and sociological studies focusing on the 'truth' of scientific knowledge claims and discussing the political strategies of climate change denial (cf. Oreskes/Conway 2010), Stehr's emphasis on a more comprehensive understanding of the social role of 'knowledge' offers a much needed fresh perspective.

Instead of putting up believers in scientific truth against climate change deniers, instead of theorizing on the basis of the dichotomy of "knowledge-based decision-making" versus the manipulative distortion of scientific truth, Nico Stehr's theoretical approach provides the context for a much more sophisticated analysis. He not only raises that ever more important question of whether one can assume that there is such a thing as "globally valid or globally present knowledge" (Adolf/Stehr 2017: 142) but also investigates the unequal distribution of knowledge and the role of global social inequality in modern knowledge societies (Stehr 1996). In consequence, anthropogenic climate change cannot be approached as a mere scientific fact that needs to be dealt with but as a crucial element of a social context. Climate change is not a problem which can be 'solved'. Stehr's work reminds us that climate change, like other fundamental changes, for example 19th-century industrialization, will have equally fundamental implications for the social structuring of modern societies and the way decisions are made and legitimized. Most importantly, placing knowledge in the center of the analysis allows for an understanding of climate change not as an external force, or something that has to be endured passively. It was therefore no surprise that Nico Stehr became engaged with thinking about alternatives to deal with the wicked problem of climate change. When the London School of Economics and Political Science under the impression of the failed 2009 climate summit in Copenhagen convened an international group of experts from the natural and social sciences at Hartwell House, Nico Stehr was a natural fit. The scholars were tasked with debating and developing alternative approaches to the political problem of climate change. The subsequently published "Hartwell Paper" (Prins et al. 2010) argues in favor of "relentlessly pragmatic options" (ibid.) and is still a much-debated contribution to international climate politics. But perhaps more important than its immediate impact is the fact that the Hartwell Paper stresses the "activist element" of knowledge. Rather than decry traditional approaches in climate politics it seeks to develop new, alternative or accompanying measures for dealing with a complex global challenge.

2.4 Conclusion

The concept of the knowledge society arguably is a 'trademark' term for Nico Stehr's work. Yet it is vital not to reduce his thinking to this one concept. It is particularly important to recall why the sociology of knowledge and knowledge as an attribute could rise to such importance. Rooting Nico Stehr's work in the long sociological tradition (of which he himself is an expert) is essential for understanding his theoretical thinking. It is important to note that he continues the work of outstanding social scientists like Peter Drucker and Daniel Bell and that he aims at reformulating their hypotheses, hence expanding and refining a pre-existing set of theories. All this is crucial for *understanding* Nico Stehr's approach.

However, the analysis of, and involvement with his works should not stop at understanding his approach. What is much more important is to *apply* the theories and concepts he developed to interesting every-day problems of modern societies. On the first pages of *The Knowledge Society*, Gernot Böhme and Nico Stehr wrote:

> Despite many insightful, sophisticated and engaged inquiries into the interrelation of science and society, particularly in the 1920s and early 1930s and again in the 1960s and early 1970s, the void of a theory of society which captures the dynamics of science, technology and society remains to a significant extent. The fundamental issues of 'the modes of interplay between society, culture and science are with us still'. (Böhme/Stehr 1986: 7)

Since then, Nico Stehr's work has picked up on insightful and sophisticated studies conducted by Karl Mannheim and others. He has significantly contributed to our understanding of the interplay between society, culture and science. Yet, as his own studies on contemporary problems such as climate change or the future of democratic decision-making have shown, the fundamental issue is "with us still". This means that Nico Stehr is not done with his pioneering work. The same is true for his readers. Rather than regarding his extensive and comprehensive writings as part of the sociological tradition, it is essential to engage with his ideas and to apply his concepts.

References

Adolf, Marian and Nico Stehr. 2017. *Knowledge. Is Knowledge Power?* 2nd Edition. London and New York: Routledge.

Bell, David. 1973. *The Coming of the Post-Industrial Society. A Venture in Social Forecasting.* New York: Basic Books.

Böhme, Gernot and Nico Stehr. 1986. *The Knowledge Society. The Growing Impact of Scientifc Knowledge on Social Relations.* Dordrecht: Reidel.

Drucker, Peter F. 1969. *The Age of Discontinuity. Guidelines to Our Changing Society.* New Brunswick & London: Transaction Publishers.

Elias, Norbert. 1987. "The Retreat of Sociologists into the Present", in: Meja Volker, Dieter Misgeld, Nico Stehr (eds.), *Modern German Sociology.* New York: Columbia University Press.

Fleck, Christian and Nico Stehr. 2007. *Paul F. Lazarsfeld. Empirische Analyse des Handelns – Ausgewählte Schriften.* Frankfurt am Main: Suhrkamp.

Garfield, Eugene. 1977. "Robert K. Merton: Among the Giants", *Essay of an Information Scientist,* 3(1977–78): 176–178.

Kettler, David, Volker Meja and Nico Stehr. 1980. *Karl Mannheim. Strukturen des Denkens.* Frankfurt am Main: Suhrkamp.

Kettler, David, Volker Meja and Nico Stehr. 1984. "Karl Mannheim and Conservatism: The Ancestry of Historical Thinking", *American Sociological Review,* 49, February 1984: 71–85.

Kettler, David, Volker Meja and Nico Stehr. 1990. "Rationalizing the Irrational: Karl Mannheim and the besetting Sin of German Intellectuals", *American Journal of Sociology,* 95(6), May 1990: 1441–1473.

Lazarsfeld, Paul F. and Nico Stehr. 1976. "Wissenschaft und und Sozialforschung. Ein Gespräch mit Paul F. Lazarsfeld", *Kölner Zeitschrift für Soziologie und Sozialpsychologie,* 28(4): 794–807.

Mannheim, Karl. 1936. *Ideology and Utopia: An Introduction to the Sociology of Knowledge.* Translated by Louis Wirth and Edward Shils. New York: Harcourt, Brace & World, Inc.

Meja, Volker and Nico Stehr. 1998. "Robert K. Merton's Structural Analysis: The Design of Modern Sociology", in: Carlo Mongardini, Simonetta Tabboni (eds.), *Robert K. Merton & Contemporary Sociology.* New Brunswick, NJ: Transaction Publishers: 21–43.

Oreskes, Naomi and Erik Conway. 2010. *Merchants of Doubt: How a Handful of Scientists Obscured the Truth on Issues from Tobacco Smoke to Global Warming.* London: Bloomsbury Publishing.

Prins, Gwyn, Isabel Galiana, Christopher Green, Reiner Grundmann, Mike Hulme, Atte Korhola, Frank Laird, Ted Nordhaus, Roger Pielke, Steve Rayner, Daniel Sarewitz, Michael Shellenberger, Nico Stehr and Hiroyuki Tezuka. 2010. *The Hartwell Paper.* London: London School of Economics.

Stehr, Nico. 1994. *Knowledge Societies.* London, Thousand Oakes, New Delhi: SAGE.

Stehr, Nico. 1996. "Social Inequality and Knowledge", *Comparative Social Research,* Supplement 2: 41–51.

Stehr, Nico. 2001. "A World Made of Knowledge", *Society,* November/December 2001: 89–92.

Stehr, Nico. 2005. *Knowledge Politics. Governing the Consequences of Science and Technology.* Boulder: Paradigm Press.

Stehr, Nico. 2008a. *Moral Markets: How Knowledge and Affluence Change Consumers and Products.* London and New York: Routledge.

Stehr, Nico. 2008b. *Knowledge & Democracy. A 21st Century Perspective.* New Brunswick, London: Transaction Publishers.

Stehr, Nico. 2015a. *Information, Power, and Democracy. Liberty is a Daughter of Knowledge.* Cambridge: Cambridge University Press.

Stehr, Nico. 2015b. "Climate Policy: Democracy is Not an Inconvenience", *Nature,* 525(7570): 449–450.

Stehr, Nico. 2016. "Exceptional Circumstances: Does Climate Change Trump Democracy?", *Issues in Science & Technology,* 32(2): 37–44.

Stehr, Nico and Hans von Storch. 2010. *Climate and Society: Climate as Resource, Climate as Risk.* Singapore: World Scientific Publishing.

Stehr, Nico and Marian Adolf. 2010. "Consumption Between Market and Morals. A Sociocultural Consideration of Moralised Markets", *European Journal of Social Theory,* 13(2) (May): 213–228.

Stehr, Nico and Reiner Grundmann. 2003. "Social Control and Knowledge in Democratic Societies", *Science and Public Policy,* 30: 183–188.

Stehr, Nico, Christoph Henning and Bernd Weiler. 2006. *The Moralization of the Markets.* New Brunswick, London: Transaction Publishers.

United Nations, The Department of Economic and Social Affairs. 2005. *Understanding Knowledge Societies in Twenty Questions and Answers with the Index of Knowledge Societies.* New York, ST/ESA/PAD/SER.E/66.

World Bank. 2002. *Constructing Knowledge Societies: New Challenges for Tertiary Education. Directions in Development.* Washington, DC: World Bank.

Nico Stehr (in Krumbach, Austria, 2015). *Source* Personal archive

Chapter 3
Reflections on Society and Knowledge: An Interview with Nico Stehr

Nico Stehr and Marian T. Adolf

This interview offers the reader the opportunity to learn about Nico Stehr's career and the background of his work and provides an accessible introduction to his thinking from a personal point of view. It focuses on the most original and central of Nico Stehr's topics: knowledge and society.

3.1 On Nico Stehr's Career and Biography

Q: You were born during the Second World War and grew up in the years of rebuilding a war-torn country. If you think back to your childhood and adolescence, can you identify any hints towards your later career as an academic?

A: I went to a high school that emphasized economics. This encouraged me, upon graduation, to search out the University of Cologne which had an excellent reputation in the fields of economics and social science, such as political science and sociology. The emphasis was not on quantitative economics but on the intellectual history of economic ideas and the impact of the economy on society.

Q: How did you end up studying with René König in Cologne? How did his legacy influence your own career as a sociologist?

A: My first student assistantship was in the research institute of a well-known law professor who encouraged me to become a law student. However, the

M. T. Adolf (ed.), *Nico Stehr: Pioneer in the Theory of Society and Knowledge*, Pioneers in Arts, Humanities, Science, Engineering, Practice 16,
https://doi.org/10.1007/978-3-319-76995-0_3

charismatic teacher René König persuaded me to major in sociology and economics, and I graduated with a master's in economics. His lectures and writings emphasized the history of ideas in sociology and the contribution of French and Anglo-Saxon historical, sociological and anthropological perspectives. He also stressed the linkage between empirical and theoretical orientations. He introduced the importance of methodological training and experience for the development of modern-day German sociology.

I spent some three years as a graduate student researcher in König's sociological research center in Cologne. My fellow students and I carried out a number of large-scale empirical studies on the economic and social preconditions of technological change. König gave us a great deal of liberty to carry out these studies, and therefore we learned a lot about empirical sociology by doing it. My fellow students all went on to hold sociology chairs in Germany and abroad. We learned more from each other than from our formal studies. By the time I graduated, it was clear to me that I wanted to become a professional social scientist. It also became evident that an intellectual apprenticeship in North America would be helpful, especially for strengthening my methodological skills.

Q: You then decided to move your career to North America. How did that come about? Were you already planning on staying there when you first moved?

A: I obtained a Fulbright Fellowship, and following the advice of Walter Schafer, a young University of Oregon faculty member, applied to their sociology program with a letter of support from René König. In the summer of 1967, I left for Eugene, Oregon, intending to stay for a year before returning to Cologne and to collective plans for a large research project. However, my experience with the professorial staff and the large graduate cohort as well as the research opportunities in the Sociology Department at the University of Oregon made me decide to stay. When I completed my doctoral studies, I realized that the Fulbright Commission rules prevented me from taking up full time employment in the United States, but with the permission of the Fulbright Commission I was able to accept an assistant professorship in sociology at the University of Alberta in Canada. I remained there for 25 years. I founded the Canadian Journal of Sociology in 1975, became a full professor in 1979 and a Fellow of the Royal Society of Canada in 1985.

Nico Stehr (in Berkeley, USA, 2017). *Source* Personal archive

Q: Rather than settling for a 'local' career, you always kept in touch with European colleagues and regularly had visiting appointments in Germany, Austria and Switzerland. Why was it important to you to keep this connection?

A: It was important to keep in close contact with colleagues, students and publishers and also to follow intellectual developments in Europe. Political changes, both long- and short-term, were of course of particular interest to a social scientist curious about the thrust of societal transformations at the macro level. Experiences in Europe and elsewhere, for example in South Africa, Latin America and Asia, were fodder for teaching and writing about the trajectory of modern societies, including their economies, political systems and civil societies.

Nico Stehr with Ulrich Beck, George Ritzer and Karin Knorr-Cetina at the ASA Annual Meeting in San Francisco 2009. *Source* Hermann Strasser

Q: Tell us about your initial research focus. What were the topics that drove your early work? Are there themes and theories that have stayed with you throughout the years?

A: My research began with a strong interest in the work of émigré scholars. My teacher in Cologne was an émigré scholar, and so was Karl Mannheim, the author of seminal work on the sociology of knowledge and science. My focus converged on examining, in the widest sense, the societal role of knowledge; that is, its production, the constructive work it performs as well as the conflicts it creates, as summarized in "Knowledge. Is Knowledge Power?" (2017).

Q: So it seems that the concept of knowledge captured your interest from early on in your career. Isn't that an unusual choice for an economist and sociologist?

A: When we move through the modern world, the phenomenon we call knowledge is always involved. Whether we talk of know-how, technology, innovation, politics or education, it is the concept of knowledge that ties all this together. And still, despite its ubiquity as a modern trope, we seldom encounter knowledge in itself, we don't pay much attention to it: How is it produced, where does it reside, and who owns it? Is knowledge always beneficial, and will we—at some point in the future—know all there is to know? And does knowledge really equal power?

My work on knowledge attempts to pursue an original approach to this concept which seems to define so many aspects of modern societies. I have always wanted to explore the topic from a distinctly sociological perspective and to trace the many ways knowledge is woven into the fabric of modern society; for example, its role in the modern economy, the globalization process, the world of politics and policies as well as the transformative power of modern scientific and technical knowledge with respect to social relations. And this interest has stayed with me throughout the years.

Seminar participants at the Inter-University Center Dubrovnik, Croatia in 2010: Thomas Brante, Helmut Willke, Nico Stehr, Gotthard Bechmann, Stephen Turner, Reiner Grundmann, Ulrich Ufer and Students of Zeppelin University. *Source* From the personal photo archive of the author

Q: Tell us a little bit about yourself. As a researcher of the social, are you a particularly sociable character?

A: As social theories can explain, I, too, enjoy the company of people who share my interests, concerns and general way of looking at the world. For example, I prefer people whose horizon extends beyond their immediate environment and who are interested in the goings-on of the world. I live what might be called a cosmopolitan life, and that has left its marks. People tell me that I seem to be good at connecting people from different geographical and intellectual backgrounds. I guess you could say that I enjoy that. When I first moved to North America, I began to go to baseball games, often with visiting colleagues, initially, to be frank, because baseball games were one of the few places where you could legally have a drink in the open air. But I soon found that the game itself is an intriguing mixture of temporary swiftness and contemplative slowness. Such games were more than a mere spectator event, they were also an opportunity for social interaction and intellectual dispute with fellow doctoral students.

At the Rockefeller Foundation Bellagio Center, Lake Como, Italy. *Source* From the personal photo archive of the author

Q: That sounds like there isn't much difference between your academic and your private self?

A: I seem to belong to that group of people who find it hard to separate work from leisure. This inability to detach scholarly observation from entertainment and discovery from recreation means that I feel uncomfortable in situations where such a distinction is constitutive.

Q: Have you ever had any role models or predecessors that you were particularly fond of?

A: Emigrants from the Nazi period who were forced to leave their homes and countries behind and start over in completely different cultures and places have always impressed me. They had to learn a new language and to write for a new audience in the social sciences and humanities, a most admirable accomplishment.

Many were immensely successful and strongly influenced the political, social and academic development in their new home countries—which also goes to show the great loss for the societies they were driven from. I have always admired these émigrés, many of whom I was lucky enough to have met throughout my career, who despite the cards they were dealt managed to assert themselves abroad, hold on to their optimism, and even find it within themselves to approach the sons and daughters of their perpetrators with an open mind, despite all that they had suffered.

Q: What is the role of social scientists today? What role do you want to play as a theorist and researcher of society?

A: I see myself as a theorist of society, which implies the role of a distanced critic of certain conditions and developments within societies. At the same time, I am also an ardent reader and critic of the social theoretical works and approaches of others. You see, social theorists deal in thought experiments, among other things. Thought experiments are the social science equivalent of laboratory experiments in the natural sciences. Both provide us with capacities for action, which does not mean that they can readily be applied or implemented. However, it's always knowledge that provides the basis for setting something in motion.

With Helga Nowotny, Vienna, October 2007. *Source* From the personal photo archive of the author

3.2 On the Knowledge Society and Knowledge Economy

Q: As already mentioned, the term of knowledge may be described as the main category of your sociological work. What is it that makes knowledge such an important notion for you? What sets it apart as a sociological concept?

A: Knowledge is indeed a very peculiar entity, distinct from other classic phenomena of social analysis, like, for example, goods and products. If knowledge is sold, it changes hands, as it were, but still stays with the producer. Knowledge is not susceptible to zero-sum phenomena; in fact everybody stands to gain from its exchange and trade. The volume of knowledge grows constantly.

Q: From your perspective, knowledge is of such great importance that you have even termed contemporary society a "knowledge society", to quote one of the concepts with which your work is most strongly associated.

A: Yes, but there are some misunderstandings in interpreting this diagnosis. What I and others who employ this term are driving at may best be understood if we look at the development of modern society as well as the development of central theoretical attempts at describing it. Let me try to explain: For the longest time, sociological inquiries into modern society were directed at the large institutions that have historically provided its foundations. But the importance of these large institutions, such as the state, the economy, the military, the educational system or the church, to name just a few, has been steadily declining as we progress through modernity. That's where the idea of the knowledge society originates: It is the immense increase in both knowledge and the ability to use what I call knowledgeability on the part of the individual and individual self-organization, which provides the basis of the argument.

Q: Can you elaborate?

A: The age of industrialization and its particular social structure, along with faculties and abilities to master life under such conditions, are on the way out and have been for some time. Knowledge society is shorthand for the observation and theorization of what comes next. The next society, to borrow a phrase from Peter Drucker, will increasingly be based on knowledge. And this applies both to the foundations of economic production and, more generally, to social organization and integration. Importantly, this is not a normative argument: We're not saying that society is becoming more rational or more just. Knowledge in a knowledge society also represents a significant source of new problems, as we can clearly observe today. Knowledge also plays a role in new patterns of stratification. What I try to elucidate with much of my work is how we have come to create our social reality on the basis of our knowledge. Or, to put it differently: Knowledge, which has for centuries been understood mainly as knowledge *of* the world, should, from my point of view, rather be understood as knowledge *for* the world.

Q: Still, some would plead that knowledge has always played a crucial role for humanity…?

A: Indeed! Knowledge is an anthropological constant, to be sure. And even knowledge societies have existed earlier, for example in ancient Egyptian civilization, which achieved a high standard of technical know-how and elaborate cultural techniques—or in ancient Israel, which relied on sophisticated knowledge of the Torah. However, my original description of *modern* society as a knowledge society refers to the classic materialist perspective, namely the source of societal production and the creation of value and wealth. I have been investigating the foundations of economic growth in modern society and how it can be explained.

Q: So should the theory of the knowledge society be understood as part of the materialist tradition of social analysis?

A: Many sociological and economic classics have emphasized that material conditions are key to understanding societies. I am willing to accept that idea, but we need to go beyond this materialist perspective. The important insight at the outset of theorizing knowledge society was to realize that all societal contexts of knowing had increasingly become affected by scientific knowledge and, importantly, by the principles of its production. In other words, not only the economy had changed but everyday life-worlds, too, had seen enormous transformations. The relationship between the large institutions and small social groups and social movements has changed, as has democracy and our understanding of democratic politics. I have since endeavored to trace the effects of knowledge on a variety of social fields or systems while always trying to describe and gauge the effect of various forms of knowledge on how sociality is established.

Q: Does this mean that knowledge has eclipsed the traditional factors of production?

A: We have witnessed that these traditional sources of growth have reached their limit when it comes to further increase in economic growth. Future growth will not spring from them but from incremental knowledge. So this is what initially convinced me to speak of knowledge societies. Still, rather than simply disappear, traditional production factors themselves are changing. They lose importance when it comes to providing the basis of social wealth. Take, for example, the pharmaceutical industry: The material cost of producing a pill is minimal but the cost of developing a new drug is staggering. Up to 95 percent of the cost of any medicine occurs during research and development. A large part of manufacturing nowadays produces goods whose cost results mainly from knowledge, research and development. The same holds true for the service and agricultural sectors. They increasingly rely on the same principles of the knowledge economy. Even they have to trade in information and knowledge today, be it seeds for farming or high-tech therapies in health care.

Q: What does that mean for classic sociological problems, for example the structure of the workforce and unemployment? How does the transition to a knowledge economy impact how a society distributes labor?

A: The most recent transformations of the knowledge-based economy represent a movement toward a growing monetization, financialization and hence expanding quantification of economic relations, as well as a whole host of social phenomena. Modern capitalism is more immaterial and financial; it witnesses and has to cope with new forms of money. New forms of money quicken the pace of life. Financial crises are a symptom of the new economic and political order. It is, for example, no longer adequate to subsume the financialization of the economy as just another manifestation of the trend toward an extension of the service sector. Yet, our knowledge about the knowledge-based economy is precarious and limited.

As for the labor market, the distinction between manual work and knowledge work does not refer to tasks that are mutually exclusive. The differentiation is not absolute. Knowledge workers are not confined to a social class formation in terms of classic political economy. There are knowledge workers engaged in manual work and there is manual work that incorporates knowledge work. Both types of knowledge workers are in immediate control of significant elements of labor as means of production. For example, knowledge workers who are physicians perform operations in close 'alliance' with machines.

The conviction that employment levels are dependent on economic cycles is not a figment of the imagination but is reflected in the macro-economic data of the time. But technical changes were seen to result in higher incomes, more employment opportunities and a lowering in the number of hours worked, despite the pains of structural adjustments. After all, as history shows, all technological advances have so far displaced workers but have ultimately led to new and better employment, though not always for everyone. In light of new technological changes, the decisive question is, of course, whether this dynamic of the labor market will be repeated. Whether or not smart technologies displace jobs and work patterns and create new work activities will be dependent on the interplay between these factors.

Q: What else characterizes knowledge societies, what are its strengths and its weaknesses?

A: Contemporary societies face countless uncertainties. The future is likely to be full of surprises, quite contrary to the once dominant belief that the future will become ever more foreseeable and projectable. The chain of events seems compressed in comparison to former times, and contingency abounds. Social, economic and political challenges that result from our heightened ability to shape our own history are also a strain on people's mental capacities. We can already witness, and I have said so for a long time, that the interplay of, or rather, the insufficient correspondence between individual knowledge resources and the demands of various social institutions, such as in professional life or political culture, raises new social problems.

At the same time, the sum of all of these societal developments represents an unprecedented, historically unique strengthening of the role of individual possibilities and the individual's place in society. An emphasis on the rights of the individual, the possibility to assert these rights and to put them to use are characteristics of what I describe as knowledge societies. But whether these freedoms are used for furthering progressive development or its opposite is difficult to anticipate. Certainly, the sum total of these changes can produce a crisis of societal control and organization, i.e. the possibility to steer society in a certain direction. The partial void that these developments cause, the waning power of the large institutions to impose their will is compensated for by new actors from within civil society whose importance grows at the cost of the power of the state. Ironically, knowledge societies are marked both by a loss of power by virtue of knowledge and by an increase in self-government owing to knowledge.

3.3 On Knowledge Politics

Q: One of your books is entitled *Knowledge Politics: Governing the Consequences of Science and Technology*, a field of research that seems to become ever more relevant. Can you name some challenges that would require such knowledge politics?

A: In a knowledge society, knowledge cannot merely be understood as something that has only beneficial qualities. Such a view is way too simple. New knowledge, especially scientific knowledge, also confronts societies with new problems both for politics and governments and people's everyday life. The central thesis is that our problem with knowledge today is no longer that we don't know enough but that we know too much.

Q: What do you mean by that?

A: The question how societies deal with incremental knowledge, new knowledge becomes more pressing. For the longest time, science and technology were seen as the harbingers of a better tomorrow. 'Progress' was closely associated with advances in engineering, the high-tech industries, medicine and biology, to name just a few. This has changed significantly as we find ourselves confronted with an explosion of knowledge that seems to have gotten out of hand. Think of genetic engineering and biotechnology. Think of the Internet and big data. Think of global financial markets and their immensely complex products that nobody seems to control anymore. Against this background, people are beginning to ask whether ignorance might indeed be bliss. Do we know too much? Should we act on everything we know and implement our knowledge as techniques and procedures? In other words, it has become necessary to police and regulate the consequences and effects of new technological inventions and science-based innovations—and, ultimately, knowledge itself. These questions play out in a field that I call knowledge politics, where such questions and measures would be discussed and weighed. Ethics counsels and policy advisory boards are merely the tip of the iceberg of knowledge societies whose problems are increasingly produced by and emerge from knowledge.

Q: But isn't the pursuit of comprehensive knowledge a defining trait of modernity as such? Isn't it inherently tied to the promise of rational conditions?

A: If I am not mistaken we have indeed set high hopes on science and technology for a long time in modern times. The scientific exploration of our world and its secrets was seen as a source for social stability and progress. Take for example Daniel Bell's book on "Post-industrial Society" (1973), for example, or Ellsworth Huntington's work on "Civilization and Climate" ([1915] 1924, see also

Huntington 1945). Such a perspective seems to have been dominant long into the 1980s. But I think we increasingly recognize that we were mistaken to trust that science and technology would be stabilizing factors in their own right, that they would somehow channel societal development and stabilize society. We have to put such simplistic notions behind us. That, by the way, also holds true for another traditional theme, namely that science and technology are inherently repressive and always serve the purposes of the already powerful, as was notoriously assumed in much of the critical theorizing of modern society. However, science is neither an element that produces social stability and security nor is it an agent of the status quo that only plays into the hands of the powers that be. Science and knowledge are also a source of instability and unrest, both for the larger societal institutions and for the average individual.

Q: Can you give us an example?

A: Take, for example, the discovery of anthropogenic climate change—a problem still denied even by a few influential people. But it won't go away simply because we choose to ignore it. For the longest time, rising global temperature averages were intangible for most people. It was a problem beyond individual experience. But once this topic, itself a product or discovery of scientific knowledge, was on the agenda politics had to attend to it, with the consequence that today climate politics consumes a great deal of political activities from mitigation to protection against changing climatic circumstances to international treaties, and so forth.

3.4 On the Fragility of Modern Society

Q: This leads us to another one of your books that deals with *The Fragility of Modern Societies*. What makes modern societies fragile? Many of the developments you describe in your work over the years have recently become more visible. Over the first one and a half decades of the new century, far right movements have garnered large support among the populace and international treaties appear to have faltered at the hands of popular protest.

A: The fragility of modern societies results from the unintended consequences of purposeful, intentional social action. There is a crucial distinction between a general fragility and a more specific and inevitable vulnerability of contemporary societies. The former is a result of some of the developments that I have described, for example the decline in importance of larger social institutions that have been so crucial for modern identities. Today, organized social action in the form of small groups or social movements can seek to confront and undermine such institutions and their efficacy and even take on large corporations as well as governments. Quite often, institutions will even steer clear of controversial plans as they already anticipate resistance.

One crucial development in this context is the increase in knowledge and knowledgeability, especially the intellectual and social capacities of even small groups of citizens, as well as the increase in the effective means of communication, including the Internet and social media. The increased ability and propensity to confront plans and policies of large national and multinational institutions, such as states, the sciences, religion and corporations, introduces greater fragility by delaying, boycotting and taking legal action against large infrastructural projects, regulatory plans and treaties. Knowledge does not automatically benefit those in control. It also increases the ability of smaller groups to resist government and other forms of power. This at the same time decreases the ability to go through with projects and policies on a more general scale.

3.5 On the Future of Modern Society

Q: So, is the crisis of the traditional party-based institutional politics permanent? What future do you see for the nation state and for our welfare systems that are rooted in it?

A: Politics and the expectations people have of politicians are still mostly based on traditional ideas. Political success is thought to consist in the ability to shape and implement top-down measures and policies. But if I am correct in my assessment of the institutional changes underway in knowledge societies, then such notions of politics as a general problem-solving mechanism are becoming less tenable. Part of the problems we see emerging is that people in general are now realizing that such hopes are bound to fail. Traditional politics seems less and less able to solve many of the notorious problems that have been relevant to generations of people. Low voter turnout as well as increasingly visible disenchantment with traditional political parties may be a symptom of this general feeling. One should interpret such developments in a political sense; i.e., we should understand that withdrawal from the political process is itself a political act. If staying home on election day were to be widely recognized as a political statement, a different picture would emerge and we wouldn't have to decry such behavior as an unfortunate lapse on the part of the citizen but as a politically rational behavior.

Q: What does the future have in store for modern societies? What will our future look like?

A: If my notion of the structural fragility of modern society is in fact an appropriate description, then the future is indeed hard to predict. We have seen, and are bound to see, further detours from social and civil progress, periods of regression to a darker past, and cruel humanitarian relapses. Yet I would argue for a stance of what might be called cautious optimism—because it would be hard to muster the will and the energy to keep on researching social phenomena without any such optimism. The reason for such an optimistic outlook lies in the same phenomenon that has rendered society fragile, namely the overall increase in the knowledge and the cultural capital of large strata of the population.

You see, the growth of knowledge and its increasing societal significance brings about at least two forms of contingency: first, the contingency of knowledge itself, and second, the expansion of the contingency of social relations as a consequence of the increased pervasion of society with knowledge. These contingencies combined lead to an acceleration of history, especially in the form of unanticipated events and occurrences. Such unforeseen occurrences always take the form of surprises; take, for example, German reunification, the attacks on the World Trade Center, the global banking crisis and other such events that have certainly left an enormous impact on global politics and modern life. Not to mention the 'Brexit' referendum or the election of Donald Trump as the president of the United States. At the same time, such an increase in the temporal evolution of modern society makes it difficult to correctly envision or anticipate the future. It seems so much harder today to predict what is to come than even Daniel Bell could have imagined in the early 1970s.

Take, for example, the anticipation of the skills required for future employment. Who can seriously and confidently predict what skills will be required 20 or 30 years down the line? How is an employer or a company supposed to foresee that? How can training programs and curricula be planned ahead if we have no reliable notion of what is going to be important? Add to that the weakening ability of the economy to impose their requirements on other fields of society, such as the educational sector. This is just one of many examples of how difficult it has become to offer any robust scenario that conforms to our future realities.

We have to give up on the idea that we are ultimately able to control the future. The notion that globalization will lead to a more homogenous world, that global development can be ascertained and that the growth of economies can be planned is becoming more and more illusory. As we have witnessed, there is no such thing as perfect public security, and modern societies will always be vulnerable. We live in unstable times, and once you have poured your finite resources into stabilizing one problem, you quickly find that another one has sprung up elsewhere. The only thing we can say for sure is that the future is highly unpredictable, contingent and full of dangers.

Q: What would your ideal society look like?

A: The ideal society is a peaceful social and cultural system in which cultural capital can be acquired and put to use unrestrictedly, in which social capital is equally distributed even if economic capital isn't necessarily equal and in which relations to its neighbors are marked by friendly reciprocity. If, however, one follows Pierre Bourdieu's assertions about the acquisition of the social resource of knowledge, one would expect that knowledge as part of the more general resource of cultural capital should primarily be found among selective strata of society who are able to take advantage of the capital controlled by their families. This is certainly still true of a more or less large sector of the population today. But it does not explain how other strata of the population, the supposedly weak strata, may acquire the attributes of knowledgeability despite a deficit in inherited cultural capital. Access to and facility with these attributes apparently is not only a function of inherited, traditional capabilities but is codetermined by the fundamental societal openness of knowledgeability. At the same time, Bourdieu's findings leave open the question of why and how the extent of available knowledge in a society, and thus the resource of knowledgeability, grows. After all, one of the unique attributes of knowledge, in comparison to other societal resources, is that it lacks restrictive zero-sum qualities. Rather, it is much more difficult to control or limit access to this resource as compared to others. We can only hope that the divisiveness of modern societies will decrease and larger access to cultural capital will be a priority of policies and social practices.

Q: How would you want to change society?

A: Based on my observations of society, what is clear to me is the great impact education and science policies—as well as knowledge politics—have on modern society and the course it takes. Consequently, we must never tire of demanding that more intellectual and material resources be made available for education, for policies, for controlling the powerful, and for coping with some of the unprecedented challenges faced by modern societies.

References

Adolf, Marian and Nico Stehr. 2017. *Knowledge. Is Knowledge Power?* Second, extended and revised edition. London & New York: Routledge.

Bell, Daniel. 1973. *The Coming of Post-industrial Society*. New York: Basic Books.

Drucker, Peter F. 1993. *Post-Capitalist Society*. New York: HarperCollins.

Huntington, Ellsworth. [1915] 1924. *Civilization and Climate*. Third Edition, Revised and rewritten with many new Chapters. New Haven: Yale University Press.

Huntington, Ellsworth. 1945. *Mainsprings of Civilization*. New York: Wiley.

Stehr, Nico. 2001. *The Fragility of Modern Society: Knowledge and Risk in the Information Age*. London & Thousand Oaks: Sage Publications Ltd.

Stehr, Nico. 2005. *Knowledge Politics: Governing the Consequences of Science and Technology*. Boulder: Paradigm Publishing.

Nico Stehr (in Hittisau, Austria, 2015). *Source* Personal archive

Chapter 4
Complete Bibliography of Nico Stehr's Writings

4.1 Books

Kettler, David, Voker Meja and Nico Stehr. 1984. *Karl Mannheim*. London: Tavistock. ISBN: 0853126879.

Kettler, David, Volker Meja and Nico Stehr. 1989. *Politisches Wissen: Studien zu Karl Mannheim*. Frankfurt am Main: Suhrkamp. ISBN: 3518282492.

Stehr, Nico. 1991. *Praktische Erkenntnis*. Frankfurt am Main: Suhrkamp. ISBN: 9783518581650.

Stehr, Nico. 1992. *Practical Knowledge: Applying Social Science Knowledge*. London: Sage. ISBN: 9780803986992.

Stehr, Nico and Hans von Storch. 1994. *The Social Construct of Climate and Climate Change*. Hamburg: Max-Planck-Institute for Meteorology.

Stehr, Nico. 1994. *Arbeit, Eigentum und Wissen. Zur Theorie von Wissensgesellschaften*. Frankfurt am Main: Suhrkamp. ISBN: 9783518581872

Stehr, Nico. 1994. *Knowledge Societies*. London: Sage. ISBN: 9780803978928.

Stehr, Nico, Hans von Storch and Moritz Flügel. 1995. *The 19th Century Discussion of Climate Variability and Climate Change: Analogies for the Present?* Report No. 157. Hamburg: Max-Planck-Institute for Meteorology.

Stehr, Nico and Hans von Storch. 1999. *Klima-Wetter-Mensch*. München: C.H. Beck. ISBN: 9783406446139.

Stehr, Nico. 2000. *Die Zerbrechlichkeit moderner Gesellschaften*. Frankfurt am Main und Göttingen: Velbrück. ISBN: 9783934730182.

Stehr, Nico. 2001. *The Fragility of Modern Societies: Knowledge and Risk in the Information Age*. London: Sage. ISBN: 9780761953487.

Stehr, Nico. 2001. *Wissen und Wirtschaften: Die gesellschaftlichen Grundlagen der modernen Ökonomie*. Frankfurt am Main: Suhrkamp. ISBN: 9783518291078.

M. T. Adolf (ed.), *Nico Stehr: Pioneer in the Theory of Society and Knowledge*, Pioneers in Arts, Humanities, Science, Engineering, Practice 16,
https://doi.org/10.1007/978-3-319-76995-0_4

Stehr, Nico. 2002. *Knowledge and Economic Conduct: The Social Foundations of the Modern Economy.* Toronto: University of Toronto Press. ISBN: 9780802078865.

Stehr, Nico. 2003. *Wissenspolitik: Die Überwachung des Wissens.* Frankfurt am Main: Suhrkamp. ISBN: 9783518292150.

Stehr, Nico. 2005. *Knowledge Politics. Governing the Consequences of Science and Technology.* Boulder, CO: Paradigm Publishers. ISBN: 9781594510861.

Stehr, Nico. 2007. *Die Moralisierung der Märkte. Eine Gesellschaftstheorie.* Frankfurt am Main: Suhrkamp. ISBN: 9783518294314.

Stehr, Nico. 2008. *Moral Markets. How Knowledge and Affluence Change Consumers and Producers.* Boulder, CO: Paradigm Publishers. ISBN: 9780765803375.

Ozolina, Žaneta, Carl Mitcham, Jack Stilgoe, Pamela Andana, Matthias Kaiser, Linda Nielsen, Nico Stehr and Ren-Zong Qui. 2009. *Global Governance of Science. Report of the Expert Group on Global Governance of Science to the Science, Economy and Society Directorate, Directorate-General for Research, European Commission.* Brussels: European Commission. ISBN: 9789279079726.

Stehr, Nico and Hans von Storch. 2009. *Klima, Wetter, Mensch.* Opladen: Budrich Verlag. ISBN: 9783866492288.

Stehr, Nico and Hans von Storch. 2010. *Climate and Society: Climate as Resource, Climate as Risk.* Singapore: World Scientific Publishers. ISBN: 978-9814280532.

Stehr, Nico and Reiner Grundmann. 2010. *Expertenwissen. Die Kultur und Macht von Experten, Beratern und Ratgebern.* Weilerswist: Velbrück Wissenschaft. ISBN: 9783938808825.

Stehr, Nico and Reiner Grundmann. 2011. *Die Macht der Erkenntnis.* Berlin: Suhrkamp. ISBN: 9783518295908.

Stehr, Nico and Reiner Grundmann. 2011. *Experts: The Knowledge and Power of Expertise.* London: Routledge. ISBN: 9780415608039.

Stehr, Nico and Reiner Grundmann. 2012. *The Power of Scientific Knowledge. From Research to Public Policy.* Cambridge: Cambridge University Press. ISBN: 9781107606722.

Adolf, Marian and Nico Stehr. 2014. *Knowledge.* London: Routledge. ISBN: 9781138014497.

Prins, Gwythian, Mark Caine, Keigo Akimoto, Paulo Calmon, John Constable, Enrico Deiaco, Martin Flack, Isabel Galiana, Reiner Grundmann, Frank Laird, Elizabeth Malone, Yuhji Matsuo, Lawrence Pitt, Mikael Román, Andrew Sleigh, Amy Sopinka, Nico Stehr, Margaret Taylor, Hiroyuki Tezuka and Masakazu Toyoda. 2013. *The Vital Spark: Innovating Clean and Affordable Energy for All.* London: LSE Academic Publishing. ISBN: 978-1909890015.

Stehr, Nico. 2015. *Information, Power and Democracy. Liberty is a Daughter of Knowledge.* Cambridge: Cambridge University Press. ISBN: 9781107120754.

Stehr, Nico. 2015. *Die Freiheit ist eine Tochter des Wissens.* Wiesbaden: Springer VS. ISBN: 9783658095192.

Stehr, Nico and Marian Adolf. 2015. *Ist Wissen Macht? Erkenntnisse über Wissen*. Weilerswist: Velbrück. ISBN: 9783958320741.

Adolf, Marian and Nico Stehr. 2017. *Knowledge. Is Knowledge Power?* 2nd Edition. London and New York: Routledge. ISBN: 9781138685680.

4.2 Edited Books

König, René and Nico Stehr (eds.). 1975. *Wissenschaftssoziologie. Studien und Materialien*. Sonderheft 18 der Kölner Zeitschrift für Soziologie und Sozialpsychologie. Opladen: Westdeutscher Verlag. ISBN: 9783531113265.

Stehr, Nico (ed.). 1978. *Theorien sozialer Strukturen*. Opladen: Westdeutscher Verlag. ISBN: 9783531114446.

Kettler, David, Volker Meja and Nico Stehr (eds.). 1980. *Karl Mannheim: Strukturen des Denkens*. 2nd Edition. Frankfurt am Main: Suhrkamp. ISBN: 9783518278987.

Meja, Volker and Nico Stehr (eds.). 1981. *Wissenssoziologie. Studien und Materialien*. Sonderheft 22 der Kölner Zeitschrift für Soziologie und Sozialpsychologie. Opladen: Westdeutscher Verlag. ISSN: 04541340.

Kettler, David, Volker Meja and Nico Stehr (eds.). 1982. *Karl Mannheim: Structures of Thinking*. London: Routledge and Kegan Paul. ISBN: 9780710009364.

Meja, Volker and Nico Stehr (eds.). 1982. *Streit um die Wissenssoziologie. Erster Band: Die Entwicklung der deutschen Wissenssoziologie*. Frankfurt am Main: Suhrkamp. ISBN: 9783518079614.

Meja, Volker and Nico Stehr (eds.). 1982. *Streit um die Wissenssoziologie. Zweiter Band: Rezeption und Kritik der deutschen Wissenssoziologie*. Frankfurt am Main: Suhrkamp. ISBN: 9783518079614.

Meja, Volker and Nico Stehr (eds.). 1984. *Knowledge and Society: Contemporary Perspectives on the Sociology of Knowledge*. New Brunswick: Transaction Books. ISBN: 9780878554935.

Kettler, David, Volker Meja and Nico Stehr (eds.). 1984. *Karl Mannheim: Konservatismus*. Frankfurt am Main: Suhrkamp. ISBN: 9783518280782.

Stehr, Nico (ed.). 1985. *Robert K. Merton. Entwicklungen und Wandel von Forschungsinteressen: Aufsätze zur Wissenschaftssoziologie*. Frankfurt am Main: Suhrkamp. ISBN: 9783518577103.

Böhme, Gernot and Nico Stehr (eds.). 1986. *The Knowledge Society: The Growing Impact of Scientific Knowledge on Social Relations*. Dordrecht: D. Reidel Publishing. ISBN: 9789027723055.

Kettler, David, Volker Meja and Nico Stehr (eds.). 1986. *Karl Mannheim: Conservatism. A Contribution to the Sociology of Knowledge*. London: Routledge and Kegan Paul. ISBN: 9780710203380.

Meja, Volker, Dieter Miesgeld and Nico Stehr (eds.). 1990. *Modern German Sociology: An Anthology.* New York: Columbia University Press. ISBN: 9780231058551.

Meja, Volker and Nico Stehr (eds.). 1990. *Knowledge and Politics. The Sociology of Knowledge Dispute.* London and New York: Routledge and Kegan Paul. ISBN: 9780415028813.

Stehr, Nico and Richard Ericson (eds.). 1992. *The Culture and Power of Knowledge. Inquiries into Modern Society.* New York and Berlin: de Gruyter. ISBN: 9783110131758.

Meja, Volker and Nico Stehr (eds.). 1995. *Robert K. Merton: Soziologische Theorie und soziale Struktur.* Berlin: de Gruyter. ISBN: 9783110130218.

Meja, Volker and Nico Stehr (eds.). 1999. *The Sociology of Knowledge. International Library of Critical Writings in Sociology.* Cheltenham: Edward Elgar. ISBN: 9781858985886.

Stehr, Nico and Hans von Storch (eds.). 2000. *Eduard Brückner: The Sources and Consequences of Climate Change and Climate Variability in Historical Times.* Dordrecht: Kluwer Academic Publishers. ISBN: 9780792361282.

Weingart, Peter and Nico Stehr (eds.). 2000. *Practising Interdisciplinarity.* Toronto: University of Toronto Press. ISBN: 9780802081398.

Ericson, Richard V. and Nico Stehr (eds.). 2000. *Governing Modern Societies.* Toronto: University of Toronto Press. ISBN: 9781442675452.

Stehr, Nico and Reiner Grundmann (eds.). 2001. *Werner Sombart: Economic Life in the Modern Age.* New Brunswick and London: Transaction Books. ISBN: 9780765800305.

Kettler, David, Volker Meja and Nico Stehr (eds.). 2003. *Karl Mannheim. Konservatismus.* 2nd Edition. Frankfurt am Main: Suhrkamp. ISBN: 9783518280782.

Stehr, Nico (ed.). 2004. *The Governance of Knowledge.* New Brunswick, New Jersey: Transaction Books. ISBN: 9770765801728.

Stehr, Nico (ed.). 2004. *Biotechnology: Between Commerce and Civil Society.* New Brunswick, New Jersey: Transaction Books. ISBN: 9780765802248.

Fleck, Christian and Nico Stehr (eds.). 2004. *Die Gesellschaft des Terrors. Innenansichten der Konzentrationslager Dachau und Buchenwald.* Frankfurt am Main: Suhrkamp. ISBN: 9783518583975.

Fleck, Christian and Nico Stehr (eds.). 2005. *The Society of Terror. Inside the Dachau and Buchenwald Concentration Camps.* Boulder, CO: Paradigm Publishers. ISBN: 9781594510953.

Grundmann, Reiner and Nico Stehr (eds.). 2005. *Knowledge. Five volumes.* London: Routledge. ISBN: 9780415317368.

Meja, Volker and Nico Stehr (eds.). 2005. *Society and Knowledge. Contemporary Perspectives on the Sociology of Knowledge and Science.* 2nd Edition. New Brunswick, New Jersey: Transaction Books. ISBN: 9780765802958.

Nico Stehr (ed.). 2005. *Niklas Luhmann. Risk. A Sociological Theory.* New Brunswick, New Jersey: Aldine. ISBN: 9780202304434.

Jansen, Stephan A., Birger Priddat and Nico Stehr (eds.). 2005. *Demographie: Bewegungen einer Gesellschaft im Ruhestand: multidisziplinäre Perspektiven zur Demographiefolgenforschung.* Opladen: VS Verlag für Sozialwissenschaften. ISBN: 9783531147802.

Stehr, Nico, Christoph Henning and Bernd Weiler (eds.). 2006. *The Moralization of the Markets.* New Brunswick, NJ: Transaction Books. ISBN: 9780765803153.

Jansen, Stephan A., Birger Priddat and Nico Stehr (eds.). 2007. *Die Zukunft des Öffentlichen. Multidisziplinäre Theorie-Perspektiven für eine Öffnung der Diskussion des Öffentlichen.* Wiesbaden: VS Verlag für Sozialwissenschaften. ISBN: 9783531152820.

Fleck, Christian and Nico Stehr (eds.). 2007. *Paul F. Lazarsfeld. Empirische Analyse des Handelns. Ausgewählte Schriften.* Frankfurt am Main: Suhrkamp. ISBN: 9783518294222.

Stehr, Nico and Bernd Weiler (eds.). 2007. *Who Owns Knowledge? Knowledge and the Law.* New Brunswick, NJ: Transaction Books. ISBN: 9780765803375.

Stehr, Nico (ed.). 2008. *Knowledge & Democracy. A 21st-Century Perspective.* New Brunswick, NJ: Transaction Books. ISBN: 9781412807067.

Von Storch, Hans and Nico Stehr (eds.). 2008. *Eduard Brückner. Die Geschichte unseres Klimas: Klimaschwankungen und Klimafolgen.* Wien: Österreichische Beiträge zu Meteorologie und Geophysik.

Jansen, Stephan A., Eckard Schröter and Nico Stehr (eds.). 2008. *Mehrwertiger Kapitalismus. Multidisziplinäre Beiträge zu Formen des Kapitalismus und seiner Kapitalien.* Wiesbaden: VS Verlag für Sozialwissenschaften. ISBN: 9783531158648.

Grundmann, Reiner and Nico Stehr (eds.). 2009. *Society. Four Volumes.* London: Routledge. ISBN: 9780415126565.

Jansen, Stephan A., Eckard Schröter and Nico Stehr (eds.). 2009. *Rationalität der Kreativität? Multidisziplinäre Beiträge zur Analyse der Produktion, Organisation und Bildung von Kreativität.* Wiesbaden: VS Verlag für Sozialwissenschaften. ISBN: 9783531166889.

Bechmann, Gotthard, Vitaly Gorokhov and Nico Stehr (eds.). 2009. *The Social Integration of Science. Institutional and Epistemological Aspects of the Transformation of Knowledge in Modern Society.* Berlin: Edition Sigma. ISBN: 9783894049423.

Moldaschl, Manfred and Nico Stehr (eds.). 2010. *Wissensökonomie und Innovation: Beiträge zur Ökonomie der Wissensgesellschaft.* Marburg an der Lahn: Metropolis Verlag. ISBN: 9783895185328.

Jansen, Stephan A., Eckard Schröter and Nico Stehr (eds.). 2010. *Transparenz. Multidisziplinäre Durchsichten durch Phänomene und Theorien des Undurchsichtigen.* Wiesbaden: Springer VS Verlag für Sozialwissenschaften. ISBN: 9783531174358.

Fleck, Christian and Nico Stehr (eds.). 2011. *Paul F. Lazarsfeld. An Empirical Theory of Action—Collected Writings*. Oxford: Bardwell. ISBN: 9781905622221.

Jansen, Stephan A., Eckard Schröter and Nico Stehr (eds.). 2012. *Positive Distanz. Multidisziplinäre Annäherungen an den wahren Abstand und das Abstandwahren in Theorie und Praxis*. Wiesbaden: Springer VS Verlag für Sozialwissenschaften. ISBN: 9783531192079.

Jansen, Stephan A., Eckard Schröter and Nico Stehr (eds.). 2012. *Bürger. Macht. Staat? Neue Formen gesellschaftlicher Teilhabe, Teilnahme und Arbeitsteilung*. Wiesbaden: Springer VS Verlag für Sozialwissenschaften. ISBN: 9783531193465.

Jansen, Stephan A., Eckard Schröter and Nico Stehr (eds.). 2013. *Fragile Stabilität – stabile Fragilität*. Wiesbaden: Springer VS Verlag für Sozialwissenschaften. ISBN: 9783658022471.

Meja, Volker and Nico Stehr (eds.). 2014. *Knowledge and Politics: The Sociology of Knowledge Dispute*. London: Routledge. ISBN: 9781138786158.

Stehr, Nico (ed.). 2015. *Das Hartwell-Papier. Eine Neuausrichtung der Klimapolitik an der Menschenwürde*. Wiesbaden: Springer VS Verlag für Sozialwissenschaften. ISBN: 9783658074609.

Machin, Amanda and Nico Stehr (eds.). 2016. *Understanding Inequality. The Social Costs and Benefits*. Wiesbaden: Springer VS Verlag für Sozialwissenschaften. ISBN: 9783658116620.

4.3 Book Chapters

Larson, Lyle E. and Nico Stehr. 1974. "Intra-family Conflict and Participation in Extra-Curricular Activities", in: *Proceedings Sports in the Modern World—Prospects and Problems*. München.

Meja, Volker and Volker Meja. 1980. "Karl Mannheim", in: Justin Winkle (ed.), *The Makers of Modern Culture*. London: Routledge and Kegan Paul.

Stehr, Nico. 1981. "Wie ich zur Soziologie kam und wo ich bin: Ein Gespräch mit Kurt H. Wolff", in: M. Rainer Lepsius (ed.), *Soziologie in Deutschland und Österreich 1918–1945. Sonderheft 23 Kölner Zeitschrift für Soziologie und Sozialpsychologie*. Opladen: Westdeutscher Verlag: 324–346.

Baldamus, Wilhelm and Nico Stehr. 1983. "Accounts and Action: The Logic(s) of Social Science and Pragmatic Knowledge", in: Burkart Holzner et al. (eds.), *Realizing Social Science Knowledge*. Würzburg: Physica Verlag: 73–78.

Meja, Volker and Nico Stehr. 1985. "Karl Mannheim", in: Adam Kuper and Jessica Kuper (eds.), *The Social Science Encyclopedia*. London: Routledge and Kegan Paul: 482–483.

Meja, Volker and Nico Stehr. 1985. "The Sociology of Knowledge", in: Adam Kuper and Jessica Kuper (eds.), *The Social Science Encyclopedia*. London: Routledge and Kegan Paul: 430–431.

Schafer, Walter E. and Nico Stehr. 1986. "Participation in Interscholastic Athletics, Academic Achievement and Social Mobility: Some Intervening Processes", in: H. Kroll (ed.), *Jugend und Sport*. Wien: Bundesverlag: 184–196.

Van den Berg and Nico Stehr. 1986. "Sociological Models", in: Karigoudar Ishawaran (ed.), *Introduction to Sociology*. Toronto: Addison-Wesley: 27–61.

Stehr, Nico. 1986. "Sociological Theory and Practical Reason: The Restriction of the Scope of Sociological Theory", in: Mark Wardell and Stephen Turner (eds.), *Sociological Theory in Transition*. London: Allen and Unwin.

Meja, Volker and Nico Stehr. 1987. "Karl Mannheim", in: Jessica Kuper (ed.), *Key Thinkers, Past and Present*. London and New York: Routledge and Kegan Paul: 152–154.

Stehr, Nico. 1987. "Conditions for the Success of Classical Social Theory", in: Helmut Steiner (ed.), *Soziologie und Sozialpolitik*. Berlin: Akademie der Wissenschaften: 392–405.

Meja, Volker and Nico Stehr. 1988. "Wissen und Soziales Handeln: Zur praktischen Effektivität der Keynes'schen Theorie", in: Akademie der Wissenschaften *Soziologie und Sozialpolitik*. Berlin: Akademie der Wissenschaften: 239–256.

Stehr, Nico. 1989. "Wissenschaftliches Wissen und soziales Handeln: Überlegungen zum Problem des Wissenstransfers", in: Karlheinz Rebel (ed.), *Wissenschaftstransfer in der Weiterbildung. Der Beitrag der Wissenssoziologie*. Weinheim und Basel: Beltz Verlag: 48–71.

Meja, Volker and Nico Stehr. 1989. "Robert K. Merton's Structural Analysis: The Design of Modern Sociology", in: Carlo Mongardini and Simonetta Tabboni (eds.), *L'Opera di Robert K. Merton e la Sociologia Contemporanea.* Genova: Edizioni Culturali Internationali: 37–60.

Meja, Volker and Nico Stehr. 1990. "The Sociology of Knowledge and the Ethos of Science", in: Eileen Leonard, Hermann Strasser and Kenneth Westhues (eds.), *In Search of Community: Essays in Memory of Werner Stark, 1909–1985*. New York: Fordham University Press: 65–83.

Stehr, Nico. 1990. "Robert K. Merton's Sociology of Science", in: Jon Clark et al. (eds.), *Masters of Sociology: Robert K. Merton*. London: Falmer: 285–294.

Meja, Volker and Nico Stehr. 1991. "Wissenschaftssoziologie", in: Gerd Reinhold (ed.), *Soziologie-Lexikon*. München: R. Oldenbourg: 662–664.

Meja, Volker and Nico Stehr. 1992. "Social Scientific Discourse and Epistemological Discourse: The Problem of Relativism", in: Diederick Raven, Lieteke van Vucht Tijssen and Jan de Wolf (eds.), *Cognitive Relativism and Social Science.* New Brunswick, NJ: Transaction Books: 1–13.

Benninghaus, Hans and Nico Stehr. 1992. "Erkenntnis und Offenheit", in: Heine von Alemann and Gerhard Kunz (eds.), *René König. Gesamtverzeichnis der Schriften. In der Spiegelung von Freunden, Schülern, Kollegen*. Opladen: Westdeutscher Verlag: 195–199.

Böhme, Gernot and Nico Stehr. 1992. "Spoleczenstwo nauki", in: *Pace Instytutu Nauk Ekonomiczno-Spolecznych*. Warszawa: Politechniki Warszawskiej.

Meja, Volker and Nico Stehr. 1993. "The Sociology of Knowledge", in: William Outhwaite and Tom Bottomore (eds.), *The Blackwell Dictionary of Twentieth-Century Social Thought*. Oxford: Basical Blackwell: 637–639.

von Storch, Hans and Nico Stehr. 1994. "The Social Construct of Climate and Climate Change", in: American Meteorological Society, *Symposium on Global Electrical Circuit, Global Change, and the Meteorological Applications of Lightning. Fifth Symposium on Global Change Studies*. Boston: American Meteorological Society: 118–125.

Meja, Volker and Nico Stehr. 1994. "The Uneasy Relation Between Scientific and Religious Codes", in: Carlo Mongardini and Marieli Ruini (eds.), *Religio. Ruolo del Sacro, Coesione Sociale e Nuove Forme di Solidarieta nella Societa Contemporanea*. Roma: Bulzoni Editore: 103–111.

Stehr, Nico. 1994. "Wissenschaftssoziologie", in: Harald Kerber and Arnold Schmieder (eds.), *Spezielle Soziologien. Problemfelder, Forschungsbereiche, Anwendungsorientierungen*. Reinbek bei Hamburg: Rowohlt: 541–555.

Stehr, Nico. 1994. "Macht und Ohnmacht wissenschaftlicher Erkenntnis", in: Horst Reimann and Hans-Peter Müller (eds.), *Probleme moderner Gesellschaften. Peter Atteslander zum 65. Geburtstag*. Opladen: Westdeutscher Verlag: 281–305.

Stehr, Nico. 1994. "Moderne gesellschaftliche Prozesse – Motor der Kulturlandschafts-bewegung?", in: Franz Moser (ed.), *Mensch und Landschaft 200. Nutzung Bedrohung Chancen: Nachhaltige Entwicklung von Kulturlandschaften im Donau- und Alpenraum*. Graz: Technische Universität Graz: 68–86.

von Storch, Hans, Moritz Flügel and Nico Stehr. 1995. "Climate Variability, Climatic Change and Societal Consequences", in: Walter Lenz and Sonnfried Streicher (eds.), *Historisch-Meereskundliches Jahrbuch. Volume 3*. Stralsund: Deutsches Museum für Meereskunde und Fischerei: 51–70.

Stehr, Nico. 1996. "Social Inequality and Knowledge", in: David Sciulli (ed.), *Normative Social Action; Cross-National and Historical Approaches*. Greenwich, CT: JAI Press: 41–51.

Stehr, Nico. 1997. "Wissensgesellschaft", in: Gerd Reinhold (ed.), *Soziologie-Lexikon*. 3rd Edition. München: R. Oldenbourg: 739–742.

Stehr, Nico. 1997. "Sapere", in: *Enciclopedia delle Scienze Sociali. Volume VII*. Roma: Enciclopedia Italiana: 598–609.

von Storch, Hans and Nico Stehr. 1998. "Das soziale Konstrukt des Klimas", in: VDI-Gesellschaft Energietechnik (ed.), *Umwelt- und Klimabeeinflussung durch den Menschen*. Düsseldorf: VDI Berichte: 187–197.

Stehr, Nico. 1998. "Wissensberufe", in: Wolfgang Schulz (ed.), *Expertenwissen. Soziologische, pädagogische und psychologische Perspektiven*. Opladen: Leske + Budrich: 17–32.

von Storch, Hans and Nico Stehr. 1998. "Climate Research: The Case for the Social Sciences", in: Henning Rodhe and Robert Charlson (eds.), *The Legacy of Svante Arrhenius: Understanding the Greenhouse Effect*. Stockholm: Royal Swedish Academy of Sciences: 143–153.

Stehr, Nico. 1998. "Die Grenzen der Macht wissenschaftlicher Erkenntnis", in: Michael Hauskeller, Christoph Rehmann-Sutter and Gregor Schiemann (eds.), *Naturerkenntnis und Natursein. Festschrift für Gernot Böhme*. Frankfurt am Main: Suhrkamp Verlag: 53–75.

Meja, Volker and Nico Stehr. 1998. "Robert K. Merton's Structural Analysis", in: Carlo Mongardini and Simotta Tabboni (eds.), *Robert K. Merton and Contemporary Sociology*. London: Transaction Books: 21–43.

Von Storch, Hans and Nico Stehr. 1999. "Climate Works: An Anatomy of a Disbanded Line of Research", in: Heidrun Kaupen-Haas and Christian Saller (eds.), *Wissenschaftlicher Rassismus. Analysen einer Kontinuität in den Human- und Naturwissenschaften*. Frankfurt am Main: Campus: 137–185.

Stehr, Nico. 1999. "Mastering the Global Commons", in: Hans von Storch and Götz Flöser (eds.), *Anthropogenic Climate Change*. Berlin: Springer VS: 261–286.

Stehr, Nico. 1999. "Technologischer Fortschritt und gesellschaftlicher Wandel", in: Gert Kaiser, Katharina Wetzel und Arne Claussen (eds.), *Technologiebedarf im 21. Jahrhundert. Technikvorschau und Technologiepolitik in Europa*. Frankfurt am Main: Campus: 91–104.

Stehr, Nico. 1999. "Wissensgesellschaften oder die Zerbrechlichkeit moderner Gesellschaft", in: Wilfried Konrad and Wilhem Schumm (eds.), *Wissen und Arbeit. Neue Konturen der Wissensgesellschaft*. Münster: W. Dampfboot: 13–23.

Stehr, Nico. 2000. "The Productivity Paradox: ICTs, Knowledge and the Labour Market", in: John de la Mothe and Gilles Paquet (eds.), *Information, Innovation and Impacts*. Kluwer Academic Publishers: Boston, MA: 255–272.

Stehr, Nico. 2000. "Wissen und Information als Problem in der modernen Gesellschaft", in: Jürgen Mittelstraß (ed.), *Die Zukunft des Wissens. XVIII. Deutscher Kongreß für Philosophie*. Berlin: Akademie Verlag: 92–102.

Stehr, Nico. 2000. "Knowledge, Markets and Biotechnology", in: John de la Mothe and Jorge Niosi (eds.), *The Economic and Social Dynamics of Biotechnology*. Boston: Kluwer Academic Publishers: 205–214.

Grundmann, Reiner and Nico Stehr. 2001. "Die Arbeitswelt in der Wissensgesellschaft", in: Thomas Kurtz (ed.), *Aspekte des Berufs in der Moderne*. Opladen: Leske + Budrich: 315–338.

Stehr, Nico. 2001. "Knowledge Politics: The Paradox of Regulating Knowledge Dynamics", in: Sabine Maasen and Matthias Winterhager (eds.), *Science Studies. Probes into the Dynamics of Scientific Knowledge*. Bielefeld: Transcript: 269–290.

Stehr, Nico. 2001. "Knowledge Societies", in: George Ritzer and Barry Smart (eds.), *Handbook of Social Theory*. London: Sage: 494–508.

Stehr, Nico and Hermann Strasser. 2001. "75 und kein bißchen unproduktiv: Wider die Vergeudung von kulturellem Kapital in der Wissensgesellschaft", in: *Gelassenheit. Festschrift für Johannes Weiß*. Kassel: 475–478.

Stehr, Nico. 2001. "Museen in Wissensgesellschaften: Unfertige Erkenntnisse oder vom Nutzen der Vergangenheit für die Zukunft", in: Landesmuseum für Technik und Arbeit (ed.), *Zauberformeln des Zeitgeistes. Erlebnis, Event, Aufklärung,*

Wissenschaft. Wohin entwickelt sich die Museumslandschaft? Mannheim: Landesmuseum für Technik und Arbeit: 14–28.

Bechmann, Gotthard and Nico Stehr. 2001. "The Uncertainty of Risk Communication in the Knowledge Society", in: Michael Matthies, Horst Malchow and Jürgen Kriz (eds.), *Integrative Systems Approaches to Natural and Social Dynamics*. Berlin: Springer VS: 289–303.

Stehr, Nico. 2001. "Von der Zukunft der Wissenschaftskulturen", in: Jörn Rüsen (ed.), *Kulturwissenschaftliches Institut: Jahrbuch 2001/2002*. Essen: Kulturwissenschaftliches Institut: 441–452.

Meja, Volker and Nico Stehr. 2001. "Modern Societies as Knowledge Societies", in: Luigi Tomasi (ed.), *New Horizons in Sociological Theory and Research. The Frontiers of Sociology at the Beginning of the Twenty-First Century*. Aldershot: Asgate: 127–146.

Stehr, Nico. 2001. "The Grammar of Productive Knowledge", in: John de la Mothe and Domique Foray (eds.), *Knowledge Management in the Innovation Process*. Boston: Kluwer: 193–204.

Stehr, Nico. 2002. "Models as Focusing Tools: Linking Nature and the Social World", in: Hans von Storch and Götz Flöser (eds.), *Models in Environmental Research*. Heidelberg: Springer VS: 1–16.

Stehr, Nico. 2002. "Die Macht des Schwachen", in: Hartmut Rupp, Christoph Scheilke and Heinz Schmidt (eds.), *Zukunftsfähige Bildung und Protestantismus*. Stuttgart: Calwer Verlag: 90–100.

Stehr, Nico. 2002. "Wissen", in: Christoph Engel, Jost Halfmann and Martin Schulte (eds.), *Wissen-Nichtwissen-Unsicheres Wissen*. Freiburg: Nomos: 17–33.

Stehr, Nico. 2002. "Biotechnology and the Governance of Knowledge", in: Kulturwissenschaftliches Institut, *Jahrbuch 2001/2002*. Essen: Kulturwissenschaftliches Institut: 304–316.

von Storch, Hans and Nico Stehr. 2002. "Towards a History of Ideas of Anthropogenic Climate Change", in: Gerold Wefer, Wolfgang H. Berger, Karl-Ernst Behre and Eystein Jansen (eds.), *Climate and History in the North Atlantic Realms*. Heidelberg: Springer VS: 17–23.

Grundmann, Reiner and Nico Stehr. 2002. "Klimawissenschaft als Akteur in der öffentlichen Arena", in: Walter Hauser (ed.), *Klima. Das Experiment mit dem Planeten Erde*. München: Deutsches Museum: 384–396.

von Storch, Hans and Nico Stehr. 2002. "Das Klima in den Köpfen der Menschen", in: Walter Hauser (ed.), *Klima. Das Experiment mit dem Planeten Erde*. München: Deutsches Museum: 180–191.

Stehr, Nico. "The Social Role of Knowledge", in: Nikolai Genov (ed.), *Advances in Sociological Knowledge*. Paris: International Social Science Council: 84–113.

von Storch, Hans and Nico Stehr. 2003. "Micro/Macro and Soft/Hard: Divergence and Converging Issues in the Physical and Social Sciences", in: Jan Rotmans and Dale S. Rothman (eds.), *Scaling in Intergrated Assessment*. Lisse: Swets & Zeitlinger: 35–49.

Stehr, Nico. 2003. "Expertise in Knowledge Societies", in: Gotthard Bechmann and Imre Hronsky (eds.), *Expertise and Its Interfaces*. Berlin: Sigma: 35–44.

Stehr, Nico. 2003. "Das Produktivitätsparadox", in: Stefan Böschen and Ingo Schulz-Schaeffer (eds.), *Wissenschaft in der Wissensgesellschaft*. Wiesbaden: Westdeutscher Verlag: 77–95.

Stehr, Nico. 2003. "Die Chancen und die Risiken der Globalisierung", in: Marcus S. Kleiner and Hermann Strasser (eds.), *Globalisierungswelten. Kultur und Gesellschaft in einer entfesselten Welt*. Köln: Herbert von Halem Verlag: 244–257.

Grundmann, Reiner and Nico Stehr. 2003. "Sozialwissenschaft ohne Natur?", in: Margarete Maurer and Otmar Höll (eds.), *Natur als Politikum*. Wien: RLI-Verlag: 47 68.

Stehr, Nico. 2003. "Wissenspolitik oder die gesellschaftliche Disziplinierung neuer Erkenntnisse", in: Herbert Kubincek, Dieter Klumpp and Alexander Roßnagel (eds.), *Next Generation Information Society? Notwendigkeit einer Neuorientierung*. Mössingen-Talheim: Talheimer Verlag: 318–330.

Stehr, Nico. 2004. "Wissensökonomie und verwandte Entwicklungen", in: Friedrich-Ebert-Stiftung (ed.), *Die neue SPD. Menschen stärken – Wege öffnen*. Bonn: Dietz: 185–196.

Stehr, Nico. 2004. "Wissensgesellschaft", in: Friedrich Jaeger und Jörn Rüsen (eds.), *Handbuch der Kulturwissenschaften. Band 3: Themen und Tendenzen*. Stuttgart: J.B. Metzler: 34–49.

Grundmann, Reiner and Nico Stehr. 2004. "Werner Sombart", in: George Ritzer (ed.), *Encyclopedia of Social Theory*. Thousand Oaks, CA: Sage: 776–777.

Grundmann, Reiner. 2004. "Die Natur und die Natur der Gesellschaft", in: Ludwig Fischer (ed.), *Projektionsfläche Natur. Zum Zusammenhang von Naturbildern und gesellschaftlichen Verhältnissen*. Hamburg: Hamburg University Press: 261–276.

Stehr, Nico. 2004. "Fragile Utopien", in: Jörn Rüsen, Michael Fehr and Annelie Ramsbrock (eds.), *Die Unruhe der Kultur. Potentiale des Utopischen*. Weilerswist: Velbrück: 327–331.

Stehr, Nico. 2004. "Einleitung", in: Jörn Rüsen, Michael Fehr and Annelie Ramsbrock (eds.), *Die Unruhe der Kultur. Potentiale des Utopischen*. Weilerswist: Velbrück: 137–142.

Stehr, Nico and Hermann Strasser. 2004. "Noch ist nichts entschieden: Chancen und Risiken der Globalisierung", in: Eckart Gottwald and Folket Rickers (eds.), *Die Zukunft des Religionsunterrichts im Horizont von Globalisierung und Multikulturalität*. Nordhausen: Verlag Traugott Bautz: 11–34.

Stehr, Nico. 2004. "The Social Role of Knowledge", in: Nikolai Genov (ed.), *Advances in Sociological Knowledge*. Wiesbaden: VS Verlag für Sozialwissenschaften: 83–105.

Stehr, Nico. 2004. "The Economic Structure of Knowledge Societies", in: Frank Webster (ed.), *The Information Society Reader*. London: Routledge: 212–236.

Stehr, Nico. 2005. "Die soziale Bedeutung des Wissens", in Nikolai Genov (ed.), *Die Entwicklung des soziologischen Wissens*. Wiesbaden: VS Verlag für Sozialwissenschaften: 101–128.

Stehr, Nico. 2005. "Die gesellschaftliche Kontrolle neuer Erkenntnisse? Zur Wissenspolitik in demokratischen Gesellschaften", in: Alfons Bora, Michael Decker, Armin Grunwald and Ortwin Renn (eds.), *Technik in einer fragilen Welt. Die Rolle der Technikfolgenabschätzung*. Berlin: editon sigma: 205–214.

Stehr, Nico. 2005. "Von der Zukunft der Wissenschaftskulturen und den Bedingungen der Transdisziplinarität", in: Martin Döring, Wolfgang Settekorn and Hans von Storch (eds.), *Küstenbilder, Bilder der Küste. Interdisziplinäre Ansichten, Ansätze und Konzepte*. Hamburg: Hamburg University Press: 351–360.

Stehr, Nico. 2005. "Noch ist nichts entschieden: Die Chancen und Risiken der Globalisierung", in: Endre Kiss (ed.), *Postmoderne und/oder Rationalität*. Szekesfehervar: Kodolányi-János Hochschule: 35–52.

Grundmann, Reiner and Nico Stehr. 2006. "Werner Sombart", in: Jens Beckert and Milan Zafirowski (eds.), *International Encyclopedia of Economic Sociology*. London: Routledge: 644–646.

Stehr, Nico. 2006. "Aktuelle Probleme der Wissensgesellschaft: Bildung, Arbeit und Wirtschaft", in: Klaus Kempter and Peter Meusburger (eds.), *Bildung und Wissensgesellschaft. Heidelberger Jahrbücher 49*. Heidelberg: Springer VS: 363–377.

Stehr, Nico. 2006. "Grenzenlose Wissenswelten", in: Gertraud Koch (ed.), *Internationalisierung von Wissen. Multidisziplinäre Beiträge zu neueren Praxen des Wissenstransfers*. St. Inbert: Röhrig Universitätsverlag: 27–58.

Stehr, Nico. 2006. "Is Democracy a Daughter of Knowledge?", in: Guy Neave (ed.), *Knowledge, Power and Dissent. Critical Perspectives on Higher Education and Research in Knowledge Society*. Paris: UNESCO: 129–132.

Stehr, Nico. 2006. "Eine Welt aus Wissen", in: Reinhard Fatke and Hans Merkens (eds.), *Bildung über die Lebenszeit*. Wiesbaden: VS-Verlag für Sozialwissenschaften: 97–110.

Stehr, Nico. 2007. "Knowledge Societies", in: George Ritzer (ed.), *The Blackwell Encyclopedia of Sociology*. Oxford: Blackwell.

Stehr, Nico. 2007. "Märkte als Gemeinschaftshandeln", in: Gerd Nollmann (ed.), *Sozialstruktur und Gesellschaftsanalyse. Sozialwissenschaftliche Forschung zwischen Daten, Methoden und Begriffen*. Wiesbaden: VS Verlag für Sozialwissenschaften: 231–251.

Grundmann, Reiner and Nico Stehr. 2007. "The Unique Complexity of Social Phenomena and the Use of Social Science Knowledge", in: Samir Dasgupta and Robyn Driskell (eds.), *Discourse on Applied Sociology. Volume 1: Theoretical Perspectives*. London: Anthem Press: 79–98.

Stehr, Nico. 2007. "Theories of the Information Age", in: John R. Bryson and Peter W. Daniels (eds.), *The Handbook of Service Industries*. Cheltenham: Edward Elgar: 62–76.

Stehr, Nico. 2007. "Modern Societies as Knowledge Societies", in: Arnaud Sales and Marcel Fournier (eds.), *Knowledge, Communication and Creativity.* London: Sage: 31–42.

Stehr, Nico. "Moralisierung der Märkte", in: Christian Eigner and Peter Weibel (eds.), *Un/Fair Trade. Die Kunst der Gerechtigkeit.* Wien: Springer: 400–407.

Stehr, Nico. 2007. "Die politische Herausforderung des globalen Klimawandels", in: Arbeitskreis KLIWA (ed.), *Klimaveränderung und Konsequenzen für die Wasserwirtschaft. 3. KLIWA Symposium am 25. und 26. Oktober 2006 in Stuttgart.* Stuttgart: Arbeitskreis KLIWA: 247–248.

Stehr, Nico. 2007. "Vorfahrt für welche Arbeit und für welchen Arbeitsmarkt?", in: Klaus Kornwachs (ed.), *Bedingungen und Triebkräfte technologischer Innovationen.* Stuttgart: acatech – Konvent für Technikwissenschaften: 187–198.

Stehr, Nico. 2007. "A Conversation with Kurt H. Wolff", in: Gary Backhaus and George Psathas (eds.), *The Sociology of Radical Commitment: Kurt H Wolff's Existential Turn.* Lexington Books: 37–62.

von Storch, Hans and Nico Stehr. 2007. "Der öffentliche Diskurs über das Klima oder – die politische Macht der Klimaforschung", in: Philipp Missfelder (ed.), *Umdenken. Für eine nachhaltige Klimapolitik.* Berlin: Weiss Verlag: 14–31.

Stehr, Nico. 2007. "Thomas Schelling und das Klimaproblem", in: Ingo Pies and Martin Leschke (eds.), *Thomas Schellings strategische Ökonomik.* Tübingen: Mohr Siebeck: 117–120.

Stehr, Nico. 2008. "Sociology of Knowledge", in: William A. Darity Jr. (ed.), *International Encyclopedia of the Social Sciences.* Vol. 7. 2nd Edition. Detroit: Macmillan Reference: 674–676.

Adolf, Marian and Nico Stehr. 2008. "Medien in der Wissensgesellschaft: Auf der Suche nach Schnittstellen", in: Johannes Raabe, Rudolf Stöber, Anna M. Theirs-Berglmair and Kristina Wolf (eds.), *Medien und Kommunikation in der Wissensgesellschaft.* Konstanz: UVK Verlagsgesellschaft: 62–73.

Stehr, Nico. 2008. "Is Civil Society a Daughter of Knowledge? Worlds of Knowledge and Democracy", in: Sven Eliaeson (ed.), *Building Civil Society and Democracy in New Europe.* Cambridge: Cambridge Scholars Publishing: 212–227.

Adolf, Marian and Nico Stehr. 2008. "Konsum zwischen Markt und Moral: Eine soziokulturelle Betrachtung moralisierter Märkte", in: Stephan Jansen, Eckhard Schroeter and Nico Stehr (eds.), *Mehrwertiger Kapitalismus. Multidisziplinäre Beiträge zu Formen des Kapitalismus und seiner Kapitalien.* Wiesbaden: VS Verlag für Sozialwissenschaften: 195–218.

Stehr, Nico. 2008. "Gestaltungsspielräume. Vom Nutzen wissenschaftlichen Wissens", in: Anette Schavan (ed.), *Keine Wissenschaft für sich.* Hamburg: Edition Körber Stiftung: 51–60.

Kaube, Jürgen, Wilhelm Krull and Nico Stehr. 2008. "Rollen in der Relevanzdebatte. Über Forschungsförderung, Anwendungsdruck und die Macht der Geisteswissenschaften", in: Anette Schavan (ed.), *Keine Wissenschaft für sich.* Hamburg: Edition Körber Stiftung: 35–50.

Stehr, Nico and Hans von Storch. 2008. "Von der Erforschung und der Politik des Machbaren", in: TU Graz (ed.), *Stadt statt Energie. Urbane Zukunftsszenarien mit wirklicher Nachhaltigkeit. Grazer Architektur Magazine. Band 5.* Wien: Springer VS: 14–21.

Stehr, Nico. 2008. "Ein Gespräch mit Paul F. Lazarsfeld", in: Wolfgang R. Langenbucher (ed.), *Paul Felix Lazarsfeld – Leben und Werk.* Wien: Braumüller: 123–142.

Stehr, Nico and Hans von Storch. 2008. "Feasibility-Based Policies: Investigation and Application", in: TU Graz (ed.), *Urbanity not Energy. Urban Future Scenarios with Genuine Sustainability. Graz Architecture Magazine. Volume 5.* Vienna: Springer VS: 14–21.

Meja, Volker and Nico Stehr. 2008. "Karl Mannheim", in: Stefan Gosepath, Wilfried Hinsch und Beate Rössler (eds.), *Handbuch der politischen Philosophie und Sozialphilosophie.* Berlin: de Gruyter: 763–765.

Adolf, Marian and Nico Stehr. 2009. "Moralisierte Märkte und postrationale Konsumenten", in: Susanne Böhler, Daniel Bongardt and Siegfried Frech (eds.), *Jahrhundertproblem Klimawandel. Forschungsstand, Perspektiven, Lösungswege.* Schwalbach: Wochenschau Verlag: 213–226.

Stehr, Nico. 2009. "Wissenswelten, Governance und Demokratie", in: Sebastian Bolzen, Janetter Hofmann, Sigrid Quack, Gunnar Folke Schuppert und Holger Straßheim (eds.), *Governance als Prozess. Koordinationsformen im Wandel.* Baden-Baden: Nomos: 479–502.

Stehr, Nico. 2009. "Research in Diverse Social Contexts: Tensions, Dynamics and Challenges", in: UNESCO (ed.), *Research in Diverse Social Contexts: Tensions, Dynamics and Challenges.* Paris: United Nations Educational, Scientific and Cultural Organization: 7–19.

Stehr, Nico. 2009. "Nothing Has Been Decided: The Chances and Risks of Feasible Globalization", in: Samir Dasgupta and Jan Nederveen Pieterse (eds.), *Politics of Globalization.* London: Sage: 334–355.

Adolf, Marian and Nico Stehr. 2009. "Die neue Macht der Kreativität. Wissensklassen in modernen Gesellschaften", in: Stephan A. Jansen, Eckard Schröter and Nico Stehr (eds.), *Rationalität der Kreativität? Multidisziplinäre Beiträge zur Analyse der Produktion, Organisation und Bildung von Kreativität.* Wiesbaden: VS Verlag für Sozialwissenschaften: 185–206.

Stehr, Nico. 2009. "Knowledge Worlds, Governance and Democracy", in: Žanetas Ozoliņas and Iveta Reinholdes (eds.), *Laba Pārvaldība.* Riga: Zinātne: 11–49.

Adolf, Marian and Nico Stehr. 2010. "Zur Dynamik des Kapitalismus", in: Ursula Pasero, Karen van den Berg and Alihn Kabalek (eds.), *Capitalism Revisited. Anmerkungen zur Zukunft des Kapitalismus.* Marburg an der Lahn: Metropolis: 15–38.

Stehr, Nico. 2010. "Five Theses on Climate Change", in: Anne Dölemeyer, Janek Zimmer and Gerd Tetzlaff (eds.), *Risk and Planet Earth. Vulnerability, Natural Hazards, Integrated Adaptation Strategies.* Stuttgart: Schweizerbart Science Publishers: 28–31.

Stehr, Nico and Ulrich Ufer. 2010. "Wissen als Ware", in: Stefan Leible, Ansgar Ohly and Herbert Zech (eds.), *Wissen – Märkte – Geistiges Eigentum.* Tübingen: Mohr Siebeck: 51–68.

Stehr, Nico. 2010. "Global Knowledge", in: Peter Meusburger, David Livingstone and Heike Jöns (eds.), *Geographies of Knowledge.* Heidelberg: Springer VS: 23–34.

Adolf, Marian and Nico Stehr. 2010. "Unternehmen, Markt und Moral: Zu einer neuen politischen Ökonomie", in: Ludger Heidbrink and Peter Seele (eds.), *Unternehmertum. Vom Nutzen und Nachteil einer riskanten Lebensform.* Frankfurt am Main: Campus: 178–198.

Stehr, Nico. 2010. "Enabling Knowledge", in: Karel Mueller, Steffen Roth and Milan Zak (eds.), *Social Dimension of Innovation.* Prague: Center for Economic Studies (College of Economics and Management): 1–9.

Stehr, Nico. 2010. "Was ist gesellschaftliche relevante Wissenschaft?", in: Klaus Kornwachs (ed.), *Technologisches Wissen. Entstehung, Methoden, Strukturen.* Wiesbaden: Springer VS: 227–240.

Stehr, Nico. 2010. "The Moralization of the Markets in Europe", in: Sven Eliaeson and Nazdezhda Gorgieva (eds.), *New Europe. Growth to Limits?* Oxford: The Bardwell Press: 149–156.

Stehr, Nico. 2010. "Theories of the Information and Knowledge Society", in: Leonid Melnyk and M.V. Byukhanov (eds.), *Social and Economic Problems of Information Society.* Kiev: Sumy University Books: 620–652.

Stehr, Nico. 2010. "The Concept of Knowledge Societies", in: Barry Smart (ed.), *Post-Industrial Society.* Volume 4. London: Sage: 79–108.

Stehr, Nico. 2010. "The Fragility of the Future", in: Barry Smart (ed.), *Post-Industrial Society.* Volume 4. London: Sage: 419–450.

Adolf, Marian, Jason Mast and Nico Stehr. 2011. "Culture and Cognition: The Foundations of Innovation in Modern Societies", in: Frane Adam and Hans Westlund (eds.), *Socio-Cultural Dimensions of Innovation Performance.* Ljubljana and Stockholm: Institute for Development and Strategic Analysis: 9–22.

Adolf, Marian and Nico Stehr. 2011. "Moralities", in: Dale Southerton (ed.), *Encyclopedia of Consumer Culture.* Thousand Oaks, CA: Sage: 999–1003.

von Storch Hans, Armin Bunde and Nico Stehr. 2011. "The Physical Sciences and Climate Politics", in: John S. Dryzek, Richard B. Norgaard and David Schlosberg (eds.), *Oxford Handbook on Climate Change and Society.* Oxford: Oxford University Press: 113–128.

Adolf, Marian and Nico Stehr. 2011. "Soziale Milieus und Konsum: Die Moralisierung der Märkte und ihre Kritik", in: Ludger Heibrink, Imke Schmidt and Björn Ahaus (eds.), *Die Verantwortung des Konsumenten. Über das Verhältnis von Markt, Moral und Konsum.* Frankfurt am Main: Campus: 245–268.

Stehr, Nico. 2012. "Education, Knowledgeability, and the Labour Market", in: Daniel Tröhler and Ragnhild Barbu (eds.), *The Future of Education Research.* Volume 1. Amsterdam: Sense Publishers: 145–162.

Stehr, Nico. 2012. "Educación, cognoscibilidad y mercado laboral", in: Daniel Tröger and Ragnhild Barbu (eds.), *Los sistemas educativos. Perspectiva histórica, cultural y sociológica.* Barcelona: Ediciones Octaedro: 143–160.

Stehr, Nico and Markus Rhomberg. 2012. "Adaptation or Mitigation. The Approach to Rapid Climate Change", in: Scott G. McNall, George Basile and James Hershauer (eds.), *The Business of Sustainability. Trends, Policies, Practices, and Stories of Success.* Volume 1. New York: Praeger: 175–192.

Stehr, Nico. 2012. "Knowledge", in: George Ritzer (ed.), *Encyclopedia on Globalization.* Volume III. Oxford: Wiley-Blackwell: 1237–1240.

Stehr, Nico. 2012. "Knowledge Societies", in: George Ritzer (ed.), *Encyclopedia on Globalization.* Volume III. Oxford: Wiley-Blackwell: 1240–1244.

Stehr, Nico and Markus Rhomberg. 2012. "Environmental Change", in: George Ritzer (ed.), *Encyclopedia on Globalization.* Volume II. Oxford: Wiley-Blackwell: 519–522.

Mast, Jason and Nico Stehr. 2012. "Knowledge Economy", in: David Rooney, Greg Hearn and Tim Kastelle (eds.), *Handbook on the Knowledge Economy.* Volume 2. Cheltenham, UK: Edward Elgar: 17–37.

Stehr, Nico and Bernd Weiler. 2012. "Social Transformations and Knowledge", in: Arnaud Sales (ed.), *Sociology Today.* London: Sage: 17–37.

Stehr, Nico. 2012. "The Knowledge of Civil Society", in: Ortwin Renn, André Reichel and Joa Bauer (eds.), *Civil Society for Sustainability: A Guidebook for Connecting Science and Society.* Bremen: Europäischer Hochschulverlag: 37–55.

Adolf, Marian and Nico Stehr. 2012. "Kommunikation. Macht. Öffentlichkeit. Zu zeitgenössischen Formen der Kommunikation", in: Stephan A. Jansen, Eckard Schröter and Nico Stehr (eds.), *Bürger. Macht. Staat?* Wiesbaden: Springer VS: 37–51.

Grundmann, Reiner, Markus Rhomberg and Nico Stehr. 2012. "Der Klimawandel und die Rolle der Sozialwissenschaften", in: Heike Eigner and Martin Schmid (eds.), *Jenseits traditioneller Wissenschaft? Zur Rolle von Wissenschaft in einer vorsorgenden Gesellschaft.* München: Oekom: 95–110.

Grundmann, Reiner, Markus Rhomberg and Nico Stehr. 2012. "Nature, Climate Change and the Culture of the Social Sciences", in: Pernille Almlund, Per Homann Jespersen and Søren Riis (eds.), *Rethinking Climate Change Research. Clean-Technology, Culture and Communication.* London: Ashgate: 133–142.

Adolf, Marian and Nico Stehr. 2012. "The Power of the Practice of Consumption: Markets, Morals, Politics", in: George O. Tsobanoglou (ed.), *The Politics of Participation and Empowerment. Current Issues and Practices.* Hildesheim: Verlag für Gesellschaftsarchitektur: 9–22.

Stehr, Nico. 2012. "Democracy and Knowledge", in: Fatos Tarifa (ed.), *Dija, universiteti dhe demokracia.* Tiranë: Ombra GVG: 63–90.

Adolf, Marian, Jason Mast and Nico Stehr. 2013. "Culture and Cognition: The Foundations of Innovation in Modern Society", in: Frane Adam und Hans Westlund (eds.), *Innovation in Socio-cultural Context.* London: Routledge: 25–39.

Adolf, Marian, Jason Mast and Nico Stehr. 2013. "Knowledge Society, Knowledge-Based Economy, and Innovation", in: Elias G. Carayannia (ed.), *Encyclopedia of Creativity, Invention, Innovation, and Entrepreneurship*. Heidelberg: Springer Science + Business Media LLC: 1186–1190.

Stehr, Nico. 2013. "Knowledge Society", in: Byron Kaldis (ed.), *Encyclopedia of Philosophy and the Social Sciences*. Volume 1. London: Sage: 520–522.

Mast, Jason and Nico Stehr. 2014. "Modernity and Post-modernity", in: Samir Daspupta and Peter Kivisto (eds.), *Postmodernism in a Global Perspective*. Los Angeles: Sage: 116–132.

Rhomberg, Markus and Nico Stehr. 2014. "Experts", in: George Ritzer (ed.), *Blackwell Encyclopedia of Sociology*. Malden, MA: Blackwell Publishing.

Stehr, Nico and Markus Rhomberg. 2014. "Environmental Change", in: George Ritzer (ed.), *Blackwell Encyclopedia of Sociology*. Malden, MA: Blackwell Publishing.

Moldaschl, Manfred and Nico Stehr. 2014. "Woran scheitert der globale Klimaschutz? Postdemokratie und andere Erklärungsmodelle", in: Robert Pfaller and Klaus Kufeld (eds.), *Arkadien oder Dschungelcamp. Leben im Einklang oder Kampf mit der Natur?* Freiburg und München: Karl Alber: 248–275.

Stehr, Nico. 2014. "Knowledge Societies", in: George Ritzer (ed.), *Wiley Blackwell Encyclopedia of Sociology*. Malden, MA: Blackwell Publishing.

Stehr, Nico. 2014. "Practical Knowledge", in: George Ritzer (ed.), *Wiley Blackwell Encyclopedia of Sociology*. Malden, MA: Blackwell Publishing.

Adolf, Marian and Nico Stehr. 2014. "Moralizace trhů: spotřeba v modern společnosti", in: Pavel Zahradka (ed.), *Spotrebni kultura: historie, teorie a vyzkum ("Consumer Culture: History, Theory and Research")*. Prague: Nakladatelství Academia: 402–418.

von Storch, Hans and Nico Stehr. 2014. "Dem Klima ausgeliefert?", in: Johannes Ebert and Andrea Zelle (eds.), *Klima Kunst Kultur. Der Klimawandel in Kunst und Kulturwissenschaften*. München: Steil Verlag: 37–42.

Grundmann, Reiner and Nico Stehr. 2014. "Social Science and the Absence of Nature: Uncertainty and the Reality of Extremes", in: Steve Rayner and Mark Caine (eds.), *The Hartwell Approach to Climate Policy*. London: Routledge: 130–146.

Adolf, Marian and Nico Stehr. 2014. "Der Konsum der Verbraucher", in: Heribert Mefferrt, Peter Kenning and Manfred Kirchgeorg (eds.), *Sustainable Marketing Management. Grundlagen und Cases*. Wiebaden: Springer Fachmedien: 55–70.

Stehr, Nico and Markus Rhomberg. 2014. "Environmental Change", in: George Ritzer (ed.), *The Wiley-Blackwell Encyclopedia of Globalization*. Malden, MA: Blackwell Publishing.

Adolf, Marian and Nico Stehr. 2015. "Die soziale Umwelt des Unternehmens. Von Konsumenten, Motiven und Kultur", in: Michael Hüther, Knut Bergman and Dominik H. Enste (eds.), *Unternehmen im öffentlichen Raum: Zwischen Marktorientierung und Mitverantwortung*. Wiesbanden: Springer VS: 171–184.

Stehr, Nico. 2015. "Knowledge Society, History of", in: James D. Wright (ed.), *International Encyclopedia of the Social & Behavioral Sciences.* 2nd edition, Vol 13. Oxford: Elsevier: 105–110.

Stehr, Nico. 2015. "Die Freiheit ist eine Tochter des Wissens", in: Gernot Böhme and Ute Gahlings (eds.), *Wie lebt es sich in unserer Gesellschaft?* Bielefeld: Aisthesis Verlag: 203–215.

Machin, Amanda and Nico Stehr. 2015. "Die politische Macht des Klimas", in: Heinrich-Böll-Stiftung (ed.), *Ökologie und Freiheit. Böll.Thema.* Berlin: Heinrich-Böll-Stiftung: 8–9.

Stehr, Nico. 2015. "Enabling Knowledge", in: Peter Meusburger, Derek Gregory and Laura Suarsana (eds.), *Geographies of Knowledge and Power.* Wiesbaden: Springer VS: 75–88.

Stehr, Nico. 2017. "Knowing and Not Knowing", in: Peter Meusburger, Benno Werlen and Laura Suarsana (eds.), *Knowledge and Action. Knowledge and Space 9.* Heidelberg: Springer VS: 113–126.

4.4 Refereed Articles

Stehr, Nico. 1968. "Status Consistency: The Theoretical Concept and Its Empirical Referent", *Pacific Sociological Review,* 11: 95–99.

Stehr, Nico. 1971. "Statuskonsistenz", *Kölner Zeitschrift für Soziologie und Sozialpsychologie,* 23: 34–54.

Stehr, Nico and Lyle E. Larson. 1972. "The Rise and Decline of Areas of Specialization", *The American Sociologist,* 7: 3, 5–6.

Larson, Lyle E. and Nico Stehr. 1973. "Family Sociology as an Area of Specialization", *International Journal of Family Sociology,* 3: 198–206.

Krishnan, P. and Nico Stehr. 1973. "A Statistic for Random Model Testing", *Quality and Quantity,* 7: 197–202.

Stehr, Nico. 1974. "Paradigmatic Crystallization: Patterns of Interrelation Among Areas of Competence in Sociology", *Social Science Information,* 13: 119–137.

Stehr, Nico. 1974. "Die Technik in der öffentlichen Meinung", *Philosophia Naturalis,* 15: 79–89.

Stehr, Nico. 1974. "Consensus and Dissensus in Occupational Prestige", *British Journal of Sociology,* 25: 410–427.

Stehr, Nico and William Avison. 1974. "Forms of Competition and Sociological Knowledge in Organized American Sociology", *Archives for Philosophy of Law and Social Philosophy,* 70: 213–229.

Stehr, Nico. 1974. "Ascriptive Career Contingencies for Sociologists: A Longitudinal Analysis", *The American Sociologist,* 9: 206–211.

Stehr, Nico. 1975. "Factors in the Development of Multi-paradigm Disciplines: The Case of Sociology", *Journal for the History of the Behavioral Sciences,* 11: 172–197.

Stehr, Nico. 1976. "Sozialforschung und Gesellschaft: Ein Gespräch mit Paul F. Lazarsfeld", *Kölner Zeitschrift für Soziologie und Sozialpsychologie,* 28: 796–807.
Stehr, Nico. 1978. "Man and the Environment: A General Perspective", *Archives for Philosophy of Law and Social Philosophy,* 74: 1–17.
Stehr, Nico and Anthony Simmons. 1979. "The Diversity of Modes of Discourse and the Development of Sociological Knowledge", *Zeitschrift für allgemeine Wissenschaftstheorie,* 10: 141–161.
Stehr, Nico. 1979. "Sociological Languages. Comment on Riggs", *The American Sociologist,* 14: 205–206.
Stehr, Nico and Anthony Simmons. 1979. "The Diversity of Discourse", *Society,* 16: 45–49.
Kettler, David, Volker Meja and Nico Stehr. 1980. "Karl Mannheim's Sociological Theory of Culture", *Canadian Journal of Sociology,* 5: 405–411.
Simmons, Anthony and Nico Stehr. 1981. "Language and the Growth of Knowledge in Sociology", *Social Science Information,* 20: 703–741.
Stehr, Nico. 1981. "The Magic Triangle: In Defense of the Strong Programme of the Sociology of Knowledge", *Philosophy of the Social Sciences,* 11: 225–229.
Stehr, Nico. 1982. "The Evolution of Meaning Systems: A Discussion with Niklas Luhmann", *Theory, Culture and Society,* 1: 33–48.
Stehr, Nico. 1982. "Sociological Languages", *Philosophy of the Social Sciences,* 12: 47–57.
Stehr, Nico. 1982. "A Conversation with Paul F. Lazarsfeld", *American Sociologist,* 17: 150–155.
Stehr, Nico. 1982. "Zwischen Resignation und Hoffnung: Erkenntnisfortschritt in der Soziologie", *Philosophische Rundschau,* 20: 178–187.
Stehr, Nico and Volker Meja. 1982. "The Classical Sociology of Knowledge Revisited", *Knowledge: Creation, Diffusion, Utilization,* 4: 33–50.
Kettler, David, Volker Meja and Nico Stehr. 1983. "Arguing for Democracy", *Times Higher Educational Supplement,* July 15.
Kettler, David, Volker Meja and Nico Stehr. 1983. "Politik als Wissenschaft. Über Theorie und Praxis bei Karl Mannheim", *Angewandte Sozialforschung,* 11: 403–417.
Kettler, David, Volker Meja and Nico Stehr. 1984. "Karl Mannheim and Conservatism", *American Sociological Review,* 49: 71–85.
Meja, Volker and Nico Stehr. 1985. "Sozialwissenschaftlicher und erkenntnistheoretischer Diskurs: Das Problem des Relativismus", *Soziale Welt,* 36: 261–270.
Böhme, Gernot and Nico Stehr. 1986. "Key to the Black Box", *Times Higher Educational Supplement,* August 1.
Stehr, Nico. 1986. "Kopfarbeit", *Soziologische Revue,* 9: 133–138.
Kettler, David, Volker Meja and Nico Stehr. 1987. "Schattenseiten einer erfolgreichen Emigration: Karl Mannheim im englischen Exil", *Exilforschung,* 5: 170–19.
Kettler, David, Volker Meja and Nico Stehr. 1987. "La Scienzia Politica di Karl Mannheim", *Mondoperaio,* 40: 76–81.

Kettler, David, Nico Stehr and Volker Meja. 1987. "Is a Science of Politics Possible? The View from Mannheim", *Society,* 24: 76–82.

Kettler, David, Volker Meja and Nico Stehr. 1987. "Karl Mannheim et 'Le Conservatisme'", *Cahiers internationaux de Sociologie,* 83: 247–256.

Meja, Volker and Nico Stehr. 1988. "Social Science, Epistemology, and the Problem of Relativism", *Social Epistemology,* 2: 263–271.

Kettler, David, Volker Meja and Nico Stehr. 1988. "Rationalizzarea l'irrationale: Karl Mannheim e il vizio degli intellectuali tedeschi", *Rassagna Italiana di Sociologia,* 29: 487–512.

Stehr, Nico. 1989. "Von den Tugenden sozialwissenschaftlichen Wissens: Max Weber und der evangelisch-soziale Kongress", *Sociologia Internationalis,* 27: 129–147.

Stehr, Nico. 1990. "L'autorité du discours et le discours de l'autorité. Les spécialistes des sciences sociales, la question sociale et la religion dans la société industrielle", *Sociétés,* 29: 45–64.

Kettler, David, Volker Meja and Nico Stehr. 1990. "Karl Mannheim und die Entmutigung der Intelligenz", *Zeitschrift für Soziologie,* 19: 117–130.

Böhme, Gernot and Nico Stehr. 1990. "Wissensgesellschaften", *Universitas,* 45: 225–231.

Kettler, David, Volker Meja and Nico Stehr. 1990. "Rationalizing the Irrational: Karl Mannheim and the Besetting sin of German Intellectuals", *American Journal of Sociology,* 95: 1141–1173.

Stehr, Nico. 1991. "The Power of Scientific Knowledge—And Its Limits", *Canadian Review of Sociology and Anthropology,* 29: 460–482.

Böhme, Gernot and Nico Stehr. 1992. "Spoleczenstwo nauki", *Pace Instytutu Nauk Ekonomicczno-Spolecznych.* Warszawa: Politechniki Warszawskiej.

Messner, Wayne, Dennis Bray, Guy C. Germain and Nico Stehr. 1993. "Climate Change and Social Order: Knowledge for Action?", *Knowledge and Policy,* 5: 82–100.

Stehr, Nico. 1994. "Om gränserna för den vetenskapliga kunskapens makt", *Vest (Tidskrift för Vetenskapsstudier),* 7: 37–64.

Stehr, Nico. 1994. "The Culture and Structure of Social Inequality", *International Journal of Group Tensions,* 24: 361–382.

Stehr, Nico. 1994. "Max Weber and the Lutheran Social Congress: The Authority of Discourse and the Discourse of Authority", *History of the Human Sciences,* 7: 21–39.

Stehr, Nico and Hans von Storch. 1995. "The Social Construct of Climate and Climate Policy", *Delta,* 5: 8–14.

Stehr, Nico and Hans von Storch. 1995. "The Social Construct of Climate and Climate Change", *Climate Research,* 5: 99–105.

Stehr, Nico. 1995. "Wissensberufe", *Vorgänge,* 34: 38–49.

Stehr, Nico, Hans von Storch and Moritz Flügel. 1995. "The 19th Century Discussion of Climate Variability and Climate Change: Analogies for the Present Debate?", *World Resources Review,* 7: 589–604.

Stehr, Nico. 1996. "The Ubiquity of Nature: Climate and Culture", *Journal for the History of the Behavioral Sciences*, 32: 151–159.
Stehr, Nico. 1996. "The Salt of Social Science", *Sociological Research Online*, 1.
Stehr, Nico and Volker Meja. 1996. "Die Zerbrechlichkeit der modernen Gesellschaft", *Vorgänge*, 35: 13-21.
von Storch, Hans and Nico Stehr. 1997. "Climate Research: The Case for the Social Sciences", *Ambio*, 26: 66–71.
Stehr, Nico and Hans von Storch. 1997. "Rückkehr des Klimadeterminismus?", *Merkur*, 51: 560–562.
Stehr, Nico. 1997. "Trust and Climate", *Climate Research*, 8: 163–169.
Stehr, Nico. 1997. "Los límites de lo posible: modernidad y postmodernidad", *Politica y Sociadad*, 24: 23–31.
Stehr, Nico. 1997. "Un sapere per la dimensione practica", *Rassegna Italiana di Sociologia*, 38: 559–600.
Grundmann, Reiner and Nico Stehr. 1997. "Klima und Gesellschaft. Soziologische Klassiker und Außenseiter. Über Weber, Durkheim, Simmel und Sombart", *Soziale Welt*, 48: 85–100.
Stehr, Nico. 1997. "Les limites du possibles: La postmodernité et les sociétés du savoir", *Sociétiés*, 58: 101–124.
Stehr, Nico. 1998. "Interdisziplinäres Arbeiten – in einer disziplinär bestimmten Welt", *SWS-Rundschau*, 37: 441–451.
Stehr, Nico. 1998. "The University in Knowledge Societies", *Social Epistemology*, 12: 33–42.
Stehr, Nico and Hans von Storch. 1998. "Soziale Naturwissenschaft. Oder: Die Zukunft der Wissenschaftskulturen", *Vorgänge*, 37: 8–12.
Stehr, Nico. 1998. "Società moderna, disuguaglianza e conoscenza: dalla disuguaglianza basata sulla classe alla disuguaglianza basat sulla conoscenza", *Studi di Sociologia*, 36: 215–231.
Weinstein, Jay and Nico Stehr. 1999. "The Power of Knowledge: Race Science, Race Policy, and the Holocaust", *Social Epistemology*, 13: 3–36.
Stehr, Nico. 1999. "The Future of Inequality", *Society*, 36: 54–59.
Stehr, Nico. 1999. "Interessen und Erkenntnis", *Merkur*, 599: 167–170.
Stehr, Nico. 2001. "Le savoir en tant que pouvoir d'action", *Sociologie & Sociétiés*, 32: 157–170.
Stehr, Nico. 2000. "Deciphering Information Technologies. Modern Societies as Networks", *European Journal of Social Theory*, 3: 84–93.
Grundmann, Reiner and Nico Stehr. 2000. "Social Science and the Absence of Nature", *Social Science Information*, 39: 155–179.
Stehr, Nico. 2000. "Da desigualdade de classe à desigualdade de conhecimento", *Revista Brasileira de Ciências Sociais*, 15: 101–112.
von Storch, Hans and Nico Stehr. 2000. "Climate Change in Perspective. Our Concerns About Global Warming Have an Age-Old Resonance", *Nature*, 405: 615.

Bechmann, Gotthard and Nico Stehr. 2000. "Risikokommunikation und die Risiken der Kommunikation wissenschaftlichen Wissens: Zum gesellschaftlichen Umgang mit Nichtwissen", *Gaia,* 9: 113–201.

Stehr, Nico and Hans von Storch. 2000. "Kultur und Klima", *Vorgänge,* 152: 100–104.

Stehr, Nico. 2000. "Die Kultur der Wissensgesellschaften", *Forum TTN,* 3: 3–14.

Stehr, Nico. 2000. "Warum es SO schwierig ist interdisziplinär zu sein: Von der Zukunft der Wissenschaftskulturen und den Bedingungen der Transdisziplinarität in den Wissenschaften", *Forum Forschung,* 100–102.

Stehr, Nico and Hans von Storch. 2000. "Von der Macht des Klimas. Ist der Klimadeterminismus nur noch Ideengeschichte oder relevanter Faktor gegenwärtiger Klimapolitik?", *Gaia,* 9: 187–195.

Grundmann, Reiner and Nico Stehr. 2001. "Sombart's Fate: From Fame to (Near) Oblivion", *Classical Sociology,* 1: 257–287.

Stehr, Nico and Reiner Grundmann. 2001. "The Authority of Complexity", *The British Journal of Sociology,* 52: 313–329.

Stehr, Nico. 2001. "A World Made of Knowledge", *Society,* 39: 89–92.

Stehr, Nico. 2001. "Economy and Ecology in an Era of Knowledge-Based Economies", *Current Sociology,* 49: 67–90.

Stehr, Nico. 2001. "Politische Arithmetik", *Merkur,* 265: 453–456.

Bechmann, Gotthard and Nico Stehr. 2001. "Niklas Luhmann", *Tempo Social,* 13: 185–200.

Stehr, Nico. 2001. "Moderne Wissensgesellschaften", *Aus Politik und Zeitgeschichte,* 36: 7–14.

Bechmann, Gotthard and Nico Stehr. 2002. "The Legacy of Niklas Luhmann", *Society,* 39: 67–75.

Stehr, Nico. 2002. "A tudástársadalmak", *Információs Társadalom* 2: 5–22.

Stehr, Nico and Hans von Storch. 2002. "Micro/Macro and Soft/Hard: Diverging and Converging Issues in the Physical and Social Sciences", *Integrated Assessment,* 3: 115–121.

Stehr, Nico. 2002. "Knowledge World", *Sotsiologicheskiy Zhurnal,* 2: 31–35.

Stehr, Nico. 2003. "Le società moderne come società della conoscenza", *Quaderni di Teoria Sociale,* 3: 267–276.

Bechmann, Gotthard and Nico Stehr. 2003. "Niklas Luhmann", *Convergencia: Revista de Ciencias Sociales,* 10: 256–279.

von Storch, Hans and Nico Stehr. 2003. "Effiziente Kommunikation", *Universitas,* 684: 608–614.

Grundmann, Reiner and Nico Stehr. 2003. "Social Control and Knowledge in Democratic Societies", *Science and Public Policy,* 30: 183–188.

Stehr, Nico. 2003. "The Social and Political Control of Knowledge in Modern Societies", *International Social Science Journal,* 55: 643–655.

Stehr, Nico. 2003. "Wissen als Handlungsvermögen", *Tumult. Schriften zur Verkehrswissenschaft,* 27: 64–68.

Stehr, Nico. 2003. "Die gesellschaftliche Kontrolle neuer Erkenntnisse", *Sociologica Internationalis,* 41: 113–138.

Stehr, Nico. 2004. "En värld as kunskap", *Utbildung & Demokrati,* 13: 11–20.

Stehr, Nico. 2004. "Knowledge, Markets and Biotechnology", *Social Epistemology,* 18: 301–314.

Stehr, Nico and Hermann Strasser. 2005. "Die Grenzen von McWorld", *Denken + Glauben,* 135: 20–22.

Pielke, Roger, Shardul Agrawala, Laurens Bouwer, Ian Burton, Stanley Changnon, Michael Glantz, William Hooke, Richard Klein, Kenneth Kunkel, Dennis Mileti, Daniel Sarewitz, Emma Thompkins, Nico Stehr and Hans von Storch. 2005. "Clarifying the Attribution of Recent Disaster Losses: A Response to Epstein and McCarthy", *Bulletin of the American Meteorological Society,* 86: 1481–1483.

von Storch, Hans and Nico Stehr. 2006. "Anthropogenic Climate Change: A Reason for Concern since the 18th Century and Earlier", *Geografiska Annaler,* 88: 107–113.

Stehr, Nico. 2007. "Societal Transformations, Globalization and the Knowledge Society", *International Journal of Knowledge and Learning,* 3: 139–153.

Stehr, Nico. 2008. "The Moralization of the Markets in Europe", *Society,* 45: 62–67.

Stehr, Nico. 2008. "Il ruolo dei consumatori nella moralizzazione dei mercati", *Sociologia del Lavoro,* 108: 43–54.

Stehr, Nico. 2008. "Globale Wissenswelten oder grenzenlose Erkenntnisse", *Rechtstheorie,* 39: 301–327.

Stehr, Nico. 2008. "Liberdade é filha do conhecimneto?", *Tempo Social,* 20: 221–234.

Stehr, Nico and Hans von Storch. 2008. "Anpassung und Vermeidung oder von der Illusion der Differenz", *Gaia,* 17: 270–273.

Stehr, Nico and Hans von Storch. 2009. "Climate Protection", *Journal of Consumer Protection and Food Safety,* 4: 1–5.

Stehr, Nico. 2009. "Useful Scientific Knowledge: What Is Relevant Science for Society?", *Applied Social Science,* 3: 18–26.

Stehr, Nico. 2009. "Damit wir wissen, was wir tun: Warum Deutschland in Bildung und Ausbildung investieren muss", *Internationale Politik,* 64: 18–23.

Stehr, Nico. 2009. "What Is Socially Relevant Science?", *Society,* 46: 262–266.

Stehr, Nico. 2009. "Enabling Knowledge", *Sociological Analysis,* 3: 69–80.

Stehr, Nico and Ulrich Ufer. 2009. "On the Global Distribution and Dissemination of Knowledge", *International Social Science Journal,* 60: 7–24.

von Storch, Hans and Nico Stehr. 2010. "Klimaforschung und Klimapolitik – Rollenverteilung und Nachhaltigkeit", *Naturwissenschaftliche Rundschau,* 63: 301–307.

Stehr, Nico and Marian Adolf. 2010. "Consumption Between Markets and Morals: A Socio-cultural Consideration of Moralized Markets", *European Journal of Social Theory,* 13: 213–228.

Stehr, Nico. 2010. "Knowledge, Democracy and Power", *Central European Journal of Public Policy,* 4: 14–35.

Grundmann, Reiner and Nico Stehr. 2010. "Climate Change: What Role for Sociology?", *Current Sociology,* 58: 897–910.

Adolf, Marian and Nico Stehr. 2010. "Die Macht der neuen Öffentlichkeit. Die Konstitution neuer Öffentlichkeit zwischen Internet und Straße", *Vorgänge,* 192: 4–15.

Stehr, Nico and Jason Mast. 2011. "The Modern Slaves: Specialized Knowledge and Democratic Governance", *Society,* 48: 36–40.

von Storch, Hans, Insa Meinke, Nico Stehr, Beate Ratter, Werner Krauss, Roger A. Pielke Jr., Reiner Grundmann, Marcus Reckermann, Ralf Weisse. 2010. "Regional Climate Services—Ilustrated with Experiences from Northern Europe", *Zeitschrift für Umweltpolitik & Umweltrecht,* 1: 1–15.

Bechmann, Gotthard and Nico Stehr. 2011. "Niklas Luhmann's Theory of the Mass Media", *Society,* 48: 62–67.

Grundmann, Reiner and Nico Stehr. 2012. "How Does Knowledge Relate to Political Action?", *Innovation: The European Journal of Social Science Research,* 25: 29–44.

Stehr, Nico. 2012. "Knowledge and Non-knowledge", *Science, Technology & Innovation Studies* 8: 313.

Stehr, Nico. 2013. "An Inconvenient Democracy: Knowledge and Climate Change", *Society,* 50: 55–60.

Stehr, Nico. 2013. "Mut zur Lücke. Zur Emanzipation des Nichtwissens in der modernen Gesellschaft", *Kursbuch,* 173: 164–178.

Stehr, Nico. 2013. "Wissen und der Mythos vom Nichtwissen", *Politik und Zeitgeschichte,* 63: 48–54.

Adolf, Marian, Jason Mast and Nico Stehr. 2013. "The Foundations of Innovation in Modern Societies: The Displacement of Concepts and Knowledgeability", *Mind & Society,* 12: 11–22.

Stehr, Nico. 2015. "Democracy is not an Inconvenience", *Nature*, 525: 449–450.

Stehr, Nico. 2016. "Exceptional Circumstances. Does Climate Change Trump Democracy?", *Issues in Science and Technology,* 32: 37–44.

Stehr, Nico and Amanda Machin. 2016. "Trusting the Climate: Catastrophe vs. Stability", *Society,* 53: 573–580.

Stehr, Nico and Marian Adolf. 2016. "The Price of Knowledge", *Social Epistemology*, 30: 483–512.

Stehr, Nico. 2016. "Too Much Democracy?", *Issues in Science and Technology,* 33: 5–7.

Stehr, Nico and Alexander Ruser. 2016. "Social Scientists as Technicians, Advisors and Meaning Producers", *Innovation: The European Journal of Social Science Research,* 30: 24–35.

Nico Stehr (in Friedrichshafen, Germany, 2005). *Source* Zeppelin University

Part II
Knowledge, Science, Society: Nico Stehr's Work from the Perspective of Colleagues

Chapter 5
The Knowledge Society: Remarks on the State of Affairs on the Occasion of Nico Stehr's 70th Birthday

Gernot Böhme

5.1 Introduction

Your birthday, Nico, is first of all an opportunity to express my gratitude: my gratitude for an academic friendship, my gratitude for a solidly united collaboration that brought us together time and again, throughout our research careers, around a common topic: the topic of knowledge and society. The first occasion was the special issue of the *Kölner Zeitschrift für Soziologie und Sozialpsychologie*, dedicated to the sociology of science and edited by you and René König in 1975; followed by the special issue of *Wissenssoziologie*, edited by you and Volker Meja in 1980.[1]

The really fruitful aspect of our collaboration consisted in the fact that the two of us, a sociologist and a philosopher, met upon the common ground of *a single* issue; that is, the social significance of knowledge. So it was that for me as a philosopher, society did not remain simply a mere marginal constraint on knowledge, its creation and effect; and so it was that for you as a sociologist, knowledge and science did not remain simply a black box whose functioning a social scientist had no choice but to accept. The high point of our cooperation was therefore the International Conference on the Knowledge Society in Darmstadt in 1984, whose results were published as a volume of the *Sociology of the Sciences Yearbook* (Böhme/Stehr 1986). With this volume we brought into the debate a concept that at first seemed alien within the scientific community; and yet, in the long term, it has had an

Prof. Dr. Gernot Böhme, Professor Emeritus, TU Darmstadt; Email: re.boehme@t-online.de.

[1] Translated from German by Paul Malone.

M. T. Adolf (ed.), *Nico Stehr: Pioneer in the Theory of Society and Knowledge*, Pioneers in Arts, Humanities, Science, Engineering, Practice 16,
https://doi.org/10.1007/978-3-319-76995-0_5

extraordinary career. Thus, for example, the concept of the knowledge society served as a keynote of state policy in the policies of Jürgen Rüttgers, Minister for Education, Science, Research and Technology in Helmut Kohl's cabinet between 1994 and 1998. Also, to name only one further example, Peter Weingart's graduate school in Bielefeld was run for years under the title of *Wissensgesellschaft*—knowledge society.

In the beginning, however, we met with incomprehension and even resistance, or suffered mockery in our respective scientific communities. For the main concepts that scholars relied on to analyze the obviously urgent problem of the relationship between knowledge and society were those of the *information society* (Webster 1995) or the *scientific society* (Kreibich 1986); or else, following Daniel Bell, they attempted to conceive theoretical knowledge within the framework of the post-industrial society as the new axial principle of society (Bell 1974); or else, again, followed the Prague Spring theorists in linking it, under the label of scientific-technological revolution, to the hope for a new, more humane society in the form of democratic socialism (Richta 1968).

The distinctive feature of the concept of knowledge society that we introduced into these debates was the perspective of the sociology of knowledge. First of all, this perspective takes account of an entire spectrum of types of knowledge (Karl Mannheim) as well as bearers of knowledge, not merely individuals but rather social circles (*Man of Knowledge*, Znaniecki 1975). By this means, we had also envisaged the possibility of linking the hierarchy of forms of knowledge to a hierarchy of bearers of knowledge in society. Against this shared background, the two of us viewed the social significance of knowledge somewhat differently. Whereas you, Nico Stehr, understood knowledge as the capacity to take social action, I, Gernot Böhme, conceived of knowledge as participation in the cultural capital of a society. Thus, I was even then allowing for the fact that knowledge in advanced societies is for the most part already symbolically represented, so that as a rule it is already second-degree knowledge—that is, not participation in the object itself but rather participation in the stock of knowledge already laid up in a society (see below).

Nico Stehr went on to investigate, in particular, the relationship among knowledge bearers, exemplified by the relationship between experts and laypeople (Stehr/Grundmann 2010), and the particular fragility of societies in which knowledge has become a constitutive impetus (Stehr 2000). In contrast, diverted by other questions, I took my leave in 1996 with my essay *Structures and Prospects of the Knowledge Society* (Böhme 1997: 447–468). In this essay, I claimed that the influence of knowledge in constituting society had already passed its peak and that the concept of the knowledge society therefore no longer described a new type of society and the hopes associated with it but represented a perspective from which certain current social problems could be articulated. In this sense and from this viewpoint, I would today again like to enumerate several contemporary problems, established by the concept of the knowledge society: What are the main issues today?

5.2 The Ongoing Self-appropriation of Society

The term "self-appropriation of society" here refers to what has traditionally been called statistics; that is, the comprehensive analysis—in the form of knowledge—of social relationships, carried out as a rule by the state administration. The oldest form of this self-appropriation of society is the census. In principle, such an enumeration can always be broadly extended beyond the mere recording of citizens' civil status within society—for example, when a person's employment, origin, place of residence and current location are also collected. In 1987, an attempt to do this had, of course, already led to widespread social protest in the Federal Republic of Germany, with the result that the census was abandoned altogether because it was felt to be an intrusion into the private sphere, and most particularly insofar as it made manifest the right of unlimited residence, a sensitive issue from the standpoint of migration and asylum. Since that time, many of the problems that originally prompted this general protest have one by one become acute again due to the creation of more specific social databases for the purpose of, for example, crime prevention and counter-terrorism, and even labor administration. These are the problems of data privacy, or of the right to *informational self-determination*, a right specifically conceived to protect the individual against the misuse of personal data.

This type of social self-analysis does not always need to be carried out at the behest of the state administration. In fact, the databases compiled by the financial sector and the insurance companies (Schufa, Germany's General Credit Protection Agency), for example, which even allow them to determine a person's financial solvency, are of comparable significance. Most of all, however, this social self-appropriation proceeds via internet-based systems or the companies behind these systems, such as Google, Facebook and so on. Thus, the individual human being as a social person is consolidated into behavioral models and user profiles by means of which he can then be further defined and even constrained in his social behavior—as it happens, for instance, in the context of job applications. While these systems are primarily focused on the statements, activities and characteristics of individual persons, systems such as Google Earth and GPS go beyond this, surveilling people's movements and determining the location of individuals and objects. By now this happens not only when the person in question wants to remain in one place or travel between places but rather as a consequence of other, quite different activities because these systems enable a person's current location to be established via the GPS-mediated APS on smart phones.

In summa: What was originally a specific state administration activity under the label of statistics—that is, taking stock of society and its activities in the form of data—is today increasingly taking place in an automated mode and in real time, as it were, because these activities are themselves largely mediated via the Internet. The result is a problem the full scale of which can hardly be understood and, likewise, hardly be managed by legal and political means: the problem of continuing to politically ensure the personal freedom and privacy guaranteed by the German Basic Law.

5.3 Knowledge as Participation in the Cultural Capital of a Society

Since its emergence in the late seventeenth century it has been a basic maxim of modern science—in contrast to the traditional hermetic sciences—that its findings are to be regarded as a public good to be placed at the disposal of the public for the general benefit of society (Francis Bacon).[2]

This principle, however, was from the very beginning in conflict with the authorial principle, namely the idea that the discoverer or creator of new knowledge possesses a certain personal right to its products. This right could initially be protected within the scientific community by means of the obligation to cite one's sources; in the course of the following centuries, however, during which the economic utility of knowledge continued to increase, this right had to be secured, step by step, by means of copyright and patent regulations. Thereafter, for a long time—today we must say: temporarily—this system remained stable, ensuring a balance between the public interest in making use of knowledge and the property rights of the creator. Here it must be emphasized that this stability has always been no more than artificial; that is, it could only be secured by means of social norms, particularly laws. For in essence knowledge, or cultural products such as music, pictorial representations or literature, cannot be privatized. A scientific finding, once published, is a published public good in the form of a printed text, in principle at anyone's disposal. A picture, as an image, is in principle accessible to any observer even if the painted canvas itself is in private hands; and literature and music can likewise be *appropriated* by any reader and any listener.

This public access to cultural goods, however, has only been fully realized in our own time, that is, by means of the Internet. In fact, it is now hardly possible to maintain the balance between the contents of cultural goods as a public commons and the private interest of authors or publishers in exploiting the use of these goods.

Therefore, the contradiction between knowledge as a public good and knowledge as private property has today become a political question of unparalleled importance. This is a typical problem of the knowledge society, and so it is no wonder that in such a society you can form a political party whose program calls for the Internet community to give open access to all cultural goods—from scientific findings to films to pieces of music (Piratenpartei Deutschland 2016).[3]

> The demand for open access arises within the scientific system itself in a particular form. In this context, this demand is aimed against the private use by huge, privately-owned scientific publishers of scientific results that have been financed by means of public funding and produced by the work of individual scientists and their teams. For individual scientists themselves, so long as they work in research institutions, these publications are freely accessible only on the basis of licensing fees paid by these institutions. Before the internet era, however, these journals had also been stored in public libraries, available to freelance

[2] On the establishment of the obligation to publish scientific results, see Kronick (1962).

[3] On the Pirate Party, see http://wiki.piratenpartei.de/Parteiprogramm (6.5.2012).

> scientists and indeed to anyone at all. This is no longer the case in the current system. The result is once again a typical problem of the knowledge society: due to the extraordinary significance of scientific knowledge for social development, scientific results ought to be accessible to everyone, particularly to policymakers and active interest groups. (Jha 2012)[4]

Besides copyright, scientific findings and constructs as well as technological solutions are also protected by patents, and thus available for private use. But for quite some time now problems have arisen even at this stage and have led to political protests, namely in those cases that have to do with knowledge in the realm of life itself or, more specifically, with the artificial alteration and/or production of organisms and human tissue. In this context, groups ranging from privately organized initiatives such as the *Genethisches Netzwerk* (i.e., the Gen-ethical Network) to the UNESCO have spared no effort to define certain artifacts of knowledge, and therefore also the contents of that knowledge, as the heritage of all humanity, and to protect them as such.

Above and beyond the problems that result from the economic exploitation of knowledge that is in itself public and ought to be common property, we must not forget the realm of military research and development. As early as in Bacon's utopian *New Atlantis*, one of the uses of the new science he described actually was its ability to offer the nation a military advantage. This led to the so-called classification of knowledge, that is, the non-disclosure by a nation of research findings in the field of military research. This is a problem not only because military research constitutes a very large part of the total volume of the research done in the areas of the natural sciences and technological development but also because, due to the competition among nations in the military sector, this research is by its very nature at the leading edge—and as a rule has research funding much more generously placed at its disposal than research in the publicly or academically controlled sector.

5.4 Education as Information

The difference between knowledge as participation in an object—

- I see a rose
- I can ride a bicycle
- I measured the depth of the water in the harbor

—and knowledge as participation in symbolically represented stores of knowledge—

- Mount Everest is 8,848 m high
- Force = mass × acceleration (Newton's Second Law of Motion)

[4] On protests and alternatives to publication see: http://www.guardian.co.uk/science/2012/apr/09/frustrated-blogpost-boycott-scientific-journals (6 May 2012).

—is crucial for the knowledge society. If it is characteristic of modern science that being a scientist means adding one further finding to an already existing stock of knowledge in which the individual researcher *believes*, then the relationship between one's own experience and the acquisition of mere information has continually developed to the advantage of the latter as a result of the prodigious accretion of symbolically represented stores of knowledge. Since placing oneself in relation to an object by means of experience always entails a further step in the development of one's own personality, this tendency can also be described as a dwindling of classical education. Knowledge is nowadays understood to be no more than an access to information, or at most as the ability to manipulate pieces of information. *Education*, then, is learning how to access information, and therefore remains external to the person. As a result, the sense of knowledge acquisition is increasingly seen to lie in the operative utilization of knowledge: the sense of knowledge lies in that which I can do with it. Of course, one could understand *information* in general as something that can and must be understood. In this sense, the classical reading of a text indeed does not consist in the assimilation of the information it contains—which could still be measured by means of information theory, if necessary—but rather in *understanding* it. This must be differentiated, however, from the operational manipulation of information.

It is characteristic that a good physicist by today's standards can calculate particular trajectories using the formulae of mechanics, for instance; or with the aid of the theory of relativity, he can determine on the basis of measurement data the motion processes on a body moving very rapidly relative to his own frame of reference. And he can do this without understanding the theory of relativity, or being able to say in regard to the Newtonian formulae what kind of laws they are, let alone what a natural law is, and in what sense natural laws exist.

A fundamental problem of the knowledge society is how to deal with the rising flood of information in the course of school-based education. This will not be achieved by either the prioritization or diversification or even abolition of entire disciplines unless the decision is made to engage in a form of paradigmatic learning that consists not in the transfer of information but rather in coming to know exactly what a specific discipline—physics, for instance—is and accomplishes; what constitutes its empirical basis; what defines its boundaries of competence; and what is defined as a proof—in mathematics, for example—or, in literary studies, an interpretation.

5.5 Reason, Science and Alternative Forms of Knowledge

The dominance of the scientific form of knowledge, primarily in regard to the natural sciences, remains unchallenged at the level of public fact-finding and institutional decision-making. This can be seen, for example, in the importance of expert opinions in the political and legal realms as well as in insurance cases. These relationships are particularly evident in the areas of medicine and health insurance:

here, as always, therapies and medicines are only approved when their efficacy is proven by means of hard scientific methods. This institutionally secured prioritization of natural-scientific knowledge, however, no longer corresponds in any way to what is "in people's heads', as they say: to what average people believe. One could also say that in our time the sciences, and especially the natural sciences, have lost their worldview-building function. In practice, what people nowadays believe they know draws on completely heterogeneous sources: on non-European cultures, resurrected remnants of European traditional knowledge, religious background knowledge, a variety of esoteric secret doctrines which, paradoxically, are so esoteric as to be anything but secret. The natural sciences, too, play a significant role in this conglomeration; though not as such but rather in the form of popularizations that—in contrast to science itself—are authoritatively deployed and credulously accepted. In the general public consciousness, quantum theory and the theory of relativity thus play a role comparable to that of other esoteric doctrines.

On the level of common sense, we have to take account of the dwindling of public regard for science. If Aristotle says in his *Metaphysics*, "All men, by their nature, strive toward knowledge" (Aristotle 1933), then it is obvious that science today is no longer in a position to satisfy this striving. From the standpoint of science, this is due to two things. For one, science, and particularly the natural and mathematical sciences notwithstanding their claim to universal validity—which also means validity for everyone—, has become practically inaccessible to laypeople. The barriers are set well ahead of what we have described as problems of open access, namely in the form of technical language and the high degree of specialization as well as the prerequisite scientific training and, in very many cases, a grounding in higher mathematics and statistical methods. The other reason for the lack of regard for the natural sciences is the fact that they have failed to live up to Bacon's promise, i.e., that scientific-technological progress is at the same time the progress of humanity (Böhme 1992). Given these experiences, a new scientific discovery or a technological innovation is by no means greeted with joy anymore, as was still the case in the nineteenth century, for instance; rather, it is received with extremely ambivalent feelings.

Furthermore, science itself—aside from the public lapses that have occurred in brain research—has become much more circumspect and self-critical in its public demeanor. This is due not only to the increase in scientific-historical reflection, which is well acquainted with many cases of scientific missteps, and the necessity of continually revising dogmatically petrified basic premises and even factual knowledge, but rather to the epistemological standpoint that began to develop with the work of Helmholtz and ended with Popper: From this standpoint, scientific knowledge is in principle hypothetical, that is, contains a wealth of presuppositions and is, thus, subject to possible revisions. This self-critical awareness has induced scientists to largely restrain themselves in their demands so as not to appear, in contrast to the popularizers of science, as purveyors of a world-view.

These two factors in concert have resulted in the fact that in our culture the very form of knowledge that enjoys the highest degree of institutional respect can no longer satisfy the natural craving for knowledge that Aristotle attributes to every human being. This in turn has provided an opening for alternative forms of knowledge—which is only to be welcomed. This is particularly sensible so long as these alternative forms of knowledge do not pass themselves off as science, as Paul Feyerabend believed he could accomplish, in his "Anything Goes," with regard to *Hexenwissenschaft*, or the science of witchcraft, and the Inquisition (Feyerabend 1979). It is the very consciousness of the limits of scientific methodologies and scientific truth claims that prompts the rehabilitation of other forms of knowledge. Thus, forms of tacit knowledge (Polanyi 1958) or personal and embodied practical knowledge have won more respect, as well as forms of knowledge that developed in completely different cultural contexts from those in which the modern natural sciences arose in Europe. In this context, the case of traditional Chinese medicine primarily comes to mind.

In general, on the one hand, globalization has served very well to expand technological civilization worldwide, and thus to disseminate the technological and natural-scientific claim to authority but, on the other hand, it has also led to an acknowledgement of forms of knowledge that developed within other cultures. Here, too, as in the economic sphere, globalization is linked to a liberalization of forms of thought and of knowledge. The two go hand in hand, as in the market-like form of the world-views on offer for the modern individual. To be sure, liberalization can lead to the very edge of the dissolution of any and all claims to knowledge if wise self-restraint and the acceptance of others lead to resignation with respect to the ability to communicate; that is, quite literally, with respect to the communicability and communality of the store of knowledge. Kierkegaard once said that truth is subjectivity (Kierkegaard 1957). With this statement, he was referring provocatively to the relevance of knowledge that consists, in religious knowledge and in the knowledge of wisdom in the unity of knowledge and person, in opposition to the claim to objectivity. For a subject, the truth of a piece of knowledge in this sense lies in the fact that he takes this knowledge *seriously* (Böhme 2005). Kierkegaard did not intend, however, to use this thesis to contest the strong possibility that there might be intersubjective agreement regarding the content of this knowledge.

The rehabilitation of alternative forms of knowledge that is linked to the decline of the claim to authority of modern European science, though welcome in itself, today threatens to degenerate into an unbounded libertarianism and—if you will forgive this oxymoron—a relativism made absolute (Böhme 1998). This could lead to the loss of everything that was accomplished by the Enlightenment. In Goethe's Faust (1855: 1851f), Mephistopheles warns: "Reason and Knowledge only thou despise, The highest strength in man that lies! […] And I shall have thee fast and sure!"

Here the New Phenomenology (Schmitz 2009) seems to be a salutary antidote. As with any form of phenomenology, the acknowledgement of experiences is its basic principle. Within what since Husserl has been called the *epoché*, it attempts to integrate theory and exegeses in order to recognize phenomena in their pure actuality. The remarkable aspect of the New Phenomenology, however, is that it is an elaborated knowledge of *subjective facts*—that is, facts that necessarily need to be sought out within the realm of one's own experience and for which the experienced subject's concern in terms of feelings, atmospheres or bodily perceptions—upon experiencing them are relevant. Thus, the New Phenomenology assimilates the existential aspect of truth posited by Kierkegaard for phenomenology as a form of knowledge and pursues, in the form of natural and body phenomenology, the insight that, as Goethe already realized, many phenomena—such as colors, for example—are phenomena not in relation to an abstract subject but rather in relation to the human being in his concrete, embodied existence. The major task for the New Phenomenology, however, now consists in developing a language in which one can make oneself intersubjectively understood with regard to these subjective facts.

References

Aristotle. 1933. *Aristotle in 23 Volumes*, Vols. 17, 18, translated by Hugh Tredennick, see http://www.perseus.tufts.edu/hopper/text?doc=Perseus%3Atext%3A1999.01.0052 (3 January 2017).

Bell, Daniel. 1974. *The Coming of Post-Industrial Society*. New York: Harper Colophon Books.

Böhme, Gernot. 1992. *Am Ende des Baconschen Zeitalters. Studien zur Wissenschaftsentwicklung*. Frankfurt am Main: Suhrkamp 1993. For a shorter English version see: Böhme, Gernot. *Coping with Science*. Boulder, CO.: Westview Press.

Böhme, Gernot. 2005. "Kierkegaard, Søren: Der Begriff Angst", in: Gerhard Gamm and Eva Schürmann (eds.), *Von Platon bis Derrida. 20 Hauptwerke der Philosophie*. Darmstadt: Primus: 207–219.

Böhme, Gernot. 1998. "Philosophie und Esoterik: Konkurrenten um die geistige Orientierung der Zukunft", in: Norbert Bolz and Willem van Reijen (eds.), *Heilsversprechen*. München: Fink: 11–24.

Böhme, Gernot. 1997. "The Structures and Prospects of Knowledge Society", *Social Sciences Information*, 36(3), September: 447–468.

Böhme, Gernot and Nico Stehr (eds.). 1986. *The Knowledge Society*. Boston and Dordrecht: Reidel.

Feyerabend, Paul K. 1979. "Against Method", in: *Minnesota Studies for the Philosophy of Science* 4, 1970; *Erkenntnis für freie Menschen*. Frankfurt am Main: Suhrkamp.

Goethe, Johann Wolfgang v. 1855. *Goethes sämtliche Werke in vierzig Bänden*. Bd. 11: Faust. Erster Teil, Stuttgart und Augsburg.

Jha, Alok. 2012. *Academic Spring: How an Angry Maths Blog Sparked a Scientific Revolution* (On Protests and Alternatives to Publication) see: http://www.guardian.co.uk/science/2012/apr/09/frustrated-blogpost-boycott-scientific-journals (6 May 2012).

Kierkegaard, Søren. 1957. Abschließenden unwissenschaftlichen Nachschrift zu den Philosophischen Brocken, *Gesammelte Werke*, Vol. 16. Abt., I. Teil, Düsseldorf: Eugen Diederichs: 190.

Kreibich, Rolf. 1986. *Die Wissenschaftsgesellschaft*. Frankfurt am Main: Suhrkamp.

Kronick, David A. 1962. *A History of Scientific and Technological Periodicals. The Origins and Development of the Scientific and Technological Press 1665 to 1790*. New York: The Scarecrow Press.

Piratenpartei Deutschland. 2016. *On the Pirate Party*, see http://wiki.piratenpartei.de/Parteiprogramm (6 May 2012).

Polanyi, Michael. 1958. *Personal Knowledge*. Chicago: The University of Chicago Press.

Richta, Redovan and Collective. 1968. *Civilization at the Crossroads: Social and Human Implications of the Scientific and Technological Revolution*. White Plains, NY: Internationals Arts and Sciences Press.

Schmitz, Hermann. 2009. *Kurze Einführung in die Neue Phänomenologie*. Freiburg: Alber 3. Auflage.
Stehr, Nico. 2000. *Die Zerbrechlichkeit moderner Gesellschaften*. Weilerswist: Velbrück Wissenschaft; *The Fragility of Modern Societies*. 2001. London: Sage.
Stehr, Nico and Reiner Grundmann. 2010. *Expertenwissen. Die Kultur und die Macht von Experten. Beratern und Ratgebern*. Weilerswist: Velbrück.
Webster, Frank. 1995. *Theories of the Information Society*. London: Routledge.
Znaniecki, Florian. 1975. *The Social Role of the Man of Knowledge*. New York: Octagon Books.

Chapter 6
Knowledge Paradigms and Intransigent Intuitions

Susan A. McDaniel

6.1 Introduction

Knowledge, as Nico Stehr reminds us, in his impressive corpus of work on knowledge societies, may indeed be the capacity for action. The concept, "knowledge as capacity for action" has become a well-known, and often echoed Stehrism, elaborated from the centuries old concept that knowledge is power, attributed, perhaps inaccurately,[1] to Sir Francis Bacon. How and whether knowledge is activated depends on the links of knowledge with social and/or political power. It is those links and the tensions among various processes and bases of knowledge in society that are the focus here.

Foucault instructs us in several of his writings that knowledge can be a weapon wielded by the powerful to the detriment of the weak. But it can also be a "weapon of the weak" or of the populace as Stehr (2010) compellingly argues. As new knowledge, whether reliable or not, is developed, it becomes disseminated ever more quickly in our increasingly wired and media connected world. And the volume of knowledge production has accelerated exponentially.

A question arising with greater frequency is about the bases of knowledge claims and beliefs, and the contestation among them. This was highly visible in the U.S. election of Fall 2016 when absurd claims were put forward and accepted by some, perhaps even by many. It is also clearly apparent in the climate change debate with scientists and a small but active group of climate change deniers seemingly contesting the issue of climate change. Even among those who agree it is a serious

Prof. Susan A. McDaniel, Ph.D., FRSC, Prentice Research Chair in Global Population & Economy, University of Lethbridge, Email: susan.mcdaniel@uleth.ca.

[1] Sometimes the concept is attributed to Thomas Hobbes.

M. T. Adolf (ed.), *Nico Stehr: Pioneer in the Theory of Society and Knowledge*,
Pioneers in Arts, Humanities, Science, Engineering, Practice 16,
https://doi.org/10.1007/978-3-319-76995-0_6

challenge, knowledge about the degree of the challenge is hotly contested. The Internet, of course, enables groups of believers in every kind of myth to actively engage in consensual validation of what they know to be true. American comedian Stephen Colbert refers to this growing phenomenon as 'truthiness' (see Walton-Roberts et al. 2013). Truthiness is the idea that when people strongly believe "in their gut" that something is true, regardless of its factual accuracy, it takes on the aura of truth for them. It is increasingly the case not only that much of what we know is not useful or outdated, but that much of it is simply false, belief with an aura of truth, supported by a group of believers. Conspiracy theories fall into this category and run rampant on the Internet. There is even a conspiracy group that suggests that the Newtown, Connecticut shootings of young school children and their teachers in fall 2012 was a hoax perpetrated by those seeking gun control! Knowledge and truth have become immensely malleable.

In one illuminating paper Stehr/Mast (2011) explore the gradient between science and democratic practices, noting structural changes in society such as historically unprecedented prosperity and scientific/technological advances that enlarge possibilities for citizen action. Scientific expertise can indeed be a significant social and political resource, but of course, expertise can also be suspect by those who see power as resting not with experts and/or science but with populism. The construction of this anti-expertise political discourse has been particularly vivid in U.S. politics where so-called elites in the mainstream media and particularly in academia are seen as biased and self-interested. Witness the way in which 2008 Vice-Presidential candidate Sarah Palin talked about experts and elites with sarcastic derision.

The valid point made by Stehr/Mast (2011) that the trend is toward more and more knowledge resting with highly trained specialists parallels the growth in populist questioning or even suspicion of such knowledge. Specialist knowledge, and knowledge about that knowledge, can indeed enhance human agency. In this respect, knowledge is indeed power for individuals and for potential citizen action. A compelling example of this could be the increasing knowledge that climate change is anthropogenic, caused by human actions and activities. "Climate change has suddenly been redefined from a natural phenomenon to a political and social conflict, making it a social issue which can be tackled through further research and political decisions" (Bechmann 2009: 105). Knowledge redefines the problem as open to human action. The conundrum of this enlargement of possibility for citizen action for the new knowledge on climate change is that if science defines the problem and its causes, then it is expected to come up with solutions as well.

The focus in this paper is on a case study of knowledge production and citizen action, the cancellation in Canada of the long-form Census of 2011. It offers an intriguing study of competing knowledge claims, of the place of knowledge in human agency and citizen action, and of which kinds of knowledge can be seen as 'actionable' and by whom. Census making and taking, like other knowledge, can really be only understood in the context of social, cultural and political dynamics (Curtis 2001), a point made in relation to knowledge about climate change as well (von Storch 2012).

6.2 The Census of Canada

Census taking has a very long history in Canada. In fact, what is now Canada claims the record for North America's first census. Jean Talon, born in Champagne, France, arrived in New France in 1665 on a mission from King Louis XIV who wished to measure the progress of the colony. One of Talon's first tasks was taking stock of the colony through a systematic census which he undertook in the winter of 1665–66 (Statistics Canada 2012). The statistics from this first census were then put into action to build the new colony by developing a tax base, a health plan, building bridges and chimney sweeping. Knowledge in the young colony of New France became action, providing empirical validation for Stehr's concept of knowledge as the capacity for action.

"Census making," as Curtis (2001: 3) argues, "involves identifying political subjects and centralizing knowledge." This is borne out in the first census of New France, and subsequently in the writing into law of the need for a census in Canada. The need was clear for reliable knowledge about the population and its characteristics, and has remained clear in Canada for centuries.

The first national census of Canada was taken in 1871. In accordance with *The Census Act* of May 12, 1870, each census was to take place no later than May 1st in every tenth year. The census was part of Section 8 of *The Constitution Act* of 1867 (formerly *The British North America Act*), and seen as vital for the proper representation in Parliament by the population. The 1871 census included questions on age, sex, religion, education, race and occupation as well as questions on land holdings, vital statistics, religion, education, administration, the military, justice, agriculture, commerce, industry and finance.

The Constitution Act of 1867 does not specify exactly what is to be on the census of Canada, with certain exceptions. Language questions were mandated by various acts related to the Census, and are seen as constitutionally required. We shall return to this in a moment.

6.3 The Cancellation of the 2011 Long-Form Mandatory Census

In June 2010, an obscurely placed announcement was made that the federal Conservative Government of Canada would be limiting the 2011 Census of Canada to a very restricted list of questions that were on the short form mandatory Census sent to all Canadians. Planning had been underway for several years in Statistics Canada for the 2011 Census. A census is the largest peace-time undertaking any country takes on, necessitating much planning, testing, advertising so that every citizen will know the census is coming and complete it. The argument put forward for the precipitous decision to cancel the mandatory long form was dual: that it was an invasion of the privacy of Canadians, and that the census ought not to be mandatory. It should be remembered, however, that 2010 was not the first time government had threatened the census. In fall 1984, the newly elected Conservative government of Canada announced its intent to cancel the 1986 Census of Canada in its entirety (Dillon 2010). It was only through a huge public outcry that the 1986 Census went forward.

Serious controversy quickly broke into the public realm over competing knowledge claims. The Minister of Industry, Tony Clement, along with other Conservative Members of Parliament made the unequivocal claim that the Chief Statistician of Canada had been consulted and had advised that statistical information acquired through a voluntary survey would be reliable and comparable with previous censuses. The Chief Statistician is a Deputy Minister level member of the civil service, and is therefore bound by his oath to keep confidential any and all advice that he provides to the Minister to whom he reports. Thus, the Chief Statistician was not able, being honour-bound, to refute the government and Minister's claims. The media discussion that ensued became one on statistics and the relative reliability of voluntary survey data. Statistical knowledge, highly technical, was being contested on the pages of Canada's newspapers, certainly an unusual occurrence. The Chief Statistician, Dr. Munir Sheikh, a respected economist who had worked for much of his career in various government departments, including holding a high-ranking position in the Department of Finance, felt he had no choice, given his oath of confidentiality about ministerial advice, but to resign his position in protest, which he did. He was then free to openly refute the Government's claim which he did succinctly in a speech in Ottawa (Canada's capital) in April 2011 with these words: "I want to take this opportunity to comment on a technical statistical issue which has become the subject of media discussion [...] the question of whether a voluntary survey can become a substitute for a mandatory census. It cannot" (Sheikh 2011).

With the Minister's knowledge claim put to rest definitively by expert statistical knowledge, other knowledge claims by the government were publicly contested. On the privacy invasion argument, one Minister (not the one responsible for Statistics Canada) reported that he had received "thousands of complaints" about the census being too invasive of privacy. On looking into the matter of privacy complaints, it was discovered that only 50 official complaints about privacy concerns with the census had been made over a twenty-year period, and only two complaints were launched about the 2006 Census (Saurette 2010). Furthermore, no prosecutions had ever occurred for failing to complete the mandatory census (Walton-Roberts et al. 2013). Indeed, it seemed that Canadians were more than willing to count themselves in.[2] The response rate on the 2011 short form Census was 98% and even higher for the reinstated 2016 long form, without much follow-up needed, indicative of Canadians' commitment to completing the Census. This is a significantly higher response rate than in other countries with population-based censuses.

The most sustained citizen engagement over the sudden decision to cancel the 2011 long form Census was from various quarters who seldom speak with a single voice. Criticism—loud and public—was raised by business, police forces, professional groups, municipalities, most provincial governments, church groups, voluntary organizations, academics and some private polling companies (McDaniel/MacDonald 2011). The latter are particularly interesting since the government, at various points, put forward the argument that private polls might replace the census in providing data. In reality, pollsters need population data from which to draw their samples and as a means by which to estimate accuracy of their polls. The argument raised over and over again by all these diverse groups was that the census provided vital information/knowledge they needed. The contest over the knowledge that the census provides, and over competing knowledge paradigms, became an open civil society debate.

Several court cases were launched over the Census long-form cancellation. One crucial one was by Francophones who argued that the Census of Canada, harking back to the Constitution Act of 1867 as well as the very early censuses of New France, was required to have questions on language. These questions had previously been on the long-form census, and were hurriedly added to the short form. The government in its enthusiasm to cancel the long-form census had neglected its constitutional obligations.

6.4 Census as a Knowledge Base

A census has multiple roles as a knowledge base. It can, given its typical population-wide approach, or by statistical sampling, make objective truth claims as knowledge. A census can additionally serve as a public policy statement, i.e. what society is or is not, and perhaps open the barn door to the gaps between what is and what could or should be. This would be the case when censuses find income inequality gaps, deepening poverty or employment challenges among subgroups of the population for example. In this sense, a census is a socio-political imaginary, which may be the place of its greatest asset to society as well as its greatest point of vulnerability.

The Census of Canada mandatory long form may have been caught up in knowledge paradigm politics. There is certainly a hunger for knowledge, in this case official statistics as the many criticisms of the government decision on the

---

[2] The public campaign for recent Censuses of Canada, issued in numerous languages, has been "Count Yourself In", which most all Canadians willingly did.

long-form census indicate. As Warren (2001: 3–4) eloquently points out with respect to the U.S., but it also applies to Canada, "[…] a vigorous market for data exists." She then adds a cautionary note. 'Ironically,' she suggests, "the power of this market [for data] threatens to crush serious, policy-directed, empirical work […]. Indeed, the market is creating an anti-market in which one study seems to contradict another, leaving policy-makers free to ignore all data and making such scholarship not only difficult but useless." In this situation, official data face a new, data suspicious knowledge paradigm.

It should be emphasized in this context that the 2011 long-form census is not the only official knowledge base to be rendered either non-existent or useless. The same averted political eyes have been used with both ending the long gun registry in Canada—and incidentally proposing to completely destroy all the data acquired in the registry through careful effort by the government and big expenditure—and the "tough on crime" policy approach in the face of statistical evidence of declining crime rates. With respect to the long-gun registry, there was strong support for it from none other than chiefs of police across Canada who found it a valuable source of knowledge for their work and safety. Knowledge as the capacity to act may not have been at the tip of the police chiefs' tongues, but it was present in their minds nonetheless. With respect to evidence from official data, or scientific knowledge, the Prime Minister of Canada Stephen Harper had said clearly that he does not make policy on the basis of statistics.

The question then arises, what knowledge paradigms and claims are made and matter to policy-making in these cases, particularly with respect to the 2011 Census long form cancellation. We know that it was not accounting or accountability since the voluntary survey that was put in place purportedly to substitute for the mandatory long-form census cost considerably more than the long-form census cost—$30 million more, to be precise. We know that expert advice had no bearing given the post-resignation words of the Chief Statistician. The question is what knowledge paradigm was operative?

In any typology of knowledge paradigms (Stehr 2005; von Storch 2011), origins, social power and dynamics are known to matter. The question of whether scientifically constructed knowledge, such as a census provides, is better than other sources of knowledge is contestable in terms of social power and dynamics. The same might be said of climate science or criminology. The issue is less what is more methodologically sound or subject to falsification by science than what fits better with an emerging socio-political narrative of knowledge, or intuition. So many disparate groups called the government decision on the long-form census not only bad science in a so-called knowledge society and economy (McDaniel/MacDonald 2012), but illogical and "undignified governance" (Saurette 2010).

So, what narrative of knowledge or knowledge paradigm could the decision on the long-form census be based upon? The distrust of knowledge based on statistics by neoliberal or conservative movements in North America is well known. This sort of knowledge is deemed to be created by elites for their presumed purposes (Ramp/Harrison 2012). Instead, the knowledge paradigm could be *epistemological populism* (Saurette 2010). This paradigm of knowledge politics holds that the most trustworthy and dependable source of knowledge emanates from anecdote, or from lived experiences of the "common people". Essentially, the premise is that statistics can lie, and indeed often do, but perceptions by one's own senses or those of reliable informants, usually constituents, neighbours or family/friends, never lie. They, after all, are seen as perceiving or experiencing 'reality' as it is. Statistical knowledge, after all, it is argued, aggregates human experience into categories which are deemed distant from observed or perceived reality. And what's more, statistics are compiled and analyzed by pointy-headed analysts or academics who generalize problems and create demands for policy changes that may well be undesired by politicos of the neoliberal persuasion. Academics are seen as colluding in this venture out of self-interest rather than in the public interest.

The logic then of the census cancellation decision may have been based on a knowledge narrative and politics that is internally consistent and even coherent. It is interesting, however, to note that the narrative descends from a political, anti-knowledge elite, a reigning government, albeit a minority government, that premises its knowledge ideology on an anti-elite plank. The Harper conservative movement grew out of the populist western Canadian Reform Party. Its roots are not those of the long-existent Progressive Conservative Party of Canada, that itself wanted to end the 1986 Census, possibly for reasons of fiscal prudence at that time. Meantime, paradoxically, the vast public was clearly against this knowledge transition in 2011 to one based on anecdote and experiential realities, and resisted it openly. In the sense of competing knowledge paradigms, the public was out in front in its preferred reliance on a scientific census and knowledge based on generalizable statistics. The same cannot be said of climate change.

Knowledge in the form of a national census, a mandatory census, has had a long and distinguished history in Canada, as we have seen. Canada not only has the historical record for the earliest population census in North America, but Statistics Canada's expertise in conducting censuses and protecting privacy and confidentiality is legendary. The Chinese, for example, sought out the expertise of Statistics Canada in expanding and fine-tuning their own statistical data collection processes and procedures. The U.S. Census Bureau has also relied on the expertise of Statistics Canada. This opens the possibility of yet another dimension of the knowledge paradigm behind the government's motives in cancelling the long-form Census of 2011. The resignation of the Chief Statistician of the time is seen by many as a regrettable event in the wake of the census decision and the politics around it. Yet, seen through the lens of a populist knowledge paradigm, this may have been less collateral and more central to the logic of knowledge politics. Indeed, the events that unfolded between the government and Statistics Canada over the sudden decision to cancel the long-form census can be seen as a kind of morality play about bringing to heel those who compile statistics and value their policy relevance. As Saurette (2010: 3) suggests, "[…] it's an ideological outreach to the conservative base at a much deeper, philosophical level […] by communicating to them that this government shares their suspicion of statistics and the […] out-of-touch academics who analyze them."

In a certain sense, the clash of knowledge paradigms comes into full visibility with the lens of populism as a socio-political imaginary. Emphasizing the objectivity and scientific value of the census as knowledge useful for policy and public action may, in fact, create a contest of different knowledge paradigms, suggesting that a census is only one way of knowing. This seems to have occurred with respect to climate change as well. The media's quest for 'balance' in reporting may give the opinions of small, outlying minorities greater credence than they deserve. Internet groups of like-minded believers or deniers also contribute to this sense of equality among various kinds of knowledge paradigms. As Warren (2010: 4) wisely remarks, this may render data useless and give policy makers free reign to ignore scientific knowledge in favour of populist experiential knowledge. She continues (2010: 17), "Good studies and bad studies are not 'mutually cancelling.' Regardless of what some advocates might claim, there are some objective facts and hence, some objective truths. Whether public policy reflects that reality is not a choice left to those in the academy, but producing and protecting the research itself is our choice and moral obligation". Clearly, the populist knowledge paradigm begs to differ.

Resort to populist knowledge paradigms knocks both the state and science off any pedestals they might have previously occupied. This may seem paradoxical that a state's top down decision on what knowledge is given credence and what is not diminishes the role of the state, but indeed that is what occurs. The shrinking of the state by the state is a key part of the neoliberalist, populist agenda. Putting a state agency such as Statistics Canada and the knowledge it produces about the state under strict control as any other policy arm of the government, and reducing its scope for independent inquiry fits well with the populist knowledge politics of neoliberalism. Providing citizens with information that is objective, and crucially, actionable (a key component of Nico Stehr's notion of knowledge) is exactly counter to the populist ideology of intransigent intuitions.

As to the knock against science, that is also a crucial component of the populist knowledge paradigm. As Ramp/Harrison (2011) contend, what they call libertarian populism as both symbolic and rhetorical, works to support and indeed encourage public skepticism about societal institutions, including government, but also science and academia in general. In defending against what are seen as elites, libertarian populist stances pit themselves against what they see as the intellectual classes. Scientists, population experts and statisticians are then redefined as special interest groups, who are seen as getting a "special deal" in obtaining data through the public purse that they should be buying in the marketplace (see Ramp/Harrison 2011: 278). It is self-interest that motivates them and therefore their findings, however compiled and whatever rigorous methodologies are used, are suspect. Climate change science may be particularly vulnerable to this framing of knowledge through populist paradigms, since it is not as directly apprehendable by individual experience and perception, making its scientific findings even more suspect.

6.5 Conclusion

Knowledge then from analysis of the case study of the cancellation of the Canadian long form census for 2011 may be neither a weapon of the weak and powerless, nor a weapon of the powerful. It may be unapparent to both, and evident only in terms of the knowledge paradigm employed. Paradoxes abound in this case study. In the name of populism, the public may be left to lobby for actionable knowledge while the populist state presumes intuitions matter more. Is the wall between scientific knowledge and society lowered or raised by the census cancellation? Difficult to tell, but in the contesting knowledge paradigms, it emerges that scientific knowledge does not have only a salutary effect for society. It can be overrun by a competing knowledge paradigm which puts scientists, technocrats, intellectuals and the public at large in a position to defend themselves and what they have long considered as in the public interest.

There is a rejection of evidence as a political resource in this new populist paradigm of knowledge. What needs to be known and by whom is reframed with the only knowledge to be trusted is that obtained by individual experience, by intuition. But—and here it gets very interesting—the populists who are in charge of the state at present *do* indeed appreciate the power and importance of statistics such as those emanating from the census. That is why they are keen to eliminate or reduce them. Their greater fear may be of the power of statisticians, particularly those working for the government who may overturn the intransigent intuitions of the populists in favour of knowledge with a decided impetus for action. "When modern science scribbles on the wall, it is no longer about emancipation from nature but about possible disaster—nuclear war, genetic manipulation, climate change" (von Storch/Stehr 2000: 615). Population diversity and inequalities may also be the scribbles on the wall that populist knowledge paradigms wish not to see. Reality, however, will not be denied, whether the issue is population diversity or climate change, no matter how intransigent the intuitions.

References

Bechmann, Gotthard. 2009. "Climate Research Between Knowledge and Organization: Problems of Transdisciplinary Science", in: Gotthard Bechmann, Vitaly Gorokhov and Nico Stehr (eds.), *The Social Integration of Science: Institutional and Epistemological Aspects of the Transformations of Knowledge in Modern Society*. Berlin: Edition Sigma: 101–120.

Curtis, Bruce. 2001. *The Politics of Population: State Formation, Statistics, and the Census of Canada, 1840–1875*. Toronto: University of Toronto Press.

Dillon, Lisa. 2010. "The Value of the Long Form Canadian Census for Long Term National and International Research", *Canadian Public Policy/Analyse de Politiques*, 36(3): 389–393.

McDaniel, Susan A. and Heidi MacDonald. 2012. "To Know Ourselves Not", *Canadian Journal of Sociology*, 37(3): 253–271.

Ramp, William and Trevor Harrison. 2012. "Libertarian Populism, NeoLiberal Rationality, and the Mandatory Long-Form Census: Implications for Sociology", *Canadian Journal of Sociology,* 37(3): 273–293.

Saurette, Paul. 2010. "When Smart Parties Make Stupid Decisions", (23 July 2010).

Sheikh, Munir. 2011. *Good Data and Good Government. Festschrift in honour of Ian Stewart.* April 20–21, 2011, Ottawa, (19 June 2011).

Statistics Canada. 2012. "New France Census of 1665–1666", E-STAT, (1 February 2013).

Stehr, Nico. 2005. *Knowledge Politics: Governing the Consequences of Science and Technology.* Boulder, CO: Paradigm Publishers.

Stehr, Nico. 2010. "Knowledge, Democracy and Power", *Central European Journal of Public Policy,* 4(1): 14–34.

Stehr, Nico and Jason T. Mast. 2011. "The Modern Slaves: Specialized Knowledge and Democratic Governance", *Society,* 48: 36–40.

von Storch, Hans. 2012. "Social Science: A Must for Climate Research", Paper presented at the Workshop in honour of Nico Stehr, Zeppelin University, Friedrichshafen, Germany, March 2012.

von Storch, Hans and Nico Stehr. 2000. "Climate Change in Perspective", *Nature,* 405(June): 615.

Walton-Roberts, Margaret, Roderic Beaujot, Daniel Hiebert, Susan McDaniel, Damaris Rose and Richard Wright. 2013. "Why Do We Still Need a Census? Views from the Age of 'Truthiness' and the 'Death of Evidence'", *Canadian Geographer,* 58(1): 34–47.

Warren, Elizabeth. 2002. "The Market for Data: The Changing Role of Social Sciences in Shaping the Law", *Wisconsin Law Review,* 1: 1–34.

Chapter 7
New Forms of Civil Society: The Social Question of the 21st Century

Hermann Strasser

7.1 Introduction

It was a great honor for me to emotionally and scientifically celebrate with many others the 70th birthday of my friend and colleague Nico Stehr. He is not only the guide of the knowledgeable society but also the visionary of the moralization of markets and the social constructionist of climate change, not to mention his expertness of experts.

By focusing on new forms of civil society I try to combine three of Nico Stehr's visions: First of all, not to restrict scientific ideas to the halls of academia. Second, to make a rather fragile society worthwhile to live in. And third, to have in mind Max Weber's goal of being a scientist, which means to strive for being surpassed by other knowledge producers. Thus, we have to ask ourselves what *society* we live in, how significant *civility* is and whether new forms of civil society are *replacing* or just *extending* traditional structures and meanings of community or are simply representing *hybrid* constructions.

7.2 What Kind of Society We Live In

In other words, we should take a look at the ways people understand community in modern society. This society is no longer characterized by 'naturally' occurring forms of belonging based on orientations provided by traditional institutions like the family, the church, school, unions, and political parties. Today, people in most modern societies live longer, are more qualified, more mobile, have higher incomes, see more women join the labor force, and less children are born.

Prof. em. Dr. Hermann Strasser, Chair for Sociology, Universität Duisburg-Essen; Email: hermann.strasser@uni-due.de.

M. T. Adolf (ed.), *Nico Stehr: Pioneer in the Theory of Society and Knowledge*,
Pioneers in Arts, Humanities, Science, Engineering, Practice 16,
https://doi.org/10.1007/978-3-319-76995-0_7

This, as you know, is the individualization story. But it is also Nico Stehr's story about the *fragile society* emerging from an increasingly knowledge-based system of economic production and social organization. In this individualized *knowledge society*, individuals have expanded their capacity to act and, thus, also their capacity to say no—with the consequence that major social institutions have lost the capacity to have their way. The other consequence is that the more we know, the less we can predict the future. This has led politicians to exercise "the sovereign handling of non-knowledge", as the political scientist Korte (2012) argues—of course, in accordance with the theory of public choice based on the individual profit-oriented thinking of the economists. However, if knowledge ages faster, new realities need a new point of view, as Stehr (2000) would say.

Not only knowledge but also *networking* has become more important, as many long-term communities have been replaced by short-term networks. This does not necessarily mean that volunteering is also becoming less sustained and more episodic. However, as Putnam (1993, 2000) argued, "plug-in volunteering", task-oriented on a short-term basis as it is, will be promoted. Nevertheless, civic or voluntary engagement is the broadest and most useful notion, with the core element of active citizens participating in organized forms of engagement. Civic activities take place in the public sphere more or less regularly, they are voluntary, based on one's own initiative, independent from the state, beyond the market and the family, and in essence not being paid for (cf. Strasser/Stricker 2011).

Moreover, the *state* is in many ways overtaxed, the globalization process makes the democratic system less controllable, and educational expansion extends both the need and the capacity of people to participate. As the Brazilian NGOs studied by Elisa Reis demonstrate, the more an NGO is linked with external actors, the better it is anchored domestically (Koslinski/Reis 2009).

The political realm has also broadened in another sense, as new forms of communication allow for new social platforms through internet-based mobilization and transnational networks. The shifting boundaries between the public and the private spheres will have consequences for civic engagement, for example with internet portals operated and supported by volunteer welfare organizations such as the *Diakonie*, that help young men and women to solve their social, financial, and mental problems. In other words, we are faced with *cultural transformations* which social scientists must capture and experts must shape by reducing complexity.

This means not only strengthening the respect for others, especially strangers and those who want to remain anonymous at first. It also implies an extension of self-regulation (cf. Billante/Saunders 2002). However, civility as learned behavior will extend social capital in terms of individual resources and collective trust only if civil engagement is enacted and acknowledged in social life. This is even more important in our media-managed attention economy, with attention as the new currency (cf. Franck 2005). As the commission of inquiry of the German Bundestag concluded in 2002: "Those who call for engagement must also promote competence and offer an appropriate context."

7.3 Volunteers: Stopgaps or Social Innovators?

Therefore, we have to look for new variations of established, as well as novel ways of organizing social solidarity beyond the state, the market, and the family, in the so-called non-profit sector. But not only do we have to think about new measures of organizing solidarity but also of ways to create a feeling of belonging, thus making society worthwhile to live in. Indeed, the *willingness* to become civically engaged seems to be on the rise in most countries. Still, the *forms* of engagement are changing as well.

In other words, what is at stake is to enable community and organize solidarity. But what *kinds of engagement* make a society work beyond the provisioning welfare state? Is it charity or associational membership, political participation or community service that makes the difference?

Undoubtedly, civic organizations of the non-profit sector are still value-oriented communities, but at the same time they are *multifunctional*. With their services in the areas of nursing, education, sports, and coaching, they are not only part of the market. They are also political actors articulating the interests of their members. With the trend toward more professionalization, these third-sector organizations seem to go through a cycle of change from value communities to service enterprises. This is not the case to the same extent everywhere, perhaps more so in the U.S., Canada, Great Britain, Ireland, Belgium and the Netherlands than in Germany, Austria, Italy, France or the Scandinavian countries.

We simply do not know enough about people and their motivations for civic engagement in leading positions and in volunteering. Is it self-realization, pleasure in voluntary work or the chance to learn more about an interesting job that motivates them?

Let me therefore briefly illustrate my postulates and assumptions by referring to the present state of civic affairs in Germany. In the 1990s, we stated a *decline* of community oriented forms of relations, which generated doubts about the continuity of solidarity and civic sense. A few years later, this opened the discussion about the consequences of demographic change. We became witnesses of talk about a "senile society" with all its nursing and pension burdens. Economic questions came to the fore. Even George W. Bush made himself heard by saying that the civic sense should above all heal the wounds of the market.

But civic sense is much more than a band-aid for these wounds, just as civic engagement is more than just a stopgap for nonprofitable niches or subsistence gaps. This discussion finally led to the contrast between the model of the welfare state and the model of civil society. In this tension between welfare state and civil society the central question arose of how to achieve a fruitful balance between state-organized solidarity and privately arranged responsibility. As civic sense and voluntary engagement cannot be decreed by law, various parts of the welfare system must adapt themselves through the power of self-control. In a society with a shrinking population there is no other alternative to master the challenge of less people having to produce more.

In both models, there is no question that older people gain in importance. They will claim more services of the welfare system. At the same time, the number of older people increases who are in good health, well educated, economically secure, and have more spare time. This has motivated social scientists and politicians to announce the age of "productive aging" for the well-being of society. And yet we know very little about these developments. We are once more confronted with the *paradox* of the knowledge society: The more knowledge is being produced, the less we know about its effects in the future.

But there are two things we know for sure: First, both quantitative changes in age groups and qualitative changes in older life phases will have a tremendous effect upon the composition, the intergenerational relations, and the quality of life of future cohorts. Second, it is also sure that older people, including even silver foxes like Nico Stehr, cannot simply be activated without adapting structural conditions and developing a new understanding of their role. In other words, a bridge must be built between entering retirement age and getting involved in spare time activities. The intellectual decline does not begin with 55, 60 or 65 years. Just as life-long learning has no age limit, paedagogic and social competences do not correlate with age.

The German academic community could actually benefit a lot more from its experienced 'oldies' equipped with lots of social and cultural capital by bringing them effectively together with the younger generation of scientists. More than ten years ago, this idea encouraged Nico and me to propose the foundation of a "University of Silver Foxes", what we ironically called a 'Greisenuniversität' ("geriatric" or "aged university"). We received a number of encouraging comments, including an exciting one from the former Austrian minister of education, Elisabeth Gehrer. She was very interested in the idea but not successful in making it a reality. Its realization then apparently happened at Lake Constance.

That an *increase* in civic engagement by older people is indeed possible may also be demonstrated by the data of the German Voluntary Survey of 1999 and 2009. In 1999, 29 percent of the age group of 65–69 years reported voluntary engagement, ten years later the percentage was 37. The main areas of older volunteers were sports, the church, and social work. They clearly prefer volunteering within an institutionalized framework.

However, these engagements also happen in a fast changing environment: On the one hand, with the opening of the market for social services such as ambulant care, this sector has been *economized* in order to increase both the efficiency and the productivity of its organizations. Higher service quality and qualification standards have led to an increased *professionalization* of welfare organizations.

On the other hand, we observe a change in the *motives* for civic engagement in general. That is to say, people, including the elderly, are increasingly confronted with their own limits, their personal and social identity when expressing their competence as well as their ways of self-realization by taking over responsible activities. Unlike in former times, belonging to a certain milieu is no longer decisive. Now, other qualities are more relevant, for example: communicative competence, the ability to organize and to negotiate and, not least, to realize one's own interests, as this is closely related to personal development. Rather than (collectively oriented) selflessness, the main motivation now is (individually oriented) self-realization. What we can recognize and what various studies confirm, is a trend *from the helper to the voluntary expert* (cf. Strasser/Stricker 2008).

7.4 The Social Question of the 21st Century

In order to master the transition from a welfare state society of the past to an active civil society of the future, two elements are of great significance: *participation* in societal actions and *appropriate forms* of voluntary engagement. How, to what extent and how quickly these challenges will be mastered is the social question of the 21st century.

If fewer young people live together with more old people, a new formation of collective trust and individual networks is not just the social cement that holds society together. These newly formed networks based on norms of reciprocity and trust also represent a kind of power station for new social relationships and new social capital. This power station will deliver the energy for the necessary bonding ability in modern society. In this sense, civic engagement will turn out to be the mediator and broker of a successful living together of future generations and institutions.

Nevertheless, a lot has still to be done if these postulates should ever become reality. The main point is that in Germany as well as in many other countries, voluntary engagement is not sufficiently made *accessible*. Studies show that the civic potential is high, since about half of those who are not volunteering are interested in getting civically engaged. Another important point is that networking with professional services must be improved. Most intermediary organizations have actually 'learned' to survive without volunteers. Their rules are pretty much determined by the needs of the funding agencies and the payed employees. There is no doubt that both the market and the state play a central role in the modern welfare mix. But these structures must be adapted in order to integrate volunteers with regard to their needs of time and content. To put it differently, there is a *discrepancy* between the *willingness* to volunteer and the *factual* engagement.

Of course, with regard to the complexity of the work process and the different motivational structures of employees and volunteers it is not easy for welfare organizations in general, or for a nursing home in particular, to direct these two groups in a parallel way. But in the long run, the increase in efficiency achieved by integrating volunteers will match and surpass by far the higher costs incurred at the beginning. The managers in charge of the structure of organizations which include volunteers must keep in mind the economic necessities of as well as sufficient margins for volunteers as co-producers.

In other words, societies like Germany are confronted with the fact that there is not a lack in engaged women and men but rather a lack in *appropriate organizational structures*, i.e. the general conditions of voluntary action. State regulations should therefore focus upon the intermediary organizations rather than upon incentives on the individual level of the volunteers.

7.5 Summary

Under these circumstances, I am convinced that there is no other alternative to the active self-organization of citizens in response to the challenge of modernizing public welfare but to adjust organizational structures and to recruit old and new volunteers on this basis. If this is accomplished, the social question of the 21st century will be successfully dealt with. It is the only way of creating a civil society, even if these developments may take on different forms in different cultures with different traditions, structural conditions, and individual challenges.

Then, the time may have come for *immaterial sources* of welfare such as education, friendship networks, neighborhood help, unhurt nature as well as humane cities and institutions to come to the fore in the age of decreasing growth. Less will then be more, as Hondrich (2007) argued. Social, that is, shared wealth is at stake, which is not to say that cooperation and participation will do away with egoism, self-interest, and stubbornness over night (cf. Bosshart 2011).

And everything that makes life worth living will then be included in the *gross national product*, as Robert Kennedy claimed as early as in 1968—the time when Nico Stehr began to explore the continent of unlimited possibilities. Good luck, Nico, in detecting new possibilities in the next 31 years since you promised to invite me to your 101st birthday party!

References

Billante, Nicole and Peter Saunders. 2002. "Why Civility Matters", *Policy,* 18: 3. St. Leonards. NSW: Center for Independent Study.

Bosshart, David. 2011. *The Age of Less. Die neue Wohlstandsformel der westlichen Welt.* Hamburg: Murmann Verlag.

Franck, Georg. 2005. *Mentaler Kapitalismus: Eine politische Ökonomie des Geistes.* München: Hanser Verlag.

Franck, Georg. 1998. *Ökonomie der Aufmerksamkeit.* München: Hanser Verlag.

Hondrich, Karl Otto. 2007. *Weniger sind mehr. Warum der Geburtenrückgang ein Glücksfall für unsere Gesellschaft ist.* Frankfurt am Main: Campus.

Korte, Karl-Rudolf. 2012. "Ich kümmere mich", *Süddeutsche Zeitung,* December 4.

Koslinski, Mariane C. and Elisa P. Reis. 2009. "Transnational and Domestic Relations of NGOs in Brazil", *World Development*, 37, 714–725.

Putnam, Robert. 1993. *Making Democracy Work: Civic Traditions in Modern Italy.* Princeton: Princeton University Press.

Putnam, Robert. 2000. *Bowling Alone: The Collapse and Revival of American Community.* New York: Simon and Schuster.

Stehr, Nico. 2000. *Die Zerbrechlichkeit moderner Gesellschaften. Die Stagnation der Macht und die Chancen des Individuums.* Weilerswist: Velbrück.

Strasser, Hermann and Michael Stricker. 2008. "Bürgerinnen und Bürger als Helfer der Nation?", *Aus Politik und Zeitgeschichte* (Beilage der Wochenzeitung *Das Parlament*), No. 12–13: 33–38.

Strasser, Hermann and Michael Stricker. 2011. "Freiwilliges Engagement älterer Mitbürger: Chancen und Hemmnisse", in: *Stimmen der Zeit,* Vol. 229, 7: 471–484.

Chapter 8
The Scientific Ethos: The Case of Climate Change

Reiner Grundmann

8.1 Cooperating with Nico

My interest in climate change dates back some 20 years when I was conducting a case study on the protection of the ozone layer. At that time climate change was seen as the "next big thing" after the problem of the depletion of the ozone layer had been 'solved'. My subsequent interest was to explore the differences and similarities between the two cases and to find out what 'learning' from one case could mean for the other.

At around that time I met Nico Stehr at a workshop in Bielefeld entitled "Climate Communications as a Research Field for the Social Sciences". It was held at the Center for Interdisciplinary Studies in Bielefeld in September 1995. In the pub discussions afterwards we discovered that we both had an interest in social theory and environmental sociology, especially in climate change. Ever since, we have co-operated on numerous projects in both fields.

I have greatly benefitted from Nico's experience and advice over the years as we embarked on different projects. One of the first papers we wrote was about sociological tradition and the natural environment. For this article we explored the relevance of Werner Sombart, an author whom I had not read in detail before. Nico prompted me to do so, and we pursued the work on Sombart with a paper for the *Journal of Classical Sociology* (Grundmann/Stehr 2001) and an edited volume of his key publications in English, for Transaction Publishers (Stehr/Grundmann 2001).

Prof. Dr. Reiner Grundmann, Professor of Science and Technology Studies, University of Nottingham; Email: reiner.grundmann@nottingham.ac.uk.

M. T. Adolf (ed.), *Nico Stehr: Pioneer in the Theory of Society and Knowledge*, Pioneers in Arts, Humanities, Science, Engineering, Practice 16,
https://doi.org/10.1007/978-3-319-76995-0_8

The chairman of the publishing house, the late Irving Horowitz, was keen to publish the book and proud of the result. I still have his letter thanking us for "making accessible the writings of one of Germany's Giants of thought".

Quite a few years ago we started a book project called *The Power of Knowledge*. It was published in 2011 in German, and in 2012 in English (Grundmann/Stehr 2012c). I think it is fair to say that we underestimated the task we were facing. In the book we are dealing with three knowledge intensive policy cases, Keynesianism, "race science", and climate change. We have a theoretical introductory and a concluding chapter which provides the intellectual framework for our argument. However, the challenge of complexity was truly daunting as we were writing in tandem, on three cases, in two languages—writing simultaneously in German and English is Nico's preferred way of publishing, but not mine. Eventually we managed to finish both manuscripts at the same time and had them accepted by the publishers without major changes. I think we are both quite pleased with the product.

You may get the impression that co-authorship with Nico is stress. Nothing could be further from the truth. What I just described in terms of challenge was self-imposed through the ambition of the book projects. Cooperating with Nico on manuscripts is uncomplicated. He trusts that the changes you make—both with regard to additions or deletions—are made for a good reason, and he never fights over petty things of style or wording.

Another recent product of our cooperation is a book on *Experts*, again published at the same time by a German and a British publisher. This book took much less time to complete, also because it is much shorter.

As we embarked on the issue of climate change and its social and sociological significance Nico introduced me to Hans von Storch, a climatologist who had become interested in the social and political issues around climate science. In the meantime, Hans von Storch together with a few other people including myself set up the blog *Klimazwiebel*[1] just after the Climategate affair and the Copenhagen summit.

8.2 Climategate

In what follows I propose to examine the scandal known as Climategate, exploring Merton's analysis of the scientific ethos. I will draw on previous publications on the matter (especially Grundmann 2012a, b).

In late 2009, several hundred emails from a server hosted by the University of East Anglia were made public. This sparked a public debate known as Climategate (see among many sources Hulme/Ravetz 2009; Pearce 2010; Grundmann 2012a, b;

[1] See: http://klimazwiebel.blogspot.de/.

Lahsen 2013; Anderegg/Goldsmith 2014; Garud et al. 2014; Maibach et al. 2014; Ramírez-i-Ollé 2015).

After the release of the emails commentators raised the question of whether or not the exposed behavior was within bounds of 'normal' scientific behavior. There were two scientists who drew most attention, Phil Jones, director of the Climatic Research Unit (CRU) at the university of East Anglia, and Michael Mann, a professor of Earth System Science at Pennsylvania State University.

In one email message, Phil Jones spoke of a 'trick' to "hide the decline" of specific temperature records (email, 16 November 1999). This referred to a solution to the problem of representing diverging temperature curves based on different datasets, one from paleoclimate reconstructions obtained through tree rings, the other from thermometer readings. Both had been spliced together, hiding the obvious divergence between the two. Just when the thermometer curve showed a steep rise, the tree ring curve showed a decline.

In their defense, Jones, Mann and their supporters said that the emails had been quoted out of context and that the word 'trick' meant something innocent, a technical procedure of representing data. In another email, Jones said he would prevent research papers by competitors from being published, even if he had to redefine what peer review means (email, 8 July 2004). Jones convinced his senior management at East Anglia not to provide data under the Freedom of Information Act because of the type of people who were making the requests. According to the government information commissioner, some refusals to comply with freedom of information requests were in breach of the law but could no longer be prosecuted. Anxiety about freedom of information requests led Jones to suggest that his collaborators delete emails. This would have covered up some of the dubious practices behind the scenes. Also, there is evidence of attempting to keep critical papers out of the peer reviewed literature while at the same time fast-tracking papers from within the circle of the likeminded. The same differential treatment has been suspected in the process of compiling assessment reports for the IPCC Fourth Assessment Report published in 2007 (AR4).

Mann was the main author of a historical temperature reconstruction known as "hockey stick", suggesting "unrivalled warming" in the late 20th century. The political implications of this claim are suggestive: if record warming occurs after industrialization, it is likely to be due to anthropogenic influences. Before that, skeptics had always pointed to warmer periods in recent history despite the absence of large-scale fossil fuel burning. This was taken as rationale for doing nothing with regard to greenhouse gas emissions. For them, the emails proved that the temperature record in the hockey stick had been manipulated. Such political dynamite was bound to create high levels of attention.

In the meantime, several investigations about this issue have been carried out (see IAC 2010; Muir Russell 2010; Oxburgh 2010). While the general tenor of commentators is that these inquiries vindicated the accused climate scientists, and because no tampering with the data could be detected and many other studies came to the same or similar conclusions, this verdict does not address the question if there was malpractice. It is no surprise that skeptics saw these inquiries as a whitewash. While none put the square blame on any climate researcher it is probably fair to say that the reviews were not probing too deeply. For example, Phil Jones was never asked if he actually deleted potentially incriminating emails (Pearce 2010).

One review pointed out that scientists at CRU seemed to have dismissed critical enquiries all too quickly, which was problematic and counterproductive. They had violated the Freedom of Information legislation in one case although the offence had lapsed and was not brought for prosecution (Pearce 2010: 147). Leaving legal issues aside, climate scientists have been accused of having developed a bunker mentality. Others call it 'tribalism', the differential treatment of the insiders and the outsiders of the dominant research paradigm (on the notion of the established and the outsiders as sociological categories, see Elias/Scotson 2008).

One review, led by Sir Muir Russell, came to the conclusion that the visual representation of the hockey stick was problematic:

> One of the allegations against CRU is that they have not been sufficiently frank in communicating uncertainties about their reconstructions into the public domain. This is particularly relevant to graphical presentation such as the "hockey stick" which has taken on iconic significance. Images have a great power to persuade, and this is particularly true when complex issues are faced by lay audiences, who may often infer a level of certainty that does not in fact exist. The danger is obviously heightened where an image is being used to support arguments for policy change. Therefore, if images are likely to be used in this way, it is essential that qualifications such as uncertainties are given a closely coupled prominence and explanation. (Muir Russell 2010: 40–1)

This statement from one of the inquiries exemplifies the unease felt with some of the practices of the inner core of climate science with regard to the most powerful, and most contested, symbolic statement of anthropogenic warming.

8.3 Merton's Norms of Science and Criticisms

This sketch looks like an invitation to the sociologist of science to apply theories of scientific ethos. Most prominent, but largely forgotten today, is Robert Merton's framework where he distinguishes between four aspects of scientific norms: universalism, communism, disinterestedness and organized skepticism (later known under the acronym CUDOS).

In a 1978 paper, "The Ethos of Science Revisited", Nico Stehr gave the following summary of Merton's norms:

> 1. Universalism prescribes that knowledge claims in science should be evaluated and accepted or rejected according to impersonal cognitive criteria rather than the 'personal or social attributes of their protagonist' and that careers and opportunities in science should be based on achievement and competence only. (The choice of words perhaps is significant; the expression 'should be' denotes a prescription aspired to but not necessarily a description of consistent scientific practice.) 2. Communism refers to the interrelated 'communal' (public) character of scientific knowledge claims; the corresponding limited 'rights' of the

> originator(s) to recognition and esteem, resulting in the distinctive and anomalous character of intellectual property in science; and the imperative not to withhold knowledge claims—an imperative reinforced by the 'incentive of recognition, which is, of course, contingent upon publication'. 3. Disinterestedness, the moral imperative at the institutional level of science, is largely self-explanatory; it points to a distinctive structure of control exercised over the individual motives of scientists. 4. Organized Skepticism is 'both a methodological and an institutional mandate'. Knowledge claims should not be accepted without (socially organized) scrutiny but should be warranted with reference to the technical norms of science. (Stehr 1978: 174)

Nico Stehr follows Merton in that the norms are seen as both moral and cognitive. As Merton (1973: 270) notes, "[t]he mores of science possess a methodologic rationale but they are binding, not only because they are procedurally efficient, but because they are believed right and good. They are moral as well as technical prescriptions."

Merton suggests two different mechanisms to make the scientific ethos work, one of them functional, the other moral. The German sociologist Luhmann (1968) affirmed the functionalist rationale, postulating social mechanisms to operate as cognitive mechanisms in science. Stehr comments approvingly:

> Thus, a definite correlation between moral imperatives and the advancement of scientific knowledge is implied. For instance, 'objectivity precludes particularism. The circumstance that scientifically verified formulations refer in that specific sense to objective sequences and correlations militates against all efforts to impose particularistic criteria of validity' (Merton 1973: 270). In other words, the norms do not merely regulate the behavior or the social relations of members of the scientific community; also, in distinct ways, they enhance the institutional goal of science, which is the continuing extension of certified knowledge claims.' It is therefore clear that, in Merton's view, the norms of science derive from the purpose of science: 'The institutional imperatives (mores) derive from the goal and the methods. The entire structure of technical [cognitive] and moral norms implements the final objective' (Merton 1973: 270). The normatively prescribed social relations of science therefore complement, if not implement, the development of knowledge in science and vice versa. The dialectics of social and cognitive norms are a part of the institution of science. (Stehr 1978: 175–6)

Merton wrote in the 1940s, and his essay received broad attention in the postwar period, which was partly due to the prospering of structural functionalism at the time. However, with its demise and the rise of the new sociology of science, Merton's scientific ethos lost visibility and plausibility. Several criticisms have been leveled against Merton's CUDOS, which I will discuss below. As I will show, Mertonians gave into their critics in one important aspect.

Merton based his analysis on the reading of scientists' biographies and autobiographies and some anecdotal evidence but never "spent any time in natural scientific or engineering settings" (Shapin 2008: 114). His basic idea was that science is a specific social institution which is governed by specific social and cognitive norms. These norms become binding for the actors operating in this institution, the scientists become socialized into it. The norms are functional for science, but also for society. As society establishes a special sphere for the production of knowledge, it can rely on the knowledge thus created, provided it is subject to the criteria listed in the norms. Science produces reliable knowledge.[2]

At the same time the norms that govern science are a kind of regulative idea for the rest of society because critical views and open and public debates are vital for a democratic society. Thus, a special role can be claimed for science. Science seems less governed by interests, and scientists appear to be less partial compared to other actors.

However, one must not make the mistake to assume that scientists as persons are somehow special, or morally superior. Merton warns that the search for noble motives or moral exceptionalism in scientists is in vain: "A passion for knowledge, idle curiosity, altruistic concern with the benefit to humanity, and a host of other special motives have been attributed to the scientist. The quest for distinctive motives appears to have been misdirected." (Merton 1973: 276). In an earlier paper Merton had indicated that "scientists may be most variously motivated—by a disinterested desire to learn, by hope of economic gain, by active (or, as Veblen calls it, by idle) curiosity, by aggression or competition, by egotism or altruism. But the same motives in different institutional settings take different social expressions [...]" (Merton 1949: 532).

What matters for Merton is the institutional setting in which the scientist operates: "It is rather a distinctive pattern of institutional control of a wide range of motives which characterizes the behavior of scientists. For once the institution enjoins disinterested activity, it is to the interest of scientists to conform on pain of sanctions and, insofar as the norm has been internalized, on pain of psychological conflict" (Merton 1973: 276). According to Merton, there are two mechanisms which align the scientist with the scientific ethos, i.e., institutional sanctions and psychological stress. We will bear these in mind and return to them later.

Merton's norms were subject to criticism from various authors who advanced different arguments. So what did Merton's critics have to say? And what implications does this have for the evaluation of Climategate?

Again, Nico Stehr's exposition is a useful starting point. In the above-mentioned article, "The Ethos of Science Revisited," he provided a summary of critical arguments, distinguishing four types of critical questions: Are the norms of science peculiar to science? Do the norms actually affect the conduct of scientists? Are moral imperatives of scientific practice indeed moral? Is the ethos of science beneficial for the development of scientific knowledge?

---

[2] Note that this does not mean that science produces 'true' knowledge in the sense this term has been understood by philosophers. Steve Shapin's comment is quite insightful. He quotes Einstein who said in the 1920s that it is difficult to attach a precise meaning to the term 'scientific truth' as its semantics are context dependent. Shapin also quotes William Baker, director of Bell Labs and science advisor to five US presidents, who insisted that public trust in science was dependent upon realizing that there were limits to the notion of scientific 'truth' and 'certainty': "Scientific findings, scientific facts, are usually thought of as symbols of certainty. But people must realize that these findings are certain only with respect to a particular frame of reference" (quoted in Shapin 2008: 30).

The literature on Merton's ethos that has emerged in the meantime can be summarized in three hypotheses, running counter to Merton's framework:

1. Norms and moral rules are universal, not special for scientists
2. Scientists are self-interested like everyone else
3. Communism has been replaced by intellectual property rights.

8.4 Climategate and Scientific Norms

In this section I will present relevant criticisms of Merton's CUDOS and relate them to the example of Climategate.

According to the first criticism (Schmaus 1983: 13), scientists' norms are not different from others. For example, cases of "scientific fraud involve the violation not of a special moral rule applying to scientists but rather a general moral rule requiring that one's job or role-related responsibilities be fulfilled honestly, whatever they may be. In the case of scientists, this general moral rule entails that they fulfill their obligations as scientists. The responsibilities of scientists, in turn, consist largely in maintaining the highest standards of intellectual rigor." In other words, there are no special moral norms operating in science. But from the fact that scientists are not different from or 'better' than other citizens it does not follow that they are exempt from general moral codes.

Applied to Climategate this would mean that hiding data and giving preferential treatment to allies is in violation of such general norms.

According to the second criticism, scientists always act out of self-interest: They assess the work of others since they depend on their recognition. They invest and accumulate credibility capital which is convertible into other resources (Bourdieu 1975; Latour/Woolgar 1986). Normative statements about scientific practice such as Merton's are ideological façades for the defense of the autonomy of science (Mulkay 1976).

In this version there would be no basis to judge the Climategate affair as unethical. Scientists get away with whatever others tolerate, and the most successful scientists will write the history of the field.

According to the third criticism, the norm of communism is being or has been replaced by the norm of private intellectual property. Since the creation of national competitiveness policies in the early to mid-1970s, the academic world has been experiencing a period of 'renormalization'.

As early as in the early 1970s it was observed (e.g. by Sklair 1973: 113) that "much, if not most, contemporary science is carried out under conditions of formal or informal secrecy—necessitated by national security matters and matters of economic interest."

The message with regard to Climategate would be that as far as data secrecy is concerned, climate scientists were acting in line with an emerging and perhaps broadly shared practice.

Other critics, such as Barry Barnes (Barnes/Dolby 1970) or Mulkay (1976) argued that it is only cognitive rather than social norms that are operative in science. Cognitive norms relate to technical aspects of scientific work and dictate specific methods and practices. Normative standards are little more than ideological façades for defending the autonomy of science.

In addition, scientists' self-interest was a topic for authors like Barnes (1971) and for French scholars Bourdieu and Latour who made the case for an interest based action theory for the analysis of science. One example is the peer review system for which scientists are not rewarded in financial terms. So what prompts their participation? According to Bourdieu and Latour they profit in various, and mostly intangible, ways. They appropriate new information which they can use to their advantage. Latour (1994) coined the memorable phrase of the "scientist as a wild capitalist" accumulating academic reputation which, then, can be exchanged into other forms of capital, thus climbing the career ladder and gaining social prestige.

Collins (1985) emphasized that career opportunities and influence in science are unequally distributed. This is an empirical argument supposed to undermine Merton's norms of universalism and disinterestedness. It is above all core groups in science that have special power and influence (see also Crane 1972). They are located in a few prestige institutions and are well connected among themselves. They control access to publications, define the important research agendas and define quality. This social process of status attribution, recognition, and *gatekeeping* runs like a *leitmotiv* through the history of science. And it is a more general social mechanism, too.

Unanswered by these criticisms is the question if (and how) one can identify scientific misconduct (assuming it is still something to be worried about) and how it can be minimized.

Another set of empirical studies in the 1970s undermined the norm of communism with the rise of *intellectual property regimes.* According to Metlay (2006) this has led to a re-normalization of science. The norm of communism is increasingly displaced by the norm of protection of intellectual property. A large proportion of modern science is produced under conditions of secrecy. Military and economic interests dictate these practices. There may be good reasons to justify the secrecy of some knowledge (albeit the rise of global social media makes this less practical). But such arguments come increasingly under close scrutiny. For example, the Dutch virologist Ron Fouchier was initially prevented by the Dutch government from publishing his work on the H5N1 virus for fear that terrorists might use the knowledge in a biological weapon. Here, two motives of Merton's analysis intertwine, the scientific ethos and the critical public discourse, with the freedom of speech being crucial to both. The case went to the courts, so science lost its autonomy in this instance (Enserink 2015).

One should think that the above comments and criticisms would have led to a substantial disagreement between Mertonians and others. However, Merton did not enter the fray, his wife Zuckerman (1988a, b) and Ben-David (1991) being the most active. Both accepted the framing of the question as an empirical question, i.e. whether scientists in reality behave ethically (in Merton's sense). Ben-David (1991:

479f) claims that scientists are norm-following ("there is little doubt [...] that scientists observe the rules of their professional ethic in the allocation of rewards and punishment of fraud and other kinds of deviance"). Ben-David admits that some studies show that this is not always the case ("The outcome of these is inconclusive: answers to some questions show acceptance of norms, while others suggest rejection").

However, the following or violating of norms presupposes that such norms exist. Key to the explanation of this difference is the special role of scientific controversies. Whenever scientists are embroiled in scientific controversies ethical norms recede into the background. In such situations scientists "are indeed willing to transgress practically all the norms enumerated by Merton [...]: they may withhold findings in order to prevent advantage from competitors, they make judgments on the basis of personal prejudice, interpret results arbitrarily in order to fit theories, and so forth." Ben-David argues that scientists participating in a scientific controversy are not to be judged according to CUDOS: "scientists at this stage act like litigants concerned more with putting together a convincing case than with ultimate truth. They are not, and are not expected to be, dispassionate" (Ben-David 1991: 480).

This indicates that the email scandal could be seen as within the scientific ethos, as defined by Ben-David. In controversies, passionate scientists use various tactics to further their goals, "like litigants putting together a convincing case".

However, Merton's defenders did not develop an argument in terms of "if the norms are not followed, so much the worse for science", or "norms get violated but we agree that they have been violated and therefore we agree that the norm exists," or "norms are changing over time". Instead, they took the norms as empirical descriptions of actual practices. Perhaps this was prompted by Merton's own emphasis "that there have been comparatively few empirical instances of deviant responses to particular norms of science" (Merton 1973: 321).

This would be supported by a comment Merton made in 2000, to Nico and myself, on a draft paper which would appear much later in print (Grundmann/Stehr 2003). Merton took aim at our somewhat sloppy dismissal of norms as mythical and self-serving. He explained that in the vast majority of cases scientists follow the norms:

> As we know, if conforming behavior far outruns deviant behavior, institutional norms are not *merely* 'mythic' or 'self-serving'. Your 'of course' about scientists 'repeatedly' deviat[ing] *de facto* from the norm strongly implies quantitative evidence that deviations occur to such a great extent that the norm is no longer operative. But is that truly the case here? After all, the 1942 paper on science as a social institution is followed (in *The Sociology Of Science)* by the 1957 paper on the institutionally distinctive reward system of science, a piece that goes on at some length to analyze the institutional roots of deviant behavior (such as faked evidence and plagiary). But it is not suggested that such deviant behavior occurs on a magnitude that renders the norms as merely mythic. (Robert Merton, email to Nico Stehr, 2 August 2000, original emphases).

In the mentioned paper (originally entitled "Priorities in Scientific Discovery: A Chapter in the Sociology of Science"), Merton says:

> To say that these frequent conflicts over priority are rooted in the egotism of human nature, then, explains next to nothing; to say that they are rooted in the contentious personalities of those recruited by science may explain part, but not enough; to say, however, that these conflicts are largely a consequence of the institutional norms of science itself comes closer, I think, to the truth. For, as I shall suggest, it is these norms that exert pressure upon scientists to assert their claims, and this goes far toward explaining the seeming paradox that even those meek and unaggressive men, ordinarily slow to press their own claims in other spheres of life, will often do so in their scientific work. (Merton 1957: 639)

Merton maintains that CUDOS impels scientists to make a forceful case for their claims and discoveries. It is not their character or personality that explains the conflictual nature of such exchanges but the institutional norm that they should do so.

Merton did not comment on the nature of scientific controversies as such, and he certainly did not say that CUDOS would impel scientists to abandon some elements of it—this would have created a paradox. But critics and supporters of Merton agree that it is unrealistic to expect scientists to behave in disinterested ways when engaged in controversies.[3] If this is the case, what yardstick remains to identify malpractice in scientific controversies?

Upon reflection it appears that Merton's defenders have embarked on a flawed strategy in order to diffuse criticism, leaving the matter a hostage to fortune. They may have assumed, or hoped, that controversies would be the exception rather than the norm in scientific research. Others have suggested that we are faced with conditions of uncertainty, scientific controversies, and therefore "post-normal science".

[3] There is an interesting change in wording in Merton's characterization of "organized scepticism". In his 1942 paper "Science and technology in a democratic order" (reprinted in *Social Theory and Social Structure,* 1957) he mentions as a distinctive character of science "the suspension of judgment until 'the facts are at hand' and the detached scrutiny of beliefs in terms of empirical and logical criteria have periodically involved science in conflict with other institutions" (Merton 1957: 560). In the final version of the paper, published in 1973 as "The Normative Structure of Science" in the book *The Sociology of Science*, Merton omits the phrase "suspension of judgment until 'the facts are at hand'." It now reads: "The temporary suspension of judgment and the detached scrutiny of beliefs in terms of empirical and logical criteria have periodically involved science in conflict with other institutions" (Merton 1973: 277). This change may have been due to Barnes/Dolby's (1970) article in which they criticise Merton as follows: "[W]hen we consider organised scepticism we must admit that a universal injunction to 'reserve judgement until the facts are at hand' combines oddly with the fact that all scientific theories go beyond the knowledge on which they are based. Further, examinations of scientific controversies show that typically scientists take sides between particular theories, often developing emotional attachment to them, and arguing their merits with great vehemence and polemic skill" (Barnes/Dolby 1970: 10). In Merton's defence it could be said that his original phrasing referred to science as a subsystem of society rather than to individual scientists. Merton does not cite Barnes and Dolby's critique.

8.5 Institutional Responses?

Together with Silvio Funtowicz, Ravetz (1993) has proposed a theory of post-normal science—that is, situations in which facts are uncertain, values in dispute, stakes high, and decisions urgent. This means that we are dealing with political controversies which are mapped onto scientific controversies.

Ravetz is one of the few social science commentators who are critical of the Climategate scientists under discussion. According to him, some leading climate scientists were induced (by skeptical interest groups) to adopt

> a simple, forcefully argued position which then had to be defended, no matter what: once the position was adopted, its proponents became invested in it, and attached to it, in all sorts of ways, institutional and personal. And I suspect that a simplified, even simplistic claim, was more comfortable for these scientists than one where complexity and uncertainty were acknowledged. (Ravetz 2010)

The nature of the political constellation called for support from the like-minded, and Ravetz admits having being lured into such an attitude for some time. But after 'Climategate' he exposed the "nature trick" as malpractice. Ravetz speaks of the corruption of climate science in this context, invoking Merton, if not in name, then certainly in spirit, as this quote exemplifies:

> We found ourselves in another crusading 'War', like those on (non-alcoholic) Drugs and on 'Terror'. This new War, on Carbon, was equally simplistic, and equally prone to corruption and failure. Global warming science became the core element of this major worldwide campaign to save the planet. Any weakening of the scientific case would have amounted to a betrayal of the good cause, as well as a disruption of the growing research effort. All critics, even those who were full members of the scientific peer community, had to be derided and dismissed. As we learned from the CRU e-mails, they were not considered to be entitled to the normal courtesies of scientific sharing and debate. Requests for information were stalled, and as one witty blogger has put it, 'peer review' was replaced by 'pal review'. (Ravetz 2011: 151)

So the question arises how to avoid morally bankrupt behavior in science. Here Ravetz draws on the distinction between normal and post-normal science and their different regimes of quality control:

> In traditional 'normal' science, the peer community, performing the functions of quality-assurance and governance, is strictly confined to the researchers who share the paradigm. In the case of 'professional consultancy', the clients and/or sponsors also participate in governance. We have argued that in the case of post-normal science, the 'extended peer community', including all affected by the policy being implemented, must be fully involved. Its particular contribution will depend on the nature of the core scientific problem, and also on the phase of investigation. Detailed technical work is a task for experts, but quality-control on even that work can be done by those with much broader expertise. (Ravetz 2011: 151)

It has to be noted that Merton, when talking about potential academic misconduct, was thinking of plagiarism and of claims to priority of a discovery. Those who could be harmed and those who had to make judgments about the quality of the science were competing scientists working in the same field. With regard to

Climategate we see different actors and problems. The competitors were not only academic colleagues working in the same field (such as Hans von Storch) but outsiders (such as Steve McIntyre). Much controversy was spun around the allegations coming from the latter's blog *Climate Audit*.[4] No priority claims were involved, nor were there charges of plagiarism. Accusations were made about misrepresentation of data and about failing to engage in a proper exchange about the issues raised. As the public dispute escalated, accusations of fraud, political bias, and corruption were added, as well.

Ravetz's suggestion about quality control indicates that scientists should have realized that there were relevant others who may have been outside of the research community, or core group, but still able to comment on their work in an informed and competent way. The British inquiry led by Alastair Muir Russell came to a similar conclusion:

> Arguably the most significant change produced by the blogosphere is a transformation in the degree of openness now required of scientists whose work directly affects policy making. Without such openness, the credibility of their work will suffer because it will always be at risk of allegations of concealment and hence mal-practice. (Muir Russell 2010: 42)

But the review stops short of calling for an "extended peer review", as envisaged by Ravetz and colleagues. Rather, the problem is seen in terms of a more effective communication of scientific results to the public: "Therefore, the Review would urge all scientists to learn to communicate their work in ways that the public can access and understand; and to be open in providing the information that will enable the debate, wherever it occurs, to be conducted objectively." (ibid.)

8.6 Sanctions?

Merton conceived of scientists as morally equivalent to other members of society. No special character or motivation was seen as crucial for the scientific endeavor in society; several contradictory motives that drive scientists were identified. The task was to create the right institutions to put these motivations to work. The scientific ethos works on the individual level only if it has been internalized. Scientific norms can be assumed to be operative only if scientists have accepted them. Merton assumed that internalization would be achieved through socialization, the process by which a novice becomes a full member of the scientific community. He also identified two mechanisms of sanctions, institutional disapproval and psychological stress. Both sanctions would enforce the scientific ethos in practice.

With regard to the Climategate affair, we can perhaps see an interesting contrast between two protagonists. Neither the University of East Anglia (Phil Jones's employer) nor Pennsylvania State (Michael Mann's employer) reprimanded their

[4] See: https://climateaudit.org/.

climate scientists. However, Jones is on record as describing his actions to the UK parliamentary inquiry somewhat self-deprecatingly: "I've obviously written some really awful emails" (*The Guardian*, 1 March 2010). He considered committing suicide in the aftermath of the scandal (*The Telegraph*, 7 February 2010). This could be seen as a manifestation of psychological conflict, as mentioned by Merton. It would indicate that Jones had internalized some relevant norms of science and that the email debates had brought to light something that troubled him.

Michael Mann seems to have reacted differently. To be fair, he was (and still is) involved in various legal battles related to his actions. According to his Wikipedia entry he regretted not objecting to Jones's request to delete emails because of FOI requests (although the source—"Penn State climate professor: I'm a skeptic". *The Morning Call*, 3 January 2010, is no longer traceable). If this may be seen as a tactical ploy, it still would show adherence to some kind of scientific ethos, if only symbolically.

In 2012, Mann published a book entitled *The Hockey Stick and the Climate Wars*, with the subtitle: Dispatches from the Front Lines. If you think you are in war, pretty much anything goes. There may be the Geneva Convention, so to speak, but certainly no scientific ethos, no matter how it is defined. There are two parties to a violent conflict and both will blame the other for provoking acts of violence. There is, of course, no real violence but the metaphor is apt as it directs our attention to the suspension of normal procedures, norms and values. This logic leads to an assimilation of the conflicting parties with regard to their repertoire of weaponry.

The war metaphor lends some legitimacy to highly questionable practices. Taking into account that there is polarization and antagonism, the metaphor of an arms race was used by Bruno Latour who wrote, long before Climategate: "In this, the 'proof race' is similar to the arms race because the feedback mechanism is the same. Once one competitor starts building up harder facts, the others have to do the same or else submit" (Latour 1986). This captures the virtuality of real conflict, stopping short of exercising violence, and the need on both sides to find more and better resources to vanquish the other side. It still leaves open the question of what means are allowed to reach this end.

8.7 Conclusion

The first sentence in Merton's essay *A note on Science and Democracy* reads: "Science, as any other activity involving social collaboration, is subject to shifting fortunes" (Merton 1942: 115). This is as true today as it was back then. Social reality has changed since the publication of this essay. Does this mean that Merton's postulated norms of science are an anachronism? And if so, what criteria do we have to evaluate scientific behavior?

Mertonians have granted the suspension of the scientific ethos during scientific controversies. This position seems not warranted. It is by no means unreasonable to expect that scientists involved in such disputes follow some kind of norms. When scientists become advocates (putting together a case) this does not mean that they are beyond any normative expectations.

The email scandal which made headlines in late 2009 cast a spotlight on a small but dedicated community of climate scientists who felt beleaguered by critics. Some of them allegedly reacted by withholding research results and trying to manipulate the peer review process and abort skeptical criticism. All of this happened because skeptics claimed they had adjusted data to make them fit their preconceived ideas and theories. If one takes Ben-David's position seriously, there would be nothing wrong with this. However, the moral outrage witnessed on the occasion is testament to the fact that the behavior was seen as deeply problematic. The incriminated scientists and their supporters also accepted the premise and tried to justify what they did in moral terms. This seems to indicate that there is a standard after all, according to which malpractice can be identified, even if the standard is not codified anywhere. Several investigations of the affair concluded that this community showed a bunker mentality and too little openness and transparency. This suggests that several, if not all of Merton's norms were touched by the affair. However, the same investigations also showed appreciation towards the scientists' behavior, as they had been under immense pressure. Again, we see the argument being made that scientists, after all, are human, all-too-human and may use unethical means to defend themselves.

It is unlikely Merton would have accepted such an argument. His norms do no longer inform debates about scientific behavior. In its place we see a range of reactions, including a justification of the alleged malpractice and an appeal to professional standards and the honesty of the researcher. In a recent study, Bray/von Storch (2014: 1) report the results of an online survey on climate scientists' attitude toward the norms of science.

> The data suggests that while Merton's CUDOS remain the overall guiding moral principles, they are not fully endorsed or present in the conduct of climate scientists: there is a tendency to withhold results until publication, there is the intention of maintaining property rights, there is external influence defining research and the tendency to assign the significance of authored work according to the status of the author rather than content of the paper. These are contrary to the norms of science as proposed by Robert K. Merton.

Empirical research of this kind faces the problem of the social desirability bias: Scientists might tell you what they think is expected from them. Nevertheless, the authors were able to distinguish between the agreement to the norms in general and the specific attitudes toward individual items of the ethos. It might be useful to compare different research fields and to take into account the degree to which controversies are being fought out.

It is perhaps ironic that the line of defense taken by Mertonians has undermined Merton's ethos as effectively as the attack launched by his critics. Perhaps it is time to re-conceptualize the ethos of science by aligning it with either professional codes of conduct or moral norms more generally. In recent years, several rules and procedures in science have been opened up for discussion, among them issues such as how to deal with conflicts of interests, how to make peer review more transparent, how to provide access to data and enable sharing of data, and perhaps most importantly, how to provide advice for decision makers. These are the issues around which a debate about the ethos of science needs to be conducted.

Especially the last aspect, the interface between knowledge and decision making (sometimes called *science for policy*) needs close attention. In knowledge societies, there is an ever-increasing demand for knowledge as justification for action yet the rules for this process are set by political power holders. Individual scientists are enrolled and enticed to play along so that their recommendations can be represented as if the chosen policy was demanded, or at least supported, by 'science'. But science can never demand any course of action, it is individual scientists who do so in the process of public will formation, which is the central process in democratic societies. Advisory boards need to follow principles of accountability, independence and transparency in order to be trustworthy. All too often scientists forget this simple truth and put a higher value on the policy goals they happen to favor, no matter how transparent or impartial the process which leads to their adoption. If scientists have nothing but an instrumental view of their knowledge they will inevitably diminish its trustworthiness. If they follow the maxim of the ends justifying the means, they should think twice. The contributions of Merton and Stehr remind us of the importance of this consideration.

References

Anderegg, William R.L. and Gregory R. Goldsmith. 2014. "Public Interest in Climate Change over the Past Decade and the Effects of the 'Climategate' Media Event", *Environmental Research Letters*. IOP Publishing, 9(5): 54005. https://doi.org/10.1088/1748-9326/9/5/054005.

Barnes, Barry and R.G.A. Dolby. 1970. "The Scientific Ethos: A Deviant Viewpoint", *European Journal of Sociology*, 11: 3–25.

Ben-David, Joseph. 1991. "Norms of Science and the Sociological Interpretation of Scientific Behavior", in: Joseph Ben-David (ed.), *Scientific Growth. Essays on the Social Organization and Ethos of Science*. Berkeley, CA: University of California Press: 469–484.

Bourdieu, Pierre. 1975. "The Specificity of the Scientific Field and the Social Conditions of the Progress of Reason", *Social Science Information*, 14(6): 19–47.

Bray, Dennis and von Storch, Hans. 2014. "The Normative Orientations of Climate Scientists", *Science and Engineering Ethics*. https://doi.org/10.1007/s11948-014-9605-1.

Collins, Harry. 1985. *Changing Order: Replication and Induction in Scientific Practice*. London: Sage.

Crane, Diana. 1972. *Invisible Colleges: Diffusion of Knowledge in Scientific Communities*. Chicago: University of Chicago Press.

Elias, Norbert and John L. Scotson. 2008. "The Established and the Outsiders", in: Cas Wouters (ed.), *The Collected Works of Norbert Elias*. Dublin: University College Dublin Press.

Enserink, Martin. 2015. "Dutch Appeals Court Dodges Decision on Hotly Debated H5N1 Papers", *Science*. https://doi.org/10.1126/science.aac8867.

Garud, Raghu, Joel Gehman and Arvind Karunakaran. 2014. "Boundaries, Breaches, and Bridges: The Case of Climategate", *Research Policy*. Elsevier B.V., 43(1): 60–73. https://doi.org/10.1016/j.respol.2013.07.007.

Grundmann, Reiner. 2012a. "Climategate and the Scientific Ethos", *Science, Technology & Human Values*, 38(1): 67–93. https://doi.org/10.1177/0162243911432318.
Grundmann, Reiner. 2012b. "The Legacy of Climategate: Revitalizing or Undermining Climate Science and Policy?", *Wiley Interdisciplinary Reviews: Climate Change*, 3: 281–288. https://doi.org/10.1002/wcc.166.
Grundmann, Reiner and Nico Stehr. 2012c. *The Power of Scientific Knowledge: From Research to Public Policy*. Cambridge: Cambridge University Press.
Grundmann, Reiner and Nico Stehr. 2001. "Why Is Werner Sombart Not Part of the Core of Classical Sociology?: From Fame to (Near) Oblivion", *Journal of Classical Sociology*, 1(2): 257.
Hulme, Mike and Jerry Ravetz. 2009. BBC News—"'*Show Your Working*': What 'ClimateGate' Means." Available at: http://news.bbc.co.uk/1/hi/8388485.stm (22 August 2011).
IAC. 2010. Review of the IPCC. *A Review of the Processes and Procedures of the Intergovernmental Panel on Climate Change*. Amsterdam: Interacademy Council.
Lahsen, Myanna. 2013. "Climategate: The Role of the Social Sciences", *Climatic Change*, 119(3–4): 547–558. https://doi.org/10.1007/s10584-013-0711-x.
Latour, Bruno. 1986. "Visualisation and Cognition: Drawing Things Together", *Knowledge and Society: Studies in the Sociology of Culture and Present*, 6: 1–40.
Latour, Bruno. 1994. "Der Biologe als wilder Kapitalist. Karrierestrategien im internationalen Wissenschaftsbetrieb", *Lettre International*, Winter(Issue 27): 77–83.
Latour, Bruno and Steve Woolgar. 1986. *Laboratory Life: The Construction of Scientific Facts*. Princeton, NJ: Princeton University Press.
Luhmann, Niklas. 1968. "Selbststeurung der Wissenschaft", in: *Soziologische Aufklärung: Aufsätze zur Theorie sozialer Systeme*. Opladen: Westdeutscher Verlag.
Maibach, Edward, Teresa Myers and Aanthony Leiserowitz. 2014. "Climate Scientists Need to Set the Record Straight: There Is a Scientific Consensus that Human-Caused Climate Change Is Happening", *Earth's Future*. https://doi.org/10.1002/2013ef000226.
Merton, Robert K. 1957. "Priorities in Scientific Discovery: A Chapter in the Sociology of Science", *American Sociological Review*, 22(6): 635–659. https://doi.org/10.2307/2089193.
Merton, Robert K. [1942] 1957. "Science and Democratic Social Structure", in: Robert K. Merton (ed.), *Social Theory and Social Structure*. Glencoe, IL: The Free Press: 531–549.
Merton, Robert K. 1973. "The Normative Structure of Science", in: Norman W. Storer. (ed.), *The Sociology of Science: Theoretical and Empirical Investigations*. Chicago: University of Chicago Press: 267–273.
Metlay, Grischa. 2006. "Reconsidering Renormalization: Stability and Change in 20th-Century Views on University Patents", *Social Studies of Science*, 36(4): 565–597. https://doi.org/10.1177/0306312706058581.
Muir Russell, Alastair. 2010. *The Independent Climate Change E-mails Review*. http://www.cce-review.org/index.php.
Mulkay, Michael J. 1976. "Norms and Ideology in Science", *Social Science Information*, 15(4–5): 637–656. https://doi.org/10.1177/053901847601500406.
Oxburgh, Ronald. 2010. *Report by Lord Oxburgh's Science Assessment Panel*; at: http://www.uea.ac.uk/mac/comm/media/press/CRUstatements/SAP.
Pearce, Fred. 2010. *The Climate Files: The Battle for the Truth About Global Warming*. London: Guardian Books.
Ramírez-i-Ollé, Meritxell. 2015. "Rhetorical Strategies for Scientific Authority: A Boundary-Work Analysis of 'Climategate'", *Science as Culture*, 5431(March): 1–28. https://doi.org/10.1080/09505431.2015.1041902.
Ravetz, Jerome. 2010. *Climategate: Plausibility and the Blogosphere in the Post-normal Age*; at: http://wattsupwiththat.com/2010/02/09/climategate-plausibility-and-the-blogosphere-in-the-post-normal-age/.
Ravetz, Jerome. 2011. "'Climategate' and the Maturing of Post-normal Science", *Futures*, 43(2): 149–157. https://doi.org/10.1016/j.futures.2010.10.003.

Schmaus, Warren. 1983. "Fraud and the Norms of Science," *Science, Technology & Human Values*, 8(4): 12–22.

Shapin, Steven. 2008. *The Scientific Life: A Moral History of a Late Modern Vocation*. Chicago: University of Chicago Press.

Sklair, Leslie. 1973. *Organized Knowledge. A Sociological View of Science and Technology*. St. Albans: Paladin Books.

Stehr, Nico. 1978. "The Ethos of Science Revisited", *Sociological Inquiry*, 48(3–4): 172–196. https://doi.org/10.1111/j.1475-682x.1978.tb00825.x.

Stehr, Nico and Reiner Grundmann. 2001. *Economic Life in the Modern Age, English*. Edited by Nico Stehr and Reiner Grundmann. New Brunswick, NJ: Transaction Publishers.

Zuckerman, Harriet A. 1988a. "Introduction: Intellectual Property and Diverse Rights of Ownership in Science", *Science, Technology, & Human Values*, 13(1/2): 7–16.

Zuckerman, Harriet A. 1988b. "The Sociology of Science", in: Neal Smelser (ed.), *Handbook of Sociology*. London: Sage: 511–574.

Chapter 9
Social Science: A Must for Climate Research

Hans von Storch

9.1 Nico and Me Wandering Through Climate…

Nico Stehr and me met first in about 1992, or so, at Cecilienhof in Potsdam. The newly founded Potsdam Institute of Climate Impact Research had convened a group of people who were thought to be able to help bring together the natural science side of climate research and the "societal dimension", whatever this might mean. So, there we were, a physical climate scientist from the Max Planck Institute for Meteorology, where we naively played around with ideas how climate change might affect societies; and a social scientist from Edmonton who knew something about the social process of science, Nico Stehr. I do not remember anything about the meeting itself, but Nico offered a good opportunity to interact with people who were not just dabbling physicists. After a while, we even found out that we could speak in German to each other. Then I invited him to come to our institute in Hamburg for a couple of months, and he accepted this invitation.

What did we natural scientists think, back in the early 1990s? None of us had any training in the social or cultural sciences; we had followed a few courses in environmental economy, had read Nordhaus' "To slow or not to slow" (1991). We thought that it would be possible to include society in our models, just another model component which would interact deterministically, possibly with some stochastic components, with the natural system that comprised the atmosphere, the ocean, vegetation and the carbon cycle. That we would be able, at least in principle, to estimate optimal measures which would balance the costs for reducing human-made climate change (mitigation) and the costs for adapting societal vulnerability to changing climate conditions. The role of the political process would mostly be to

Prof. Dr. Hans von Storch, Director emeritus, Institute for Coastal Research of the Helmholtz-Zentrum Geesthacht, Professor at the University of Hamburg; Email: hvonstorch@web.de.

M. T. Adolf (ed.), *Nico Stehr: Pioneer in the Theory of Society and Knowledge*,
Pioneers in Arts, Humanities, Science, Engineering, Practice 16,
https://doi.org/10.1007/978-3-319-76995-0_9

determine an appropriate metric, apart from avoiding free-rider problems. At its core, this argument simply represented climatic determinism. It assumed implicitly that societal development would be governed by certain laws. The role we assigned to social and cultural sciences in these days was to formulate these laws and to provide parameters for modeling, such as demographic data and the usage of resources.

Then came Nico. He gave a talk on…—I am not sure. I mostly remember the set-up of the talk. He had asked for an overhead projector, as was common for us natural scientists at the Max Planck Institute. We were sitting and waiting, when he came in—and then… he *sat down*. An unheard behavior; we give our presentations standing, in front of the screen, going through our transparencies. Nico had a few slides, but they only served as decoration on the wall. And then he read his manuscript. He *read* it—and we had to listen.

Unfortunately, we were not used to that—we were used to looking at diagrams, thinking about them, having our thoughts wander around, and eventually returning to the diagrams, compact hypotheses and conclusions on the transparencies. But Nico read and read, and we were lost, at least I was lost, I guess almost all of us were. Two cultures met for the first time, and it needed some time to get the interaction going.

I do not really remember what his talk was about, but he did not provide us with laws of social development and responses to climatic change, there were no parameters to estimate emissions or vulnerability, but very likely he spoke about Emmanuel de Roy Ladurie's book "*Times of Feast, Times of Famine*" (1971), the case of the wet summers in England in 1315–17, when harvests failed and the religious authorities instituted a climate protection policy of the time, a successful policy as it had to be perceived. This case was an eye-opener to me, and my apprenticeship with Professor Stehr began then, back in 1993. That was more than 20 years ago, and I am deeply grateful to Nico for what he has taught me.

We started with using this case of the Great Famine 1315–17 as an example illustrating that the issue of man-made climate change was not a new concept, that the idea of adopting measures to end the catastrophic consequences of our mis-doings—manifest in an adverse climate—had been thought of before but in a framework entirely meaningless for our enlightened times. My question was—would coming generations find our explanations and measures as strange as the activities of the Archbishop of Canterbury some 700 years earlier had appeared to us?

The first thing we wrote was a piece in the German weekly *Die Zeit*. The title was "*Genarrt vom Wettergott*" ("Fooled by the weather god", 1993), and we wanted to see if others had thought about this link before. The response to this article was meager, and apart from Wolfgang Settekorn, a media scientist at the University of Hamburg, no significant contact emerged. A full article was published two years later in the journal Climate Research (Stehr/von Storch 1995), and we continued with other work, such as climatic determinism and the case of Eduard

Brückner (Stehr/von Storch 2008). Moreover, we continued to publish shorter articles in newspapers and weeklies to encourage exchange with other segments of the public and the sciences.

Brückner's case was an important step in our understanding because it showed us that the present perception of man-made climate change and the need for mitigation was something recurrent in history and that the issue of deforestation in the 19th century was in many aspects similar, albeit smaller in extent, to the current concern about ongoing man-made climate change. We turned this into a short article in *Nature* in 2000, "*Climate change in perspective. Our concerns about global warming have an age-old resonance*".

Another important line of research was related to climatic determinism—with a focus on the concepts of Huntington (1915). Ellsworth Huntington was a US geographer from Yale who tried to present the classical climatic determinism in modern scientific clothes, with maps, statistics and the like. He demonstrated the similarity between global maps of presumed levels of civilization, on the one hand, and the utility of regional climate for creating favorable working conditions, on the other. Not surprisingly, the two maps both pointed to Western Europe and parts of North America, while Africa and most of the rest of the world obviously suffered from unfavorable climatic conditions and needed support from the regions with favorable conditions—in short, colonialism. I was surprised to find 'my' good climatology in the neighborhood of racism.

The case of Brückner, the climatic determinism, as well as the case of the Bishop of Canterbury's mitigation policy in the 14th century, demonstrated to me very clearly that social construction processes really matter when talking about climate, climate change, climate impact and climate policy. Scientific construction processes are intertwined with these social construction processes—if you allow me, for the time being, to separate these two types of processes. Some would claim that all such construction processes would be cultural. So they would, but there are differences. I will return to this later. A significant element of the parallel existence of such constructions is the parallel existence of terminology—climate means something different in its social understanding and its physical understanding.

For Nico, this was of course evident right away, and he insisted from the beginning of our cooperation on Gernot Böhme's concept of "social natural sciences" (Böhme/Schramm 1985), i.e., that any climate science needs to be embedded in an understanding of social and cultural dynamics, that the understanding of the physical dynamics would to some extent be conditioned by our cultural world. For social scientists this seems trivial, for us natural scientists it represents an insult since, as we claim, we are objective. But Nico's early work from 1978 on the social reality of Merton's norms showed us otherwise.

We condensed all this into our book *Wetter, Klima, Mensch* which first came out in German in a small volume in 1999. Today, it has been translated into Chinese, Russian, Slovenian, Romanian, Czech, and English (*Climate and Society*, 2010), and a revised and extended 2nd edition was published in 2010.

9.2 Climate Science as a Physical Approach?

As a first educated look tells us, the scientific analysis of climate is mostly a physical science discipline. The actors are predominantly physicists, meteorologists, oceanographers, and geographers; ecologists and economists have been joining in the last decades, and the older concept of a climate model is more and more replaced by an Earth System model.

The traditional object of climate science is not a social construct but part of 'Wirklichkeit' ('reality'), of the real world that is governed by physical principles, such as conservation of energy, momentum and mass in hydrodynamics. Thus, the climate science field contains the "physics of the climate system". When we add terrestrial and marine cycles of matter, of ecosystems, other natural science concepts are getting involved.

All this is good, conventional natural science—but constrained by schools of thought, by paradigms, and ubiquitously challenged by alternative knowledge claims which employ pre-scientific or outdated views or concepts molded by world views and particular interests. We natural scientists consider ourselves to be objective, independent of our history, equally open to all ideas, always prepared to accept that our view has been falsified by new observations and findings. Except that we aren't, as Fleck (1983) and STS—science, technology and society—studies have taught us.

For one of our later papers, Nico and I teamed up with a theoretical physicist, Armin Bunde from Giessen, to analyze whether climate science would be mostly a kind of subfield of physics, and we concluded that—yes, it includes many elements of physics, albeit not necessarily in a conventional way, and, no, there is more to it, in particular on the part of social and cultural studies. This paper became a chapter in *The Oxford Handbook of Climate Change and Society* (von Storch et al. 2011).

The situation becomes more complex when the product of the knowledge construction, the "wissenschaftliches Wissen" ("scientific knowledge"), becomes an asset in political debates: when stakes are high, decisions urgent, uncertainty remains high and values are in dispute. We call this situation, following Funtovicz/ Ravetz (1993), *post-normal*. In this situation, science becomes politicized, and policy 'scienticized'. It is no longer only the scientific method which qualifies a result but also its utility as an argument in a political debate, a debate about values; on the other hand, political decisions are perceived not as mostly a matter of cultural preferences and values but as a mere 'must' provided by the natural sciences. Symptomatically, there are arguments discussing whether autocratic systems are better in dealing with environmental crises than the slow democratic system is which is built to balance different worldviews, assessments of risks and possibilities, preferences and values.

Post-normal situations are common in the case of environmental issues, among them climate. The *uncertainty* is large and will remain so for an extended time; whether the sensitivity of the climate system (the equilibrium temperature increase caused by a doubling of CO_2 concentrations) is 2K, 4K or 11K will be known only in a few decades—and it is the same with the development of the big ice sheets of Greenland and Antarctica, or with trends related to tropical cyclones, the spreading of diseases and the fate of the polar bear. If we want to limit climate change and assume that the sensitivity is high, then we need to massively reduce emissions now; if we think that the sensitivity is lower, we may have more time to do so. But the decision to implement significant measures must be taken today—think of the rhetoric of COP-15 in Copenhagen (2009). The decisions are *urgent*. Obviously, such decisions have costly consequences, the *stakes are high*: if we do not mitigate enough, the damages may be prohibitively high, while unneeded mitigation may limit other urgently needed investments, say in providing electrical power for the poor. Do we need to act now to prevent being challenged by future generations, or do we want to create wealth now as a basis for future prosperity? This is a *value* decision, closely related to the question of how to deal with risks.

The post-normal situation goes mostly unnoticed by natural scientists who become actors in a political conflict while believing to be objective; they get desperate because the political system does not follow the suggestions and requests of 'science', which have recently become more and more publicly voiced orders by an elite of climate scientists. But the policy processes do not follow these grand calls for re-organizing global economics, for the great transformation, for efficient global mitigation, whereas science seems to have a profound effect only on the more piecemeal activities of regional mitigation and adaptation.

To conclude: A significant part of the set of issues represented by 'climate' is deeply embedded in social and cultural processes, in particular the field of society-climate interaction. Thus, climate science is a genuinely trans-disciplinary scientific field which poses special challenges and approaches that require the skills of both physical and other natural science professionals as well as social and cultural scientists.

While climate science began its ascent as a conventional natural science quest for understanding real world dynamics, it is now dealing with both "physical reality" and social constructions thereof.

9.3 What Is Needed from Social and Cultural Sciences?

There are a number of issues which the social and cultural sciences should clarify and examine to help society make better use of the natural sciences and to help the natural sciences make scientifically constructed knowledge more readily available. A bit idealistic, I admit.

(1) The topology of competing knowledge claims, their origins, their social power and their dynamics; natural science knowledge is one of these claims. Ideally, it differs from others by its methodology of permanent skepticism and willingness to incorporate falsification. This includes the dynamics of what is named "skeptical climate science".

(2) Is scientifically constructed knowledge 'better' than other explanation systems? In what respect 'better'? As a better guide to practical implementation or as a system less likely to be falsified?
(3) Empirical analysis of the extent to which climate science quarters are implementing the ideals of Merton's CUDOS principles (Merton 1973 [1942]). Such an analysis requires studying specific cases—instead of broad general claims.
(4) An ethnology of the different 'tribes' in climate science—how do they determine dominant narratives and knowledge systems, how are speakers chosen, how is the interaction with other tribes and stakeholders taking place?
(5) What expectations do the public and stakeholders have concerning the function, and possibly service, of climate science? Is this consistent with Robert Merton's norms?
(6) Are STS findings relevant for the practice of climate science and of science-stakeholder interaction, and are such studies taken into account by natural scientists?

This list can certainly be extended, as well as better organized. A key element not explicitly included is the role of scientists' subjectivity. I have often heard very general assertions that all knowledge is subjective and that, thus, there is no qualitative difference between all of them; a kind of arbitrariness. While subjectivity is all over the place, and in that sense no objectivity exists, we have to face the fact that there is a 'Wirklichkeit' ('reality'), for example the timing of the next high tide at the St. Pauli gauge in Hamburg, which we can predict—and determine the skill of this forecast in an objective frequentist manner. Scientifically constructed knowledge is often of superior utility. But this *subjectivity*—we natural scientists want to learn how it manifests itself, not in general terms but concretely, in our own work.

My own answer includes two manifestations: (1) In the process of choosing issues for our scientific endeavor, we select topics which are of interest to us. Many scientists have entered the field of climate research because they are concerned about the state of the environment, some because they want to protect the climate from human interferences.

In the (2) process of accepting an explanation as valid, we employ alternative knowledge systems; when we already 'know' that climate change is manifest in our environment, we need less evidence before we accept that man-made climate change is causing, for instance, an increase in Hamburg storm surge heights. On the other hand, we would ask for more evidence if the data did not show the expected pattern of change. Our 'backpack' of culturally constructed understanding guides us in the analysis, acts as an efficient filter when looking for consistent explanations.

In a sense, all this is "nice to have", but does it matter? I believe so. I believe that better knowledge, more robust against falsification, will emerge in this way. My subjective assessment!

But there is one field where it certainly matters, and that is the practice of what is nowadays called "climate service". In this practice, the idea is to make scientifically constructed knowledge 'useful' in practical societal contexts—by providing options and perspectives that are valuable in societal and economic decision processes. This has something to do with the competition of knowledge claims, and it is by no means clear that scientific knowledge will be accepted as superior. An example is the smart physicist who tells me that our storms are getting stronger because the increased warmth of the air leads to more water vapor and thus more kinetic energy. This sounds plausible but is easily falsified by the observation that our summers are warmer than our winters, even in the North Sea area, and that strong storms are an attribute of our cold season.

The issue of regional climate servicing is the most recent topic Nico and I, plus seven other colleagues, have written about in the *Journal for Environmental Law and Policy* (von Storch et al. 2011).

9.4 Epilogue

In retrospect, after some 25 years of working with and learning from Nico Stehr, I find myself greatly enriched. He has opened various windows and doors to look or walk through, to approach relevant issues and exciting problems in other ways, with fresh ideas and from different perspectives. Maybe my own practice has converged toward his "social natural science" concept.

Nico, your presence and your teaching has had a positive, useful effect, but what's maybe more important—working with you has been fun. Thanks, Nico.

References

Böhme, Gernot and Engelbert Schramm (eds.). 1985. *Soziale Naturwissenschaft. Wege zu einer Erweiterung der Ökologie*. Frankfurt am Main: Fischer.

Fleck, Ludwik. 1983. *Erfahrung und Tatsache. Gesammelte Aufsätze*. Frankfurt am Main: Suhrkamp.

Funtowicz, Silvio O. and Jerome R. Ravetz. 1993. "Science for the Post-normal Age", *Futures,* 25 (7): 739–755.

Huntington, Ellsworth. 1915. *Civilization and Climate*. New Haven and London: Yale University Press.

Le Roy Ladurie, Emmanuel. 1971. *Times of Feast, Times of Famine*. New York: Doubleday.

Merton, Robert K. [1942] 1973. "The Normative Structure of Science", in: *The Sociology of Science. Theoretical and Empirical Investigations*. Chicago: University of Chicago Press.

Nordhaus, William D. 1991: "To Slow or Not to Slow: The Economics of the Greenhouse Effect", *The Economic Journal*, 101(407): 920–937.

Stehr, Nico. 1978. "The Ethos of Science Revisited", in: Jerry Gaston (ed.), *The Sociology of Science: Problems, Approaches and Research*. San Francisco: Jossey-Bass: 172–196.

Stehr, Nico and Hans von Storch. 1993. "Genarrt vom Wettergott." ZEIT online: http://www.zeit.de/1993/37/genarrt-vom-wettergott (5 July 2017).

Stehr, Nico and Hans von Storch. 1995. "The Social Construct of Climate and Climate Change", *Climate Research,* 5(2): 99–105.

Stehr, Nico and Hans von Storch. 2008. *Eduard Brückner: Die Geschichte unseres Klimas. Klimaschwankungen und Klimafolgen.* Österreichische Beiträge zu Meterologie und Geophysik). Wien: Zentralanstalt für Meteorologie und Geodynamik Wien.

Stehr, Nico and Hans von Storch. 2010 [1999]. *Wetter, Klima, Mensch.* München (English edition: *Climate and Society. Climate as Resource, Climate as Risk*). Singapore: World Scientific Publishing).

von Storch, Hans and Nico Stehr. 2000. "Climate Change in Perspective. Our Concerns About Global Warming Have an Age-Old Resonance", *Nature,* 405(615): 615–617.

von Storch, Hans, Armin Bunde and Nico Stehr. 2011. "The Physical Sciences and Climate Politics", in: *The Oxford Handbook of Climate Change and Society.* Oxford: Oxford University Press: 113–128.

von Storch, Hans, Insa Meinke, Nico Stehr, Beate Ratter, Werner Krauss, Roger Pielke Jr., Reiner Grundmann, Marcus Beckermann and Ralph Weisse. 2011. "Regional Climate Services Illustrated with Experience from Northern Europe", *Journal for Environmental Law and Policy,* 1/2011: 1–15.

Part III
Selected Key Texts by Nico Stehr

Chapter 10
The Ethos of Science Revisited: Social and Cognitive Norms

Nico Stehr

This chapter is a critical examination of the growing literature dealing with the normative structure of scientific practice. More specifically, my intention is to analyze the presuppositions of Robert Merton's original formulation of the ethos of science and—the critiques of his position, thereby arriving at a provisional assessment of the current status of the theory of the normative structure of science. The critiques may be divided into four major issues; (1) the extent to which the norms of science are peculiar to science; (2) the actual conduct of scientists in various organizational and historical contexts and the extent to which their conduct is governed by the norms of science; (3) the 'morality' of the moral imperatives of science; and (4) the relation between the scientific ethos and the development of scientific knowledge.[1]

10.1 The Normative Structure of Science

Critics and supporters of Merton's theoretical formulation of the moral imperatives of science seem to agree that the formulation provided one of the most significant theoretical foundations for a sociology of science (see Barnes/Dolby 1970: 3; Spiegel-Rosing 1973: 37; Storer 1973: 226; Mitroff 1974a: 12–13). A 'first'

This text was first published as Stehr, Nico. 1978. "The Ethos of Science Ethos of Science Revisited", *Sociological Inquiry*, 48: 172–196. The permission was granted on 17 July 2017 on behalf of Wiley by Ms. Kelly Hoff, Permissions Coordinator, Copyright and Permissions.

[1] I would like to acknowledge a critical reading of the first draft of the essay by Michael Kozlowslu and the helpful editorial assistance of Laura Hargrave.

M. T. Adolf (ed.), *Nico Stehr: Pioneer in the Theory of Society and Knowledge*, Pioneers in Arts, Humanities, Science, Engineering, Practice 16,
https://doi.org/10.1007/978-3-319-76995-0_10

delineation of the normative structure of science by Merton appears in the context of his essay “Science and the Social Order,” initially presented as a paper to the American Sociological Society in 1937 and published in 1938 in *Philosophy of Science.* In the context of this essay, Merton ([1938] 1973d: 259) observes, “The sentiments embodied in the ethos of science [are] characterized by such terms as intellectual honesty, integrity, organized skepticism, disinterestedness, impersonality.” In the same essay, one also finds references to psychological attributes that scientists share with non-scientists, attributes that also affect their scientific works. Merton’s first systematic exposition, however, is found in a 1942 essay, originally published under the title “A Note on Science and Democracy” in the *Journal of Legal Political Sociology.* Subsequently, the same essay appeared as “Science and Democratic Social Structure” in Merton’s *Social Structure and Social Theory* (1949). Still later, the essay was reprinted as “The Normative Structure of Science” in Merton’s *The Sociology of Science: Theoretical and Empirical Investigations* (1973). In the course of these several reprintings, the essay has not remained unchanged. Most of the changes, however, involve minor matters, such as stylistic improvements.

The ethos of science generally is conceptualized by Merton ([1942] 1973b: 268–269) as that “affectively toned complex of values and norms which is *held to be binding* on the man of science. The norms are expressed in the form of prescriptions, proscriptions, preferences, and permissions. They are legitimized in terms of institutional values. These imperatives, transmitted by precept and example and *reinforced by sanctions* are in *varying degrees internalized by the scientist.* Although the ethos of science has not been codified, it can be *inferred* from the moral consensus of scientists as expressed in use and wont, *in countless writings on the scientific spirit and in* moral indignation directed toward contraventions of the ethos.” (emphasis added).

Merton specifies four basic moral imperatives, which provide a foundation for the social relations of science, and the professional identity of individual scientists and therefore constitute important elements of the sociocultural structure of science.

1. *Universalism* prescribes that knowledge claims in science should be evaluated and accepted or rejected according to impersonal cognitive criteria rather than the “personal or social attributes of their protagonist” (Merton [1942] 1973b: 270) and that careers and opportunities in science should be based on achievement and competence only. (The choice of words perhaps is significant; the expression “should be” denotes a prescription aspired to but not necessarily a description of consistent scientific practice.)
2. *Communism*[2] refers to the interrelated ‘communal’ (public) character of scientific knowledge claims; the corresponding limited ‘rights’ of the originator(s) to recognition and esteem, resulting in the distinctive and anomalous character of

[2] I intend to retain the formulation *communism* since the *Zeitgeist* has undergone change once again and does not at the present time, so it seems, require a substitute term such as *communality* (see Barber 1952: 91).

intellectual *property* in science[3]; and the imperative not to withhold knowledge claims—an imperative reinforced by the "incentive of recognition, which is, of course, contingent upon publication" (Merton [1942] 1973b: 274).
3. *Disinterestedness,* the moral imperative at the institutional level of science, is largely self-explanatory (see Parsons 1939; Wunderlich 1974); it points to a distinctive structure of control exercised over the individual motives of scientists.
4. *Organized Skepticism* is "both a methodological and an institutional mandate" (Merton [1942] 1973b: 277). Knowledge claims should not be accepted without (socially organized) scrutiny but should be warranted with reference to the technical norms of science.

In addition, Merton ([1942] 1973b: 269–270) refers to certain technical (or methodological) norms, certified knowledge (the institutional goal of science), and institutional values as other elements of the culture of science. Certain other norms —for instance, 'individualism' and 'rationality' (Barber 1952: 86–90; Barnes/Dolby 1970: 9) 'objectivity' and 'generalization' (Storer 1966: 81), and various other formulations (see, for example, Cournand/Zuckerman 1970; Cournand 1977) —have been suggested as a part of the ethos of science; these norms were intended to be largely compatible with the original formulation advanced by Merton.

Another norm mentioned by Merton ([1957] 1973c: 294) is the "emphasis upon originality on the institutional plane" in the scientific community; as emphasized in the context of the same essay, originality counterbalances humility, thereby creating potential normative conflict ([1957] 1973c: 303, 305; see also Merton 1976). It has often been observed, therefore, that science as a social institution is, like any other societal institution, characterized by potentially incompatible normative demands (norms and counternorms). (A discussion of the sociostructural antecedents and consequences of norms and counternorms in social organizations may be found in Merton 1976.) As long as potentially conflicting norms are not 'compartmentalized' (Deutscher 1972), they may generate 'ambivalence'; for example, ambivalence toward claims of priority, particularly in the context of multiple independent

[3] A rather similar observation is made by Ravetz (1971: 299): "One essential feature of the scientist's intellectual property, as distinguished from the 'real' property of commerce, is that it exists only by being available for use by others." The anomalous and distinctive character of intellectual property in science, as an essential context and expression of the norm of communism, also is emphasized in several of Merton's subsequent writings; more recently, for example, Merton (1977: 48–49) notes that the "distinctively anomalous character of intellectual property, becoming fully established in the domain of science only by being openly given away (published), is linked with the normative requirement for scientists making use of that property to acknowledge (publish) the source, past or contemporaneous" (see also Merton [1957] 1973c: 294–295). Moreover, in direct contrast to commercial relations, the greater the 'use' of the property by others, the greater the value of the property (see also Price 1978: 80–81). The availability of knowledge claims is the condition for the possibility of a "continued appraisal of work and recognition for work judged well done *by the standards of the time* [which] constitute a mechanism for maintaining the processes of falsification and confirmation of ideas that are required for the cognitive *development* of science" (Merton 1976: 45; emphasis added).

discoveries. Finally, Merton emphasizes that there have been comparatively few empirical instances of deviant responses to particular norms of science (Merton [1957] 1973c: 321), and he stresses his theory of social structure and anomie applies also to the institution of science (Merton [1957] 1973c: 308, note 51).

The four basic moral imperatives of science are more than mere moral principles. They are linked in distinct ways to the cognitive development of science. For, as Merton ([1942] 1973b: 270) notes, "The mores of science possess a methodologic rationale but they are binding, not only because they are procedurally efficient, but because they are believed right and good. They are moral as well as technical prescriptions." (Luhmann [1968] 1970, writing from a functional-structural perspective, has described the functional necessity for *social* mechanisms to operate as *cognitive* mechanisms in science.) Thus, a definite correlation between moral imperatives and the advancement of scientific knowledge is implied. For instance, "objectivity precludes particularism. The circumstance that scientifically verified formulations refer in that specific sense to objective sequences and correlations militates against all efforts to impose particularistic criteria of validity" (Merton [1942] 1973b: 270.) In other words, the norms do not merely regulate the behavior or the social relations of members of the scientific community; also, in distinct ways, they enhance the institutional goal of science, which is the continuing extension of certified knowledge claims.[4]

It is therefore clear that, in Merton's view, the norms of science derive from the purpose of science: "The institutional imperatives (mores) derive from the goal and the methods. The entire structure of technical [cognitive] and moral norms implements the final objective" (Merton [1942] 1973b: 270). The normatively prescribed social relations of science therefore complement, if not implement, the development of knowledge in science and vice versa. The dialectics of social and cognitive norms are a part of the institution of science (for a different view, see Bourdieu 1975; Knorr 1977). However, an implicit bifurcation of social and cognitive norms is equally evident when Merton outlines the dialectics of social and cognitive norms as constitutive of science. His emphasis then shifts to a more detailed inquiry into the normative basis of the social relations of science and therefore to a dissociation of the interdependent realms of intellectual and social domains of science. The former emphasis may be called the *theoretical* or *programmatic* formulation of the ethos of science; the latter, more restrictive, formulation may be called the *research* formulation of the ethos of science (followed subsequently by many sociologists of science; see Cole/Zuckerman 1975: 157). The two formulations—differing in scope as they do—seem, on the one hand, to represent a significant *theoretical break* in the work of Merton and therefore in the sociology of science. But, on the other

---

[4] These distinctions suggest that the ethos of science (norms) pertains directly to what could be called both methodological (cognitive) and social means, which are simultaneously subsidiary to the goal of science (values). Such categorical distinctions indeed are made at times (for instance, Storer 1966: 76), but I shall not attempt to maintain such a distinction here. That would presuppose a much more explicit and developed theory of the social relations of science than actually is available at this time.

hand, they are often seen as a theoretical basis and justification for empirical research into the social relations of science. The observed theoretical break cannot be seen, in a historical context, to represent the sole condition for such categorical bifurcation; the then dominant philosophies and histories of science, as well as the dominant societal legitimations of scientific practice also support it (King 1971).[5]

In addition to the thesis that social relations of science based on these norms support the advancement of scientific knowledge, the 1942 essay also refers to potential interinstitutional normative support and conflict.[6] For example, a societal legitimation of ascriptive standards for careers would contradict the norm of universalism in the ethos of science.

Merton's normative structure of science and the subsequently developed theory of the reward system in science represent the structural-functional account of the social relations of science. In the course of their professional socialization into science (Merton [1942] 1973b: 269), scientists acquire the ethos of science; it then becomes a salient part of their identity and consciousness and is reinforced through mechanisms of social control, mediated by the reward system of science. Thus, "rewards are to be meted out in accord with the measure of accomplishment. When the institution of science operates effectively, the augmenting of knowledge and the augmenting of personal fame go hand in hand" (Merton [1957] 1973c: 323).

Merton's formulation obviously contains a number of presuppositions, many of them noted later by critics. Merton assumed a particular historiography of science and a particular theory of the institutional setting of scientific practice. The ethos of science is assimilated, on the one hand, to a particular theory and history of knowledge and, on the other hand, to a particular model of the social relations of science—a model that emphasizes the normative basis and control of social action. In each instance, the relative compatibility of cognitive and social elements within science is emphasized. Threats to such compatibility primarily come from sources

[5] The decision to concentrate attention on the normative basis of the social relations of science presumably also rests on a number of more nearly immanent meta-theoretical considerations. For example, as emphasized at various times (Merton 1948, 1975), a general theory of social action is not a realistic prospect in sociology at the present stage of its cognitive development; thus, more delimited theoretical structures have to be developed, and research strategies have to be adjusted accordingly. The scope of theories is therefore limited by the cognitive state of the discipline. At the same time, one of the central features of Mertonian discourse, as described by Coser (1975: 5), might well be relevant: "Merton's world is composed of multiple ambiguities, of conflicting and contradictory demands and requirements that need to be articulated and made accessible by the sociologist."

[6] However, I will exclude a more detailed discussion of the possible interrelation between societal (including cognitive) norms and the ethos of science (see also Merton [19738] 1973d: 259; Barber 1952; Blume 1974: 45–50), including the views advanced by Downey (1967: 253) that 'scientism' assures the obsolescence of a number of norms which are part of the ethos of science, or that the autonomy of scientific institutions is in effect the outcome of cognitive progress in science.

external to science such as the state. Cognitive divisions therefore are deemphasized,[7] because the theory of the institutional basis of science and its cognitive development reinforce each other. Yet, one problem that causes considerable discussion is the precise methodological status of Merton's formulation: Is it to be understood as an "ideal-type" or is it perhaps a "general conception" that claims to have captured the actual, observable conduct of scientists?[8] But the most significant controversial legacy of the Mertonian conception of the ethos of science is linked to the theoretical break. Critics of the ethos of science have reproduced the break, although often on the basis of different intellectual traditions and presuppositions.

10.2 Critiques of the Ethos of Science

What, then, have been the serious objections to Merton's formulation of the ethos of science? Most of the criticisms can be placed in one of several categories.

First, and perhaps most often raised because it directly pertains to both a particular historiography of science and a theory of the social relations of science, is the question: Are the norms of science peculiar to science? A related question is the extent to which the norms actually affect the conduct of scientists. A third question is whether the moral imperatives of scientific practice are indeed moral. Finally, questions have been raised about the claim of a correlation between the ethos of science and the development of scientific knowledge.

Most of the critiques in each of the categories can be classified as post-Kuhnian critiques; that is, they reflect the immense influence generated by Kuhn's *The Structure of Scientific Revolutions* (1962). Kuhn provides critics with an alternative historiography of scientific knowledge and, tacitly, an alternative theory of the social relations of science. Many of the criticisms to be discussed were made possible by a particular reconstruction of Kuhn's account of scientific practice (Barnes/Dolby 1970: 3; Spiegel-Rosing 1973). Some examples to illustrate the presuppositions are appropriate: "The paradigm is a source of *social control*" (Barnes/Dolby 1970: 197; emphasis added). "The cohesion, solidarity and commitment within [paradigm-sharing communities] stem from the technical norms of paradigms, not from an overall scientific 'ethos'" (Barnes/Dolby 1970: 23; see also Mulkay 1972b: 15). These researchers deal with Merton's formulation primarily with reference to its cognitive functions (Barnes/Dolby 1970: 4; King 1971: 15–16;

---

[7] Often, as I have tried to show elsewhere (Stehr 1975), these domain assumptions become much more explicit in the writings of authors who attempt to follow Merton in their analysis of science as a social institution; for example, Storer (1966: 82) emphasizes the point that there is but *one* science.

[8] I refer, of course, to Rickert and Weber's theory of concept formation in the empirical sciences, particularly Weber's notion of ideal types as conceptual tools in the social sciences. Such ideal types were not conceived simply as descriptions of observable conduct (Rickert 1902; Weber 1922; Burger 1976).

Mitroff 1974a: 10, 13). In contrast, researchers who do not make any explicit reference to Kuhn's historiography of scientific knowledge typically describe the ethos of science as a set of *social* norms. For instance, Hagstrom (1965: 1) writes, "I am concerned with the operation of social control within the scientific community, with the problem of discovering the social influences that produce conformity to scientific norms and values" (see also West 1960: 54; Rothman 1972). The resulting shift of domain assumptions, however, does not produce a negation of the theoretical break observed earlier. The emphasis of research shifts to an analysis of cognitive processes in science and therefore represents a separation of social and intellectual factors in scientific practice.

The general structure of these critiques is as follows. Inasmuch as (the predominant reconstruction of) Kuhn's historiography of the development of scientific knowledge provides an important foundation for such critiques, the criticism shifts to the *cognitive* development of science, including such questions as: To what extent is the ethos of science governed by the cognitive development of science? Or, even stronger: To what degree are the actual norms of science cognitive (technical) norms? Not infrequently, therefore, the cognitive norms are conceptualized, in some sense, as prior to the internalization of the social norms of science. The acquisition of social norms is seen as mediated by the conceptual structure of science (Kuhn 1962: 46; Downey 1967: 251). Obviously, such considerations have implications for a theory of the social relations of science. In contrast, those who choose not to emphasize the theoretical consequences of Kuhn's historiography of scientific knowledge for the ethos of science typically inquire into the *social* relations of science in light of the normative structure of science and are therefore not bound to an analysis of the 'rational' structure and content of scientific discourse (see, for example, Gouldner 1976). The overall result has been a perpetuation of the bifurcation of theory and inquiry into the social and cognitive norms of science. Few analyses have systematically investigated the questions raised by the observation that "methodological canons are often *both* technical expedients and moral compulsives" (Merton [1942] 1973b: 268; emphasis added).

10.2.1 Are the Norms of Science Peculiar to Science?

How are we to differentiate science as a social and cultural institution from other social and cultural institutions? Any criterion or set of criteria found acceptable depends on an implicit or explicit theory of social and cultural differentiation. One could, for example, emphasize the differentiation of modes of production, relations of production, technology, social functions, norms and values, types of knowledge, modes of discourse, or a combination of these. By virtue of its focus on "a set of cultural values and mores governing the activities termed scientific", Merton's ([1942] 1973b: 268) formulation of the ethos of science links his implicit theory of social differentiation to considerations that emphasize the distinctiveness (but not necessarily radical discontinuity) of norms in different institutions.

Most of Merton's critics do not deny that norms, in some sense, are fundamental to the differentiation of social institutions; but many of them believe that Merton's formulation of the normative structure of science does not adequately set science, as an institution, apart from other social institutions. Barnes/Dolby (1970: 8–12), for example, observe that the norms of universalism, rationality, and organized skepticism are not peculiar to science but are also observable in everyday life. Similarly, Mulkay (1969: 22, 27) questions the adequacy of the ethos of science as a valid means of differentiating science. He suggests instead that differentiation must be linked to a theory of the development of specialized knowledge claims: "The theoretical and methodological norms are more central to the structure of the scientific community than are the Mertonian social norms" (Mulkay 1969: 36; see also Mulkay 1972a). Evidently, the dichotomy of social and cognitive norms is crucial to the argument, and the claim is advanced that the shared type of knowledge as well as the procedures (the technical-methodological rules) used to generate and warrant knowledge claims are constitutive of science. Accordingly, the identity of the scientific community is a cognitive identity (see also Parsons 1951: 336–337; Boehme 1974). Mulkay's approach therefore is based on a theory of social differentiation that stresses the emergence of specialized cognitive procedures and their consequences.[9]

Thus, following Kuhn's historiography of scientific knowledge and his tacit theory of the social relations of science mediated by cognitive processes, the criticisms for the most part amount to an 'alternative' delineation of a set of norms as constitutive of scientific practice. Kuhn's approach therefore is perceived as a reversal of Merton's approach (King 1971: 18, 30). In the introduction to his most recent publication, *The Essential Tension,* however, Kuhn (1977) appears to suggest "Je ne suis pas Kuhniste," when, for example, he insists that "scientific communities must be discovered by examining patterns of education and communication before asking which particular research problems engage each group" (xvi) and when he more generally stresses the continuity between his concerns and

[9] The position that Mulkay advances is not, it seems, peculiar for a social scientist to adopt (Gaston 1978: 163–166) but rather is the result of underlying Kuhnian presuppositions. Gaston, in his critique of Mulkay's view, discusses the norms of science primarily as they manifest themselves in the social relations of science (the research formulation), whereas Mulkay discusses the ethos of science primarily in terms of cognitive processes and consequences in science. Kuhn's program for paradigm closure (Simmons/Stehr 1978) and logical empiricism as well as critical rationalism (Hempel 1965: 3–5; Popper [1935] 1968: 41, [1962] 1969: 156) also is based on a differentiation of science from nonscience; namely, the idea that demarcation must be based on cognitive attributes of systems of discourse. As far as social relations in general are concerned, positivism emphasizes cognitive factors of social action, while functionalism stresses norms, values, motives, and the consequences of social conduct (King 1971: 8; see also Parsons [1937] 1949: 387, 439–440). The emphasis of cognitive norms to the exclusion of other processes in science, of course, results in an image of a rather 'rational' scientific practice. It may therefore fail to do justice to the immense complexity of the social organization of science. That is, the reduction (and retention) of complexity in science, to use Luhmann's terminology, requires social mechanisms other than cognitive norms—for example, reputation operating *both* as a medium of communication (selection) and as motivation (Luhmann [1968] 1970).

those of Merton (xxi). Nonetheless, most of the critiques of Merton's formulation remain based on a particular reconstruction of Kuhn's theory of scientific practice.

Because these critiques view scientific practice primarily as 'conceptual', 'cognitivist', or 'discursive', they cannot cope successfully with the 'material' conditions of scientific practice, particularly as such conditions affect the cognitive development of scientific knowledge. As a result, relative to the theoretical formulation of the ethos of science, they seem to represent a retrogressive step. With respect to the research formulation of the ethos of science, however, the critiques may be seen as progressive, since they emphasize the intellectual basis of scientific practice. Yet, the criticisms lack a sense of historicity (Boehme et al. 1972), but historical changes are difficult to detect once the issue is confined to cognitive norms.[10] Thus, the fundamental issue of the historicity and interdependence of social and cognitive domain in science (Fleck 1935; Bourdieu 1975: 22; Weingart 1976: 33–92; Baldamus 1977; Knorr 1977) remains a challenge to a comprehensive theory of the normative structure of science (Barnes/Dolby 1970: 14–18).

In this context other questions about the ethos of science derive from differing theories of social institutions (see Mitroff 1974a: 77), particularly questions about the exact character of norms in institutions. As noted earlier, Merton has repeatedly stressed the presence, structural origin, and import of 'contradictory' norms in social institutions, including science.[11] Mitroff (1974a: 17) also emphasizes the importance of such conflicting sets of norms as conditions for the possibility of rationality and growth in science.[12] Mulkay (1976: 641), however, has advanced the thesis that norms and counternorms, as discussed by Mitroff or Merton or a combination of the two, should not be seen as rules governing scientific practice—in

---

[10] In an essay written some thirty years after the original formulation of the ethos of science, Merton [1968] 1973a: 327–328) appears to acknowledge the historicity of the social norms and with it, of course, the possibility that the norms of science are neither entirely autonomous nor unresponsive to basic sociostructural changes in science and society. However, the emphasis, in this essay at least, is on the continuity of certain institutional relations of science, such as the intensity and the degree of competition among scientists.

[11] In his essay "The Ambivalence of Scientists" Merton (1976: 41–42) refers to the doubling of ambivalence—scientists' ambivalence toward feeling ambivalent—as a possible explanation of the lack of attention paid to conflicts and modes of conflict in science as conflicts over claims to priority. One of the interesting issues raised by the conception of conflicting norms as a 'normal', structurally generated attribute of social organization concerns the problem of deviant conduct under such circumstances. For deviant conduct traditionally is defined as behavior at variance with specific dominant norms typifying social positions. Given the particular formulation of norms and counternorms (their scope and range, for instance), deviant conduct might well be denied *a priori*. In the case of science, however, deviant conduct also requires a set of rules signifying both social and cognitive distinctiveness of scientific practice (demarcation criteria). (For a discussion of deviant behavior from both cognitive and social norms in science, see Zuckerman 1977.)

[12] The empirical evidence Mitroff reports refers to research about the moon, which at the time was not intellectually homogeneous (in the sense of a dominant paradigm, for instance). As Mitroff (1974b: 594) therefore suggests, "whereas the conventional norms of science are dominant for well-structured problems, the counternorms proposed here appear to be dominant for ill-structured problems."

particular, the construction of specialized knowledge claims. (For a critique of Mulkay's views, see Zuckerman 1977: 126–128; Gaston 1978: 180.)

Rather, Mulkay argues, specific intellectual commitments are a part of scientific practice. These commitments mediate sociostructural processes such as inequality and communications networks (Mulkay 1976: 643; see also Kuhn 1977: ix–xxiii). Norms and counternorms, Mulkay suggests, are not institutionalized norms; they are not related, for example, to the exchange of rewards in science (Mulkay 1976: 641–645): "Conformity to most of the supposed norms and counternorms of science is largely irrelevant to the institutional processes whereby professional rewards are distributed" (Mulkay 1976: 642). Instead, and as Mulkay has argued elsewhere, operative control mechanisms are intellectual commitments.[13] According to Mulkay, scientists may refer to norms and counternorms when describing their practice—for instance, in conversations—but that does not prove the existence of a functioning normative structure in science, which, for example, may be said to contribute to the advance of scientific knowledge.

10.2.2 Are the Norms Prescriptive?

The question of the extent to which the various norms (and counternorms) actually determine or influence social and cognitive processes in science has been addressed in two different ways. Some critics relativize or limit the importance of the ethos of science. Others seek to determine the extent to which the norms of science operate empirically in different scientific communities.

Among those who attempt to limit the importance of the ethos of science, Mitroff (1974b: 585–586, 594) argues that its applicability may vary with the cognitive institutionalization of a research area, specialty, or discipline. A lack of cognitive institutionalization, (which may, of course, be a more or less permanent attribute of a research area or specialty) reduces the functional importance of the norms. Or, as Storer (1973: xiv), assimilating the historiography of scientific knowledge as developed by Kuhn, suggests, "It is when [...] a universe of discourse is only

[13] Mulkay's underlying theory of institutions largely rests on a notion of social control (and norms) based on externally imposed sanctions (or sanctioned norms) but not on what may be called 'primary' social control. The internalization of norms does not require an 'exchange' or reward (Berger/Luckmann 1966). A critique of such a restrictive conceptualization of social norms also may be found in Luhmann (1969), while Dahrendorf (1964) and Popitz (1961) advocate the restriction of the concept of norm to sanctioned conduct. The somewhat retrogressive perspective argued by Mulkay, at least in the context of his discussion of the status of the norms of science, can also be illustrated with respect to the theoretical status of "intellectual commitments." Mulkay largely abstains from linking commitments to social mechanisms. Social mechanisms may explain the homogeneity (within specific social boundaries), heterogeneity, distribution, and reproduction (or origin and alteration) of intellectual commitments. Such an explanation requires a theoretical analysis that transcends either a cognitive approach or a sociology of scientific ideas and procedures.

slightly developed (as during the Kuhnian 'preparadigm' stage in the development of a new discipline or during a 'scientific revolution'), or when group loyalties outside the domain of science take over, that violations of the norms become more frequent, leading some to reject the norms entirely."

A similar conclusion may be found in Ben-David (1977) and Zuckerman (1977), who both indicate that the norms operate relative to the progress made in resolving a particular research problem; that is, the ethos of science is seen as predominantly the normative structure of the context of justification, not the context of discovery.

Similarly, critics of Merton's formulation have asked whether one should distinguish between two not necessarily congruent sets of norms; namely, 'professed' and 'statistical' norms. The latter type of norm actually regulates conduct among scientists, while the former is ritualistically "enjoined or celebrated in tract or speech" and "may be directed mainly to out-groups in situations of justification and conflict" (Barnes/Dolby 1970: 8; see also Merton 1976: 40–41). In short, professed norms are described as an occupational ideology (Mulkay 1976: 646). The scientist is someone in need of a dual morality: an idealized morality for the 'external,' legitimating, social relations and a realistic morality for the internal social relations of science (see Spinner 1977: 565). Moreover, the idealized norms may have originated in everyday life as characteristics of popular science (Fleck 1935: 123; Spinner 1977: 563). In the context of these objections, therefore, an important question concerns the extent of the congruence between the two sets of norms.

The validity of the ethos of science is also relativized by those who emphasize another social context: the social relations of applied versus pure science. These critics claim that the ethos of science is primarily the ethos of pure science (Barber 1952: 95; Marcson 1960; Krohn 1961; Komhauser 1963; Storer 1966: 165–166; Box/Cotgrove 1966; Barnes/Dolby 1970: 7; Kowalewski 1974: 279). Other observations designed to limit the ethos of science, or particular aspects of it, make reference to social constraints. Sklair (1973: 113), for example, observes that "much, if not most, contemporary science is carried out under conditions of formal or informal secrecy"—necessitated by national security matters and matters of economic interest (see also Buehl 1974: 117). Finally, some arguments are directed toward a historical relativization of the ethos of science. These critics generally allege that the ethos of science may have been operative in the past, particularly in very early phases of the development of scientific institutions (Ravetz 1971: 310), but is no longer operative.

However, those who attempt to relative the ethos of science agree, in the last analysis, that scientific practice is indeed based on certain rules and standards after all. As Barnes/Dolby (1970: 10) therefore stress, "The scientist has a specific viewpoint which makes him skeptical of some results whilst uncritically accepting others; this is the norm." The social and cognitive relations of science, that is, do follow certain standards. At issue therefore are different formulations regarding exactly how and when the norms apply. A growing number of empirical investigations into the normative structure of science, for the most part influenced by the research formulation of the ethos of science, have dealt, at least in part, with this question. Gaston's (1978: 186) conclusion is perhaps representative of many of these empirical investigations: "The normative imperatives, universalism, communality, disinterestedness, and organized skepticism, are strongly institutionalized in the British and American scientific communities studied" (see also Cole/Cole 1973).

10.2.3 Are the Moral Imperatives Moral?

Much less frequently discussed is the question of the morality of the ethos of science, Merton's original formulation did not address this question explicitly. In general, however, critics seem to have heeded Durkheim's warning that a discussion of the morality of morals will indicate only the morals of the moralist, not those practiced by others. Nevertheless, some critics refer to this issue in passing. Mok/Westerdiep (1974: 216) observe that scientists may use, perhaps abuse, the norm of organized skepticism and disinterestedness in order to stay aloof from public controversies. Similarly, Sklair (1973: 111–112) claims that the norm of universalism, taken to the extreme, would ensure that "no squeamish, unscientific rules prohibiting crucial experiments on human subjects" would obstruct medical science. Both objections make evident that science cannot insulate its moral relations from society's moral views. Although the norms of science need not be the norms of society, society's norms affect those of science.

10.2.4 Are the Functional Imperatives Functional?

A final major issue about the norms of science concerns their influence on the growth of scientific knowledge. Practically no empirical research has been directly related to this question.[14] Any assessment of the degree to which the ethos of science claims to facilitate the production of scientific knowledge presupposes an explication of a historiography of scientific knowledge, and a specification of criteria for the growth of knowledge. Claims made about the effect of the social relations of science on the cognitive development of science obviously require some notion of the patterned development of knowledge. Moreover, the 'functional' effects of the norms on the cognitive development of science must be either direct or mediated. In either case it is necessary to have a theory of the relation between social and cognitive processes within science. The ethos of science may have an indirect effect, if it provides one of the main conditions for organizing scientific discourse in a manner which allows for the advancement of knowledge.

---

[14] Using information gathered for a small and heterogeneous sample of university scientists, West (1960: 61) reports that neither their rate of publication nor the strength of their motivation as reported by peers appears to be associated with an endorsement of the ethos of science.

The effect may be more direct, for example, if the ethos of science is formulated in such a way that it already incorporates cognitive conditions for the possibility of growth of scientific knowledge claims.

Mitroff (1974a, b) has explicitly questioned the validity of the notion that the ethos of science is functional for the advancement of scientific knowledge. Specifically, Mitroff (1974a: 75–78) suggests that certain counternorms, such as the degree of secrecy or emotional commitment of scientists, may indeed further the progress of scientific knowledge.[15] But, in spite of Mitroff's view, a comprehensive examination of the relation between the normative structure of science and the development of scientific knowledge requires the explication of a comprehensive theory of the normative (social and cognitive) structure of science, a task yet to be accomplished.

10.3 Conclusion

One of the salient characteristics of the debate about the ethos of science is the categorical bifurcation of social norms and cognitive norms and the restrictive formulation of a theory of the normative structure of science. Merton's theoretical formulation of the ethos of science, in contrast to the research formulation and the cognitivist formulation of the critiques of the research conception of the ethos of science, is best suited to transcend "the difficulty of uniting into a single vision the perspectives of science as conceptual structure and as human activity" (Gutting 1973: 209). The dichotomy between social and cognitive norms of science, and therefore the dichotomy between scientific ideas and scientific practice, clearly requires a return to the original theoretical formulation of the ethos of science.

References

Baldamus, Wilhelm. 1977. "Ludwig Fleck and the Development of the Sociology of Science", in: *Human Figurations: Essays for Norbert Elias*. Amsterdam: Amsterdam Sociologist Tijdschrift.

Barber, Benjamin. 1952. *Science and the Social Order*. New York: Free Press.

Barber, Benjamin. 1961. "The Resistance by Scientists to Scientific Discovery", *Science*, 134: 596–602.

Barnes, Barry and Dolby, Robert G.A. 1970. "The Scientific Ethos: A Deviant Viewpoint", *European Journal of Sociology*, 11: 3–25.

---

[15] Following Feyerabend (1975) and Churchman (1971), Mitroff (for example, 1974b: 590) subscribes to a historiography of scientific knowledge based on the theory of paradigm proliferation. Subsequently, Merton (1976: 59) has criticized Mitroff for exaggerating the inevitably 'subjective' attributes of the production of scientific knowledge to the detriment of its objective aspects and for ignoring the interaction of subjective and objective attributes, which results in a "storybook version of scientific inquiry".

Barnes, Barry. 1973. "The Comparison of Belief-Systems Anomaly Versus Falsehood", in: Robin Horton and R. Finnegan (eds.), *Modes of Thought. Essays on Thinking in Western and Non-Western Societies.* London: Faber: 182–198.

Ben-David, Joseph. 1977. "Organization, Social Control and Cognitive Change in Science", in: Joseph Ben-David and T. Clark (eds.), *Culture and Its Creators.* Chicago: University of Chicago Press.

Berger, Peter L. and Thomas Luckmann. 1966. *The Social Construction of Reality. A Treatise in the Sociology of Knowledge.* New York: Doubleday.

Blume, Stuart S. 1974. *Toward a Political Sociology of Science.* New York: Free Press.

Boehme, Gernot. 1974. "Die Bedeutung von Experimentalregeln für die Wissenschaft", *Zeitschrift für Soziologie,* 3: 5–17.

Boehme, Gernot, Wolfgang van den Daele and Wolfgang Krohn. 1972. "Alternativen in der Wissenschaft", *Zeitschrift für Soziologie,* 1: 302–316.

Bourdieu, Pierre. 1975. "The Specificity of the Scientific Field and the Social Conditions of the Progress of Reason", *Social Science Information,* 14: 19–47.

Box, Steven, and Stephen Cotgrove. 1966. "Scientific Identity, Occupational Selection and Role Strain", *British Journal of Sociology,* 17: 20–28.

Buehl, Walter L. 1974. *Einführung in die Wissenschaftssoziologie*. München: Beck.

Burger, Thomas. 1976. *Max Weber's Theory of Concept Formation.* Durham, NC: Duke University Press.

Churchman, C. West. 1971. *The Design of Inquiring System.* New York: Basic Books.

Cole, Jonathan R. and Stephen Cole. 1973. *Social Stratification* in *Science.* Chicago: University of Chicago Press.

Cole, Jonathan R. and Harriet Zuckerman. 1975. "The Emergence of a Scientific Specialty: The Self-exemplifying Case of the Sociology of Science", in: Lewis A. Coser (ed.), *The Idea of Social Structure.* New York: Harcourt Brace Jovanovich.

Coser, Lewis A. (ed.). 1975. *The Idea if Social Structure.* New York: Harcourt Brace Jovanovich.

Coser, Lewis A. and Robert Nisbet. 1975. "Merton and the Contemporary Mind: An Affectionate Dialogue", in: Lewis A. Coser (ed.), *The Idea of Social Structure.* New York: Harcourt Brace Jovanovich.

Cournand, Andre F. 1977. "The Code of the Scientist and Its Relationship to Ethics", *Science,* 198: 699–705.

Cournand, Andre F. and Harriet A. Zuckerman. 1970. "The Code of Science", *Studium Generale, 23*: 941–962.

Dahrendorf, Ralf. 1964. *Homo Sociologicus.* Opladen: Westdeutscher Verlag.

Deutscher, Irwin. 1972. "Public and Private Opinions: Social Situations and Multiple Realities", in: Saad Z. Nagi and Ronald G. Corwin (eds.), *The Social Contexts if Research.* New York: Wiley.

Downey, Kenneth J. 1967. "Sociology and the Modern Scientific Revolution", *Sociological Quarterly,* 8: 239–254.

Feyerabend, Paul. 1975. *Against Method.* London: New Left Books.

Fleck, Ludwik. 1935. *Entstehung und Entwicklung einer wissenschaftlichen Tatsache: Einführung* in *die Lehre vom Denkstil und Denkkollektiv [Origin and Evolution of a Scientific Fact: Introduction to the Theory of Styles and Collective Carriers of Thought].* Basel, Switzerland: Schwab.

Gaston, Jerry. 1975. "Soziale Organisation, Kodifizierung des Wissens, und das Belohnungssystem der Wissenschaft" ["Social Organization, Codification of Knowledge, and the Reward System of Science"], in: Nico Stehr and René König (eds.), *Wissenschafissoziologie*. Opladen: Westdeutscher Verlag.

Gaston, Jerry. 1978. *The Reward System in British and American Science.* New York: Wiley-Interscience.

Gouldner, Alvin W. 1976. *The Dialectic of Ideology and Technology: The Origins, Grammar, and Future of Ideology.* New York: Seabury Press.

Gutting, Gary. 1973. "Conceptual Structures and Scientific Change", *Studies in History and Philosophy of Science,* 4: 209–230.

Hagstrom, Warren O. 1965. *The Scientific Community.* New York: Basic Books.

Hempel, Carl G. 1965. *Aspects of Scientific Explanation.* New York: Free Press.

King, M. D. 1971. "Reason, Tradition, and the Progressiveness of Science", *History and Theory,* 10: 3–32.

Knorr, Karin. 1977. "Producing and Reproducing Knowledge: Descriptive or Constructive? Toward a Model of Research Production", *Social Science Information,* 16: 669–696.

Kornhauser, William. 1963. (with the assistance of W. O. Hagstrom). *Scientists in Industry.* Berkeley: University of California Press.

Kowalewski, Stephen A. 1974. "Bureaucratic Trends in the Organization and Institutionalisation of Scientific Activity", in: R. Whitley (ed.), *Social Processes of Scientific Development.* London: Routledge & Kegan Paul.

Krohn, Roger G. 1961. "The Institutional Location of the Scientist and His Scientific Values", *IRE Transactions on Engineering Management,* EM 8: 133 138.

Kuhn, Thomas S. 1962. *The Structure of Scientific Revolutions.* Chicago: University of Chicago Press.

Kuhn, Thomas S. 1977. *The Essential Tension: Selected Studies in Scientific Tradition and Change.* Chicago: University of Chicago Press.

Lammers, Cornelius. 1974. "Mono- and Poly-paradigmatic Developments in Natural and Social Sciences", in: Richard Whitley (ed.), *Social Processes of Scientific Development.* London: Routledge & Kegan Paul.

Lautmann, Rüdiger. 1971. *Wert und Norm: Begriffsanalysen für die Soziologie [Value and Norm: Concepts of Analysis for Sociology].* Opladen: Westdeutscher Verlag.

Luhmann, Niklas. [1968] 1970. "Selbststeuerung der Wissenschaft", in: *Soziologische Aufklärung: Aufsätze zur Theorie sozialer Systeme. [Sociological Enlightenment: Essays on the Theory of Social Systems*]. Opladen: Westdeutscher Verlag.

Luhmann, Niklas. 1969. "Normen in soziologischer Perspektive" [Norms in the Sociological Perspective], *Soziale Welt,* 20: 28–48.

Marcson, Simon. 1960. *The Scientist* in *American Industry.* New York: Harper & Row.

Merton, Robert K. 1948. "On the Position of Sociological Theory", *American Sociological Review,* 13: 164–168.

Merton, Robert K. [1968] 1973a. "Behavior Patterns of Scientists", in: *The Sociology of Science.* Chicago: University of Chicago Press.

Merton, Robert K. [1942] 1973b. "The Normative Structure of Science", in: *The Sociology of Science.* Chicago: University of Chicago Press.

Merton, Robert K. [1957] 1973c. "Priorities in Scientific Discovery", in: *The Sociology of Science.* Chicago: University of Chicago Press.

Merton, Robert K. [1938] 1973d. "Science and the Social Order", in: *The Sociology of Science.* Chicago: University of Chicago Press.

Merton, Robert K. 1973e. *The Sociology of Science.* Chicago: University of Chicago Press.

Merton, Robert K. 1975. "Structural Analysis in Sociology", in: Peter M. Blau (ed.), *Approaches to the Study of Social Structure.* New York: Free Press.

Merton, Robert K. 1976. *Sociological Ambivalence and Other Essays.* New York: Free Press.

Merton, Robert K. 1977. "The Sociology of Science: An Episodic Memoir", in: Robert K. Merton and Jerry Gaston (eds.), *The Sociology of Science in Europe.* Carbondale: Southern Illinois University Press.

Mitroff, Ian I. 1974a. *The Subjective Side of Science: A Philosophical Inquiry into the Psychology of the Apollo Moon Scientists.* Amsterdam: Elsevier.

Mitroff, Ian I. 1974b. "Norms and Counter-Norms in a Select Group of the Apollo Moon Scientists: A Case Study of the Ambivalence of Scientists", *American Sociological Review,* 39: 579–595.

Mok, Albert, and Anne Westerdiep. 1974. "Societal Influences on the Choice of Research Topics of Biologists", in: Richard Whitley (ed.), *Social Processes of Scientific Development.* London: Routledge & Kegan Paul, 1974.

Mulkay, Michael. J. 1969. "Some Aspects of Cultural Growth in the Natural Sciences", *Social Research*, 36: 22–52.

Mulkay, Michael J. 1972a. "Conformity and Innovation in Science", in: Paul Halmos (ed.), *The Sociology of Science.* Sociological Review Monograph, No. 18. Keele, England: University of Keele.

Mulkay, Michael J. 1972b. *The Social Process of Innovation: A Study in the Sociology of Science.* London: Macmillan.

Mulkay, Michael J. 1975. "Drei Modelle der Wissenschaftsentwicklung" ["Three Models of the Evolution of Science"], in: Nico Stehr and René König (eds.), *Wissenschaftssoziologie.* Opladen: Westdeutscher Verlag.

Mulkay, Michael J. 1976. "Norms and Ideology in Science", *Social Science Information,* 15: 627–656.

Parsons, Talcott. 1939. "The Professions and Social Structure", *Social Forces*, 17: 457–467.

Parsons, Talcott. [1937] 1949. *The Structure of Social Action.* Vol. 1. New York: Free Press.

Parsons, Talcott. 1951. *The Social System.* New York: Free Press.

Parsons, Talcott. 1970. "On building Social Systems Theory: A Personal History", *Daedalus,* 99: 826–881.

Popitz, Heinrich. 1961. "Soziale Normen", *European Journal of Sociology,* 2: 185–198.

Popper, Karl R. [1935] 1968. *The Logic of Scientific Discovery.* New York: Basic Books, (enlarged ed. 1959). London: Hutchinson.

Popper, Karl R. [1962] 1969. *Conjectures and Refutations.* London: Routledge & Kegan Paul.

Price, D. de Solla. 1978. "Toward a Model for Science Indicators", in: Y. Elkana et al. (eds.), *Toward a Metric of Science: The Advent of Science Indicators.* New York: Wiley-Interscience.

Ravetz, Jerome R. 1971. *Scientific Knowledge and Its Social Problems.* New York: Oxford University Press.

Rickert, Heinrich. 1902. *Die Grenzen der naturwissenschaftlichen Begriffsbildung [The Limits of Concept Formation in Natural Science*]. Tübingen: Mohr.

Rothman, R.A. 1972. "A Dissenting View on the Scientific Ethos", *British Journal of Sociology,* 23: 102–108.

Scherhorn, Gerhard. 1968. "Der Wettbewerb in der Erfahrungswissenschaft" ["Competition in Empirical Science"], *Hamburger Jahrbuch fur Wirtschafts- und Gesellschaftspolitik [Hamburg Yearbook of Economic and Social Policy],* 14: 63–86.

Shephard, Herbert. A. 1954. "The Value System of a University Research Group", *American Sociological Review*, 19: 456–462.

Simmons, Anthony and Nico Stehr. 1978. "Language and the Growth of Knowledge in Sociology", Unpublished Paper.

Sklair, Leslie. 1973. *Organized Knowledge. A Sociological View of Science and Technology.* St. Albans, England: Paladin Books.

Spiegel-Rosing, Ina-Susanne. 1973. *Wissenschaftsentwicklung und Wissenschaftssteuerung: Einführung und Material zur Wissenschaftsforschung [Evolution and the Control of Science: An Introduction and Materials on Science Research].* Frankfurt am Main: Athenäum.

Spinner, Helmut F. 1977. "Thesen zum Thema Reichweite und Relevanz der Wissenschaftstheorie für die Einzelwissenschaften – Analytische Philosophie Versus Marxismus", in: Karl-Heinz Braun and Klaus Holzkamp (eds.), *Kritische Psychologie [Critical Psychology].* Vol 2. Cologne: Pahl-Rugenstein.

Stehr, Nico. 1975. "Zur Soziologie der Wissenschaftssoziologie", in: Nico Stehr and René König (eds.), *Wissenschaftssoziologie [Sociology of Science].* Opladen: Westdeutscher Verlag.

Storer, Norman W. 1966. *The Social System of Science.* New York: Holt, Rinehart and Winston.

Storer, Norman W. 1973. "Introduction", in: Robert K. Merton (ed.), *The Sociology of Science: Theoretical and Empirical Investigations.* Chicago: University of Chicago Press.

Useem, Michael. 1968. "Scientific Normative Orientations and Research Methodologies: A Comparative Study of European Scientists and Engineers at Euratom", Unpublished Report.

Weber, Max. 1922. *Gesammelte Aufsätze* zur *Wissenschaftslehre [Collected Essays on the Theory of Knowledge]*. Tübingen: Mohr.

Weingart, Peter. 1972. "Wissenschaftsforschung und Wissenschaftssoziologie", in: Peter Weingart (ed.), *Wissenschaftssoziologie.* 1: *Wissenschaftliche Entwicklung als sozialer Prozess [Sociology of Science: The Evolution of Science as a Social Process]*. Frankfurt am Main: Athenäum.

Weingart, Peter. 1974. "On a Sociological Theory of Scientific Change", in: Richard Whitley (ed.), Social *Processes of Scientific Development.* London: Routledge & Kegan Paul.

Weingart, Peter. 1976. *Wissensproduktion und Soziale Struktur [The Production of Knowledge and Social Structure]*. Frankfurt am Main: Suhrkamp.

West, S. Stewart. 1960. "The Ideology of Academic Scientists", *IRE Transactions on Engineering Management,* EM-1: 54–62.

Whitley, Richard D. 1972. "Black Boxism and the Sociology of Science", in: Paul Halmos (ed.), *The Sociology of Science*. Sociological Review Monograph No. 18. Keele, England: University of Keele.

Wunderlich, Richard. 1974. "The Scientific Ethos: A Clarification", *British Journal of Sociology*, 25: 373–377.

Young, Robert. 1973. "The Historiographic and Ideological Contexts of the Nineteenth-Century Debate on Man's Place in Nature", in: M. Teich and Robert Young (eds.), *Changing Perspectives in the History of Science*. London: Heinemann.

Ziman, John M. 1968. *Public Knowledge.* Cambridge, England: Cambridge University Press.

Zuckerman, Harriet. 1977. "Deviant Behavior and Social Control in Science", in: Edward Sagarin (ed.), *Deviance and Social Change.* Beverly Hills: Sage.

Zuckerman, H., and Merton, R.K. 1971. "Patterns of Evaluation in Science: Institutionalization, Structure and Functions of the Referee System", *Minerva,* 9: 66–100.

Chapter 11
The Magic Triangle: In Defense of a General Sociology of Knowledge

Nico Stehr

> *It has become extremely questionable whether, in the flux of life, it is a genuinely worthwhile intellectual problem to seek to discover fixed and immutable ideas or absolutes.*
>
> Karl Mannheim (1936: 77)

Karl Mannheim's at times rather tentative and at times most ambivalent commitment to a programme of merging philosophical (especially epistemological), sociological and moral discourse could only emerge after he severed strong dependence on a Hegelian philosophy of history. For it was on the basis of such a philosophy that Mannheim, for many years, remained convinced that most if not all intellectual and social divisions in society were mere transitory phenomena, soon to be transcended by the logic of history. Mannheim's commitment to joining separate, narrow and unequal modes of discourse, however, is also tentative and ambivalent because such a programme strongly violates contemporary prohibitions and taboos against mixing, for example, sociological and moral issues. It would not violate the same set of norms, however, if one were to view sociological discourse as dependent and subsequent to epistemological discourse. It is my contention that a non-restrictive and self-reflexive sociology of knowledge requires a set of arguments which are based on the assumption that moral, sociological and epistemological modes of discourse are interdependent.

What is required for a general sociology of knowledge, a sociology of knowledge not limited a priori to the presumably ever contracting domain of 'irrational' knowledge claims, are arguments which emphasize the interconnection and mutual

This text was first published as: Stehr, Nico. 1981. "The Magic Triangle: In Defense of a General Sociology of Knowledge", *Philosophy of the Social Sciences*, 11: 225–229. The permission to republish this text was granted on behalf of SAGE by Ms. Yvonne McDuffee, Rights Coordinator, SAGE Thousand Oaks, CA 91320.

M. T. Adolf (ed.), *Nico Stehr: Pioneer in the Theory of Society and Knowledge*, Pioneers in Arts, Humanities, Science, Engineering, Practice 16,
https://doi.org/10.1007/978-3-319-76995-0_11

dependence of problems heretofore assigned to separate disciplines. It is evident that such a programme continues to encounter vigorous opposition, as it did some fifty years ago when Mannheim made tentative proposals in this direction. A transformation in the entrenched relations of moral, sociological and philosophical discourse, it is suggested, invariably involves a serious dilemma, perhaps even the fallacy of the magic triangle.

Economists choose the term magic triangle to refer to a dilemma of economic policy which attempts to realize three distinct but interdependent economic goals simultaneously. The goals in question are forms of economic stability, namely a balanced balance of payments, full employment and stable prices. In theory at least, all three goals cannot be attained simultaneously (cf. Veit 1965).

Dilemmas such as these of course are not restricted to economic affairs. Analogous difficulties are perceived to exist in social science generally. For social scientists almost always face the dilemma of having to exclude one, or even two, of three concerns held to be part of the essential nature of social scientific considerations. A strictly enforced division of intellectual labor therefore gives rise to such a magic triangle in social science. Moral, philosophical and sociological issues form such a magic triangle since we often see them as interdependent concerns, although one is not supposed to practice all three modes of discourse simultaneously. What may well have been initially a provisional separation has become a permanent fixture of the academic division of labor.

Mannheim as indicated displays but a cautious commitment towards a programme aimed at developing a theory which disregards some of the presumed dilemmas of the magic triangle. He, too, it would seem, succumbs to the tendency to focus on one of the corners of the triangle to the exclusion of the others. As Mannheim's interest and commitment to sociology grew stronger, he increasingly abandoned philosophical concerns and his moral commitments became more and more explicit.

In a similar way, discussions about a general sociology of knowledge exemplify the operation and the strength of the prohibitions associated with the magic triangle: both proponents and opponents of a strong programme in the sociology of knowledge seem to act in conformity with these prohibitions. That is, the frequent rejection of the possibility of a general sociology of knowledge, for example, is almost exclusively based on logico-philosophical grounds taken to be self-sufficient. Similarly, many arguments in defense of a general sociology of knowledge remain within the boundaries of epistemological discourse and rarely consider sociological or even moral considerations to be relevant.

It is impossible to examine many of the ramifications of a foundation of a general sociology of knowledge which takes the prohibitions of the magic triangle less seriously. My remarks will have to be extremely limited. I will therefore confine myself to certain aspects of the debate on relativism.[1]

[1] Despite the arguments advanced here, charges of relativism will of course persist. From a hermeneutic-linguistic perspective Apel (1972: 10) suggests that such a charge is consequential

Such a limitation is perhaps justified once one considers the importance this issue has in the history of the sociology of knowledge dispute. Moreover, the legitimacy of a general sociology of knowledge most certainly rests to a large extent on a satisfactory answer to the very charge of relativism levelled against the strong programme. However, such a response remains elusive. It is my view that this, to a considerable extent, stems from the widespread acceptance of the prohibitions of the magic triangle. The discussion of relativism should also be used to demonstrate some of the advantages of revising entrenched divisions of intellectual labor.

Given the lack of a separate cognitive and institutional identity of sociology at the time, and given the relatively important role of philosophy in German universities as a *Grundlagenwissenschaft* (foundational science), it does not come as a surprise that the sociologists whose pioneering efforts led to the sociology of knowledge, all had a strong interest and background in philosophy and epistemology. These strong initial links between philosophical and sociological concerns, one might expect, would lead to continued fruitful reciprocal relations between philosophy and the sociology of knowledge. This has not been the case.

For most North American social scientists, the sociology of knowledge was quickly transformed and absorbed into the pre-existing division of academic labor, and justifiably so. That is, a most significant aspect of the entrenched conception of the sociology of knowledge, even in Europe today, is the conviction that a divorce from philosophical or epistemological concerns is the *sine qua non* for the possibility of a 'positive' or 'substantive' sociology of knowledge (cf. Parsons 1936: 680; Speier 1936: 682; Aron 1953: 85; Coombs 1966: 232; Barber 1975: 104; Walter 1967: 342). The sociology of knowledge is, in the words of Stark (1958: 152), "a positive-descriptive-historical discipline rather than a philosophical and epistemological one". And, reflecting an essential criterion of critical rationalism, e.g., Reichenbach's demarcation between the context of discovery and the context of justification, Stark (1958: 152) notes in agreement with most sociologists and philosophers that the "sociology of knowledge is concerned in the first place with the origin of ideas, and not with their validity".

Given the nature of the reconstructed sociology of knowledge today, sociologists have succeeded in immunizing their speciality from the charge of relativism by declaring it to be irrelevant to the *sociology* of knowledge. The problem of relativism is therefore a concern which has to do with the *logic* of the sociology of knowledge and is thus a matter with which logicians or epistemologists might be concerned. In other words, as Barnes (1974: 180, note 5) correctly observed, "despite Mannheim, relativism has never really been an issue in sociology".

because "der Mensch [erwirbt] mit der kommunikativen Erlernung der Sprache zugleich ein wie immer unausdrückliches Bewusstsein davon, was es heisst, *sich* nach einer Regel zu richten. D.h. er wird nicht nur faktisch in eine bestimmte Lebensform eingeübt, sondem erwirbt ein effektives Reflexionsverhältnis zur Form des Lebens überhaupt. Z.B. hat er mit der Erlernung der Sprache zugleich ein Verständnis von Sprachbegrauch überhaupt erlernt, das ihn in den Stand setzt, aus einer Sprache in eine andere zu *übersetzen*, und d.h. zugleich *fremde* Lebensformen zu *verstehen*".

For the most part then, the post-Mannheim debate about relativism and the self-referential problem of the sociology of knowledge has been carried out on the basis of the premise that these questions are solely epistemological issues. Sociological or, for that matter, moral considerations are irrelevant by definition. Efforts to reopen the debate have, in many ways, affirmed traditional divisions, and, not infrequently, have turned into celebrations of orthodox positions.

The magic triangle referred to earlier would suggest that the issue of relativism too may be viewed from a number of points of view. For the sake of simplicity, I shall concentrate on the claim of the interdependence of epistemological and sociological (existential) issues which rarely have been made a topic of investigation in the relativism debate. Of course I shall not attempt to claim that sociology is a kind of epistemological wonderweapon. However, I shall claim that presumably purely epistemological arguments advanced against a general sociology of knowledge have a number of sociological implications which make them appear far less convincing.

Among the more important implications of the charge of relativism levelled against a general sociology of knowledge is that such a programme somehow must admit all forms of knowledge claims as equally possible or sensible and that the programme therefore fails to recognize the existence of ahistorical rules or methods as final arbiters among competing claims: a general sociology of knowledge implies anything goes. On the basis of purely formal considerations, such a charge appears eminently reasonable and makes a general sociology of knowledge a most irresponsible programme. An examination of the tacit sociological assumptions which underly such conclusions, however, might well show that this is a most implausible conclusion.

A reference to one of Chomsky's methodological devices may illustrate the point initially. As is well known, one of the foundations of Chomsky's account of human language is the observation that the languages of the world are by no means as diverse as they might be. As a matter of fact, Chomsky argues that there are linguistic universals which have to do with the syntax of language, rules which are constitutive for meaningful utterances and which have to do with the hierarchical organization of sentences. Chomsky further argues that the invariant syntactic properties of human languages are genetically determined. Linguistic variations are limited. Without going into further details and without accepting Chomsky's controversial conclusions, one lesson relevant to the debate on relativism appears to be that the empirically observable diversity of knowledge claims, too, is limited. A limitless variation and diversity of knowledge claims of course requires that such unrestructured transformations are possible in the first place. But this clearly is not the case, for what this would *require* are precisely those social conditions which are said to *follow* from relativism.

In other words, the formal possibility that everything goes, that everything passes as a knowledge claim, that no limits exist, that proliferation is endless, etc., actually demands a society which in fact does not deserve the term 'society'. It is most implausible that such social conditions will ever be realized. Paradoxically perhaps, the very ability to advance knowledge claims requires that we must lose, to use the terminology of philosophical anthropology, our "world-openness", a condition which presumably allows for almost unrestricted and limitless intellectual developments.

With analogy to Chomsky then, one could point out that contending knowledge forms in a variety of institutions and societies are not nearly as diverse as they might be. There are distinct social limits to the proliferation of ideas.

Similarly, the charge of relativism may be said to imply—from a sociological point of view—"methodological solipsism" (cf. Apel 1973: 233–45). Among the transcendental conditions for the possibility of an almost limitless proliferation of knowledge as charged by the critics of a general sociology of knowledge, is the assumption that a single person is capable of knowing and, therefore, for example, of doing science. However, the presence and importance of conventions (or language games) for scientists as a precondition for doing science would indicate that a single scientist could not produce knowledge claims in splendid isolation.

A further sociological criticism of the critique of the strong programme in the sociology of knowledge could point to the highly implausible assumption of a radical discontinuity among knowledge claims or cognitive systems generally. As Mannheim (1924, 1964: 325) observed, different knowledge claims, in all likelihood, do not develop in a vacuum but are oriented toward each other, they learn from each other, they are directed toward each other, without necessarily losing their identity.

In short, a sociological critique and analysis of the charge of relativism could come to the conclusion that the orthodox criticism of a general sociology of knowledge exaggerates the dangers of an unchecked proliferation of knowledge claims and underestimates (or, is silent on) the threats to the very development of knowledge which may well come from a monopolization of discourse.

References

Apel, Karl-Otto. 1972. "Sprache als Medium der transzendentalen Reflexion", in: Heinrich Starke (ed.), *Sprache und Erkenntnis*. Meisenhain am Glan: Anton Hain Verlag: 1–15.

Apel, Karl-Otto. 1973. *Transformation der Philosophie*, Volume II. Frankfurt am Main: Suhrkamp.

Aron, Raymond. 1953. *Die deutsche Soziologie der Gegenwart*. Stuttgart: Kröner.

Barber, Bernard. 1975. "Toward a New View of the Sociology of Knowledge", in: Lewis A. Coser (ed.), *The Idea of Social Structure*. New York: Harcourt Brace Jovanovich: 103–116.

Barnes, Barry. 1974. *Scientific Knowledge and Sociological Theory*. London and Boston: Routledge and Kegan Paul.

Coombs, Robert H. 1966. "Karl Mannheim, Epistemology and the Sociology of Knowledge", *Sociological Quarterly*, 7: 229–233.

Mannheim, Karl. [1924] 1952. "Historicism", in: *Essays on the Sociology of Knowledge*. London: Routledge and Kegan Paul: 84–133.

Mannheim, Karl. 1936. *Ideology and Utopia: An Introduction to the Sociology of Knowledge*. New York: Harcourt.

Parsons, Talcott. 1936. "Review of Alexander von Schelting's Max Weber's Wissenschaftslehre", *American Sociological Review,* 1: 675–681.
Speier, Hans. 1936. "Review of Ernst Gruenwald's Das Problem einer Soziologie des Wissens", *American Sociological Review,* 1: 681–682.
Stark, Werner. 1958. *The Sociology of Knowledge.* London: Routledge & Kegan Paul.
Veit, Otto. 1965. "Zentralbankpolitik", in: Erwin Beckerath et al. (eds.), *Handwörterbuch der Sozialwissenschaften.* Stuttgart: Gustav Fischer: 423–434.
Walter, Benjamin. 1967. "The Sociology of Knowledge and the Problem of Objectivity", in: Llewellyn Gross (ed.), *Sociological Theory: Inquiries and Paradigms.* New York: Harper & Row: 335–357.

Chapter 12
Karl Mannheim

Nico Stehr (with Volker Meja)

Karl Mannheim, one of the founders of the sociology of knowledge (*Wissensoziologie*), was born in Budapest in 1893, and held academic posts at the universities of Heidelberg and Frankfurt and The London School of Economics and Political Science. He died in London in 1947. His biography, which is one of intellectual and forced geographical migration, falls into three phases: Hungarian (to 1920), German (1920–33) and British (1933–47). Among the important intellectual influences on Mannheim are Georg Lukács, Georg Simmel, Edmund Husserl, Karl Marx, Alfred and Max Weber, Max Scheler and Wilhelm Dilthey. Through these and other writers, German historicism, Marxism, phenomenology, sociology and, later, Anglo-Saxon pragmatism became decisive influences on his work.

The writings of Mannheim's Hungarian phase—primarily on literary and philosophical themes—demonstrate a first attempt to go beyond the German idealist view of history and society.

The German phase was Mannheim's most productive; he gradually turned from philosophy to sociology (although he never completely abandoned philosophical questions), inquiring into the possible social roots of culture and knowledge. Many of his essays on the sociology of knowledge have become classics. Mannheim's most influential work, *Ideologie und Utopie*, Bonn, 1929 (English translation, *Utopia,* London, 1936), was also written during this period. These writings became the focus of a vigorous intellectual dispute in Germany towards the end of the Weimar Republic, in part because of what many critics regarded as the 'relativistic'

This text was first published as: Meja, Volker and Nico Stehr. 1987. "Karl Mannheim", in: Jessica Kuper (ed.), *Key Thinkers, Past and Present*. London and New York: Routledge and Kegan Paul: 152–154. The permission to republish this text was granted on 26 October 2017 by UK Book Permissions of Taylor and Francis.

M. T. Adolf (ed.), *Nico Stehr: Pioneer in the Theory of Society and Knowledge*,
Pioneers in Arts, Humanities, Science, Engineering, Practice 16,
https://doi.org/10.1007/978-3-319-76995-0_12

implications of Mannheim's sociology. Mannheim, however, claimed that his ideas prepared the ground for a new comprehensive perspective capable of transcending the fragmented and partial social and political views held up till then. He maintained that the "socially unattached" intelligentsia had an instrumental role in developing such a synthesis.

Mannheim's British phase was in some ways foreshadowed by the more practical orientation already evident in his work prior to his emigration from Germany. Applied sociology should be concerned with the comprehensive analysis of the structure of modern society, especially through democratic social planning, in which education should occupy a central role.

The original themes of the sociology of knowledge were formulated in Germany during a period of major social crisis, and may be seen, as Mannheim himself saw them, as the product of one of the greatest social, political and economic dissolutions and transformations, accompanied by the highest form of reflexivity, self-consciousness and self-criticism. The renewed interest in the problems posed by the sociology of knowledge reflects a similar crisis in our own period and may therefore be said to owe more to the course of events than to analytical progress.

Further Reading

Other Works by Mannheim in translation include:

Man and Society in an Age of Reconstruction, London, 1941.
Essays on Sociology and Social Psychology, London, 1953.
Essays on the Sociology of Culture, London, 1956.
Systematic Sociology: An Introduction to the Study of Society, London.

Secondary Literature on Mannheim Includes:

Kettler, David, Volker Meja and Nico Stehr (eds.). 1984. *Karl Mannheim.* London: Tavistock Publications.

Loader, Colin. 1985. *Culture, Politics and Planning: The Intellectual Development of Karl Mannheim.* New York: Cambridge University Press.

Simonds, Arthur P. 1978. *Karl Mannheim's Sociology of Knowledge.* Oxford: Clarendon Press.

Wolff, Kurt H. (ed.). 1971. *From Karl Mannheim.* London Transaction.

Chapter 13
The Sociology of Knowledge and the Ethos of Science

Nico Stehr (with Volker Meja)

The Critical Assessment of the classical sociology of knowledge, especially of its Mannheimian version,[1] which has been in evidence for nearly sixty years even among otherwise radically different theoretical traditions in the social sciences and philosophy, is at once surprising and significant. If we take the widespread cognitive disagreement among sociological traditions (see Merton 1978) as an initial criterion, then it might be expected that assessments of the basic claims advanced by the sociology of knowledge would vary greatly. What we find instead is an unusual degree of convergence in most evaluations of the sociology of knowledge —an extremely uncommon event in the history of sociological ideas. The conclusion may therefore be warranted that this is in fact a first indication that the degree of dissensus in sociological traditions (cf. Friedrichs 1970; Boudon 1971; Ritzer 1975; Sztompka 1979; Ultee 1980) is by no means as extensive, definite, or dysfunctional as is sometimes assumed (cf. Stehr 1982). The widespread consensus in regard to the classical sociology of knowledge raises at the same time the question of the grounds for the agreement among very varied critics of the major sociology of knowledge propositions.

This text was first published as: Meja, Volker and Nico Stehr. 1993. "The Sociology of Knowledge and the Ethos of Science", in: Eileen Leonard, Hermann Strasser & Kenneth Westhues (eds.), *In Search of Community: Essays in Memory of Werner Stark (1909–1985)*. New York: Fordham University Press: 65–83. The permission to republish this text was granted on by Will Cerbone, Fordham University Press, New York.

[1] As is well known, the sociology of knowledge tradition is by no means confined to Karl Mannheim's particular conception, even if his approach remains in many respects the most important exemplar. There are a variety and a considerable range of theoretical positions, including relatively recent ones (see Stehr/Meja 1984; Meja/Stehr 1990).

M. T. Adolf (ed.), *Nico Stehr: Pioneer in the Theory of Society and Knowledge*, Pioneers in Arts, Humanities, Science, Engineering, Practice 16,
https://doi.org/10.1007/978-3-319-76995-0_13

Our first task therefore will be to summarize these shared assessments of the classical sociology of knowledge and to inquire into some of the reasons for them.

Two basic flaws of the Mannheimian sociology of knowledge (cf. Luhmann 1980: 195–207) are usually identified in the literature and designated there either as logical flaws or as fatal errors of reasoning. There is, first, the claim that the sociology of knowledge fails to restrict its cognitive interest to its proper *sociological* problematic, i.e., to the analysis of the genesis and development of knowledge. In an essay on the sociology of knowledge which decisively influenced the reception of this new field of inquiry in North America (cf. Wolff 1978: 372), Merton (1957: 508) argues that

> Mannheim has sketched the broad contours of the sociology of knowledge with remarkable skill and insight. *Shorn of their epistemological impedimenta,* with their concepts modified by the lessons of further *empirical* inquiry and with occasional *logical* inconsistencies eliminated, Mannheim's procedures and substantive findings clarify the relations between knowledge and structure which have hitherto remained obscure [emphasis added].

Merton insists that Mannheim's approach, in spite of its 'speculative' brilliance, suffers from inadequate empirical grounding and, even more important, from its failure to delimit the cognitive ambitions of its *sociological* inquiry into knowledge. Mannheim's many critics have never tired of pointing out that the separation of philosophical or epistemological issues from 'proper' sociological questions is a *conditio sine qua non* for the emergence and development of a 'positive' (or 'existential') sociology of knowledge (see especially Parsons 1936: 680; Speier 1936: 682; Aron 1953: 85; Coombs 1966: 232; Walter 1967: 342; Lukács [1954] 1974: 82; Barber 1975: 104).[2]

Mannheim's cautious claim that the sociology of knowledge might have epistemological relevance has been almost universally rejected in favor of a persistent affirmation of the prevailing division of labor among disciplines[3] and the conviction

[2] Berger/Luckmann (1967: 14) argue that although the sociology of knowledge ought to inquire into all ideas that pass for knowledge, epistemological questions cannot be part of a sociology of knowledge agenda because they "properly belong to the methodology of the social sciences, an enterprise that belongs to philosophy and is by definition other than sociology, which is indeed an object of its inquiries."

[3] As a study by Köhnke (1985: 58–105) has shown, epistemology emerges in the first part of the nineteenth century and then begins to assume a separate identity within philosophy.

that these cognitive boundaries, including those of fairly recent origin, should not be broken down.[4] Mannheim's program has clear implications for a variety of scientific disciplines and runs counter to the historical trend toward specialization in science which began in the nineteenth century and has greatly accelerated ever since. It remains to be seen, however, whether trends that may be functional in the natural sciences do in fact have the same desirable consequences in the social sciences. In any case, it is unlikely that Mannheim's ambitions went beyond wanting to cast doubt on the cognitive advantage of excessive specialization.

The objections to Mannheim's effort to reduce the impact of cognitive demarcations on sociological inquiry would have confirmed his own realistic expectations. But this is not so in regard to what is widely perceived as a second flaw of the classical sociology of knowledge: its relativistic implications. Mannheim never analyzed this issue in an entirely convincing manner. Yet the widespread interpretation of the relativistic implications of the sociology of knowledge clearly runs counter to Mannheim's conception of the kind of sociological approach to knowledge he wanted to develop. Numerous passages in his writings indicate that he did not advocate a doctrine of a complete lack of standards and criticism, a *Lehre der Maßtabs- und Kritiklosigkeit.*[5]

But as we know only too well from the secondary literature (cf. Stehr/Meja 1982), there is a widely held perception that Mannheim suffers from false self-consciousness, as it were. The virtually unanimous conclusion has been that Mannheim's efforts to chart a non-relativistic sociology of knowledge ended in failure. The reception of his sociology of knowledge shows that his repeated attempts to locate the sociology of knowledge somewhere between "irrelevance to

[4] Mannheim's at least implicit conviction that the sociology of knowledge may well be a foundational science and therefore of greater use than an ordinary sociological speciality (see Lieber 1965. 82–83)—which some scholars have interpreted to mean that the sociology of knowledge aspires to replace philosophy (e.g., Salomon 1947: 350–64)—has encountered strong opposition (cf. Horkheimer 1930: 38; Rüschemeyer 1958: 4; Lenk 1961: 313; Neusüss 1968: 26). It is tempting to ask whether the precarious intellectual and social identity and legitimacy of sociology in the past (and to some extent even today) have not invariably generated strong opposition to any efforts that discourage specialization. Disagreement about cognitive demarcations is by now a nearly traditional feature of intellectual disputes in sociology and in the other social sciences (see Merton 1973: 10). Opposition to a general sociology of knowledge with implications beyond the discipline of sociology also results from disagreements about which is the leading social science discipline. While in the 1930s support for such an interdisciplinary program that would reduce the rigid division of labor in the social sciences and the humanities was broad, opinions about the 'leading' discipline or speciality differed. Horkheimer (1937: 3–16), for example, in his inaugural lecture at the University of Frankfurt, supported similar cognitive aims but suggested philosophy as the leading perspective.

[5] Mannheim uses this expression in describing the position he vainly attempted to avoid in a draft of his survey essay on the sociology of knowledge published in the *Handwörterbuch der Soziologie.* This draft is among the Mannheim papers deposited in the University Library of the University of Keele (Staffordshire, England). In his essay on historicism, Mannheim (1952: 127) similarly refers to the doctrine that "all action and decision is relative and lacks a standard" as one of the typical interpretations of historicism, namely, one conflating historicism and relativism.

the establishment of truth on the one hand, and entire adequacy for determining truth on the other" (Mannheim 1936: 285) have been rejected as unconvincing. Mannheim's notion of "relational objectivity" which he proposed in order to avoid the very charge to which he was subjected, has been widely interpreted as a merely reformulated relativism. Whatever distance Mannheim may have been able to establish between relativism and the location he desired for the sociology of knowledge, in the end his 'relationism' proved to be a platform that failed to attract much sympathetic assent. Mannheim, in the words of Toulmin (1972a: 495), did not manage to escape the invidious choice between the "arbitrariness of the absolutist and the defeatism of the relativist."[6] The history of the reception of Mannheim's ideas about the nature and function of the sociology of knowledge is therefore well suited to serve as a reminder of the typical static and dichotomous character of epistemological categories and traditions. Werner Stark's reasoning (1958: 160) is perhaps most representative of this kind of argument: "We must insist that we do not see it as a problem of relativity, so far as the truth is concerned. The truth is always absolute, otherwise it does not deserve to be called the truth at all. We utterly repudiate that anarchical historicism which holds that the terms 'relative' and 'true' are reconcilable."[7]

The more serious of the two major objections advanced against the classical sociology of knowledge is the accusation that it is essentially relativistic. Relativism raises the possibility of context-independent criteria such as truth, logic, and reason, and it invariably becomes entangled in a logical contradiction (see Grünwald 1934; von Schelting 1934: 99; Nagel 1961: 500; Bergmann 1968: 132–33; Merton 1971: 196; Lukes 1977: 137).[8] Proponents of critical theory charge, furthermore, that relativism implies a general inability to develop a critique of ideology or to generate any serious resistance to ideologies in the first place.[9] But relativism also implies opposition to the prevailing ethos of science.

The view of Mannheim's sociology of knowledge as relativistic is contradicted, as Merton (1957) has observed, by Mannheim's own recognition of "the logical fallacy and intellectual nihilism implicit in such a position." He repeatedly objected

[6] Some of those who interpret Mannheim's sociology of knowledge as an exemplar of a relativistically inclined epistemology merely want to stress that they disagree with the means by which he attempted to avoid the charge of relativism (for example, Merton 1957: 502–508). At the same time, other critics assume a much stronger intellectual and political affinity between a relativistic position and Mannheim's position. This applies especially to those who stress that Mannheim's 'relationism' is 'relativism' in disguise (see von Schelting 1936: 664; Mandelbaum 1938: 67; Aron 1953: 85; Maquet 1951: 84; Lukács [1954] 1974; Louch 1966: 205; Hamilton 1974: 128).

[7] But compare Wilhelm Baldamus' instructive discussion (1976) of the function of dichotomous notions in sociological theory-building.

[8] We have attempted to demonstrate elsewhere (Meja/Stehr 1988) that such a critique is usually linked with efforts to show that the issue of relativism itself is not a *sociological* matter, that it cannot be addressed from a sociological perspective. The upshot, as Barnes (1974: 180) has observed, is that "despite Mannheim, relativism has never really been an *issue* in sociology."

[9] Lukács ([1954] 1974: 89) pointedly argued that Mannheim's sociology of knowledge, in the face of fascism and imperialism, represents an "ideology of helpless surrender".

to a doctrine of a complete lack of standards and criticism. This puzzling discrepancy can be better understood by an immanent interpretation of the classical sociology of knowledge. A more adequate interpretation of Mannheim's views, furthermore, may well prove to be of more than merely historical significance.

Mannheim persistently attempted to show that neither his investigation of the social roots of ideas nor the theoretical perspective guiding his sociology of knowledge can in any serious sense be regarded as implying a relativistic perspective. In his early work on the sociology of knowledge, in particular in his 1924 essay on 'Historicism' (Mannheim 1952), which still reflects the influence of Georg Lukács and Hegel's philosophy of history, Mannheim suggests that the observable multiplicity of intellectual positions will in the long run be transcended through the distinctive *logic of reason* built into the evolution of human thought. This presupposes that a 'synthesis' of formerly contending intellectual perspectives is historically inevitable.[10] In the 1920s, as Mannheim increasingly saw himself as a sociologist (cf. Kettler et al. 1984), he became convinced that the *logic of history* will help overcome rival world-views. The existential connectedness of thought, and even the sociology of knowledge itself, Mannheim thought, will be transcended by the efforts of a unique social group, the socially unattached intellectuals. The process of mediation and the overcoming of competing ideas is set in motion by specific forms of social action subject to the control of a particular social group, and not by the iron necessity of an impersonal evolutionary principle. The competition of world views, the plurality of intellectual perspectives of every conceivable kind, and the fact that ideas come to serve a particular function in the social process are constitutive of a particular historical epoch and belong to a historically limited phase in the development of human self-reflection and knowledge. This period will be followed by an "age of conciliation", an intellectual synthesis on a higher plane. As a result of such dramatic changes, a period of greater social harmony and equality will ensue. Although Mannheim now believes that the conditions leading to a collapse of the competition of world views should be analyzed in terms of a particular logic of history, he nonetheless retains his basic convictions about the ultimate outcome of these intellectual developments. Although he provides only occasional indications about the future state of affairs he anticipates, Mannheim

---

[10] Very much in congruence with such a conception, Mannheim (1952: 178) writes in his 1925 essay on "The Problem of a Sociology of Knowledge": "Thus, one can at most arrive at the belief —by extrapolating from the structural position observed today—that the present rivalry of antagonistic systems and standpoints, and their attempts at incorporating the rival positions within themselves, indicate an inherent tendency of all human thought to account for the whole of reality, a tendency which falls short of achieving its goal as long as a fully comprehensive systematic principle is not yet discovered. This, then, will be reflected in the 'finiteness', the limitation to partial perspectives, of actual thinking. So far as we can see, reality is always more comprehensive than any of the partial standpoints it brings forth. Then, if we extrapolate, we may believe that a central systematic idea will eventually be found which will in fact permit a synthesis of the entire process."

remains convinced of the real possibility of a transcendence of the currently prevailing multiplicity of intellectual perspectives.[11]

[11] Already in 1924, Mannheim writes in his essay on 'Historicism': "We notice in all spheres (the 'ideological' is most conspicuous) that in contrast to these tendencies toward autonomy, atomization, and analysis (three fundamentally different tendencies, which nevertheless have something in common) there is taking place a movement toward synthesis. What historicism undertakes in the individual historico-cultural spheres, in art history, in the history of religion, in sociology, etc., in that it exhibits these different spheres of culture, not in their immanent exclusiveness, but as an integrative part of a totality—what historicism accomplishes here, is attempted also—to give one example out of many—in modern psychology. Here too, for example, the principle holds, that we should not only investigate the various sensory fields in isolation from one another, but should also explore the problems of the solidarity and unity of sensory experience. Here too, that analysing, atomizing, isolating tendency which dominated the other sciences as well, and which led to the endeavour to build up the most complex structures out of the most simple elements, is being supplanted by the recognition of 'complexes' and 'totalities' as primary and irreducible data, as given, for example, in perceptions of *Gestalt.* All these examples may be regarded as symptomatic of the fact that on the reflective ('ideological') side of the total process, one can find a number of parallel trends. This raises the question whether these phenomena do not represent a counterpart, at the level of scientific method, to the transformation process which is taking place in the social structure. If the atomizing, sectionalizing mode of thought may be regarded as corresponding to a social structure which allowed a maximum dissolution of the social bonds and which produced an economy consisting of liberalistically independent, atom-like units, then the present trend toward synthesis, toward the investigation of totalities may be regarded as the emergence, at the level of reflection, of a force which is pushing social reality into more collectivistic channels. It may very well be, indeed, that this newly developing impulse to restore a psychic and intellectual unity in the place of the separation of spheres brought about by the previous epoch, the levelling down of the sharp boundaries between them, corresponds to a general change in practical attitudes" (Mannheim 1952: 95–96). In 1928 Mannheim writes in his essay "Competition as a Cultural Phenomenon" (and similarly, a year later, in *Ideologie und Utopie*): "There are periods in modern history during which a representative generation becomes free to achieve a synthesis. Such generations take a fresh approach in that they are able to envisage from the higher platform of a synthesis those alternatives and antagonisms which their fathers had interpreted in a dogmatic, absolute sense. Then, if there are existential problems not yet ripe for a solution, such a generation will experience them in entirely different contexts; the old antagonisms, however, become less sharp, and it will be possible to find a point, so to speak, farther back, from which partisan positions can be seen as merely partial and relative, and thus transcended. (It seems, by the way, that the sociology of knowledge itself provides just such a viewpoint 'farther back' from which theoretical philosophical differences which cannot be reconciled on the level of manifest content, can be seen through in all their partiality and therewith made amenable to a synthesis. The existence of this continually receding viewpoint—which one might be tempted to interpret inaccurately as a sign of an ever-increasing reflexiveness—presents us with a hitherto untouched but nevertheless important problem of the sociology of knowledge.) The problem of synthesis is far too complicated to allow us even to approach its solution here. It must suffice for us to see that syntheses do exist, and that the history of thought in modern times provides instances not only of polarization, but of association, crossing-over, and synthesis. One thing, however, we must not lose sight of: the syntheses are not confined to purely intellectual currents; they also represent interpretations of social forces" (Mannheim 1952: 224–25).

Nearly twenty years later, in his last comment on the sociology of knowledge, in a letter to Kurt H. Wolff (April 15, 1946) a few months before his death, Mannheim calls upon our whole generation to complete the work of synthesis "as nothing is more obvious than that we transcended in every field the idea that man's mind is equal to an absolute Ratio in favour of a theory that we

Mannheim therefore argues that the social conditions which give rise to a sociology of knowledge are historically specific and that significantly reduced social divisions will eventually make a sociology of knowledge unnecessary. The political function of the sociology of knowledge is an instrumental one: to serve as a mediating force in reorienting all vital participants in the political process toward a transcendence of dissensus, and to prepare for a synthesis of perspectives. The elimination of communicative misunderstandings and a reduction in societal conflicts are regarded as salient contributions of the sociology of knowledge to political life.

In our view, Mannheim's ideas on the ways in which the sociology of knowledge might become a force for mediating conflicts between competing ideas, and the persistence with which he pursued the possibility that a sociological inquiry into thought might help produce a rapprochement between politics and reason, clearly indicate that his rejection of relativism does not originate primarily in epistemological reflections. While Mannheim strongly and consistently opposed an absolutist position in epistemology,[12] his opposition to relativism was equally strongly grounded in his convictions concerning philosophy of history, theory of society, and ethics. But because he did not consistently and explicitly link these views with those passages in his texts in which he proclaimed himself an opponent of relativism, he was unable to prevent his critics from imposing upon him the norms of their own discourse. Nonetheless, an immanent interpretation of Mannheim's theoretical program for a sociology of knowledge must not overlook the fact that his tolerance for diversity was hardly unlimited. On the contrary, Mannheim's moral and political views strengthened his conviction that the diversity of ideas is subject to human control, can be restricted, and can be reduced in scope. The alleged dangers of an unrestrained relativism, in consequence, were for Mannheim matters to be negotiated and dealt with through political effort, and not primarily in the arena of pure intellectual effort. Consistent with these convictions, Mannheim (1952: 126–133) regarded the growing preoccupation with relativism as an expression of static conceptions of history and of inflexible standards of thought.

---

think on the basis of changing frames of reference, the elaboration of which is one of the most exciting tasks of the near future" (Wolff 1983: 204).

[12] Mannheim might easily have been tempted to employ a fairly straightforward strategy to ward off objections to a thoroughgoing relativism by arguing that there are no absolute truths, except one. As Gellner (1974: 49) notes, for example, it is a frequent strategy to make "an exception on one's behalf; having difficulty in accounting for oneself is the professional ailment of philosophers, and is virtually written into the terms of reference under which they work." However, Mannheim clearly was not primarily concerned with epistemological discourse. As a matter of fact, he considered a merely 'logical' resolution of the issue of relativism unsatisfactory and incomplete. He consequently writes (Mannheim 1952: 128f.): "Only a mode of thought, only a philosophy which is able to give a concrete answer to the question 'what shall we do?' [and not 'what shall we know?'] can put forward the claim to have overcome relativism."

Mannheim's mode of presentation and reasoning—for example, the frequently intended ambiguity of his formulations—have provided fertile grounds for misinterpretation and misrepresentation of his theses. He methodically employed ambiguity and multiple meanings, and he was quite prepared to leave alternative possibilities in suspension and not let difficulties deter him from pressing on with the task at hand or moving on to other questions. This willingness to experiment with openness has not always been sufficiently recognized and accepted. At the same time, Mannheim was not immune to pressures to formulate his sociology of knowledge approach in an ethically and philosophically *neutral* manner. The prohibition against violating the entrenched cognitive division of labor in science was already evident fifty years ago, but today it has practically become a taboo against forms of discourse which threaten to challenge these boundaries. Work at the boundaries of disciplines has its own inherent logic and dangers; it sometimes inhibits a scholar from revealing the premises and aims of such work. Mannheim himself was painfully conscious of these difficulties, having become aware of some of the penalties for work of this kind.

Mannheim's critics employ criteria and standards of judgment in their assessment of the sociology of knowledge that they evidently believe to be beyond reproach. That these criteria are indeed context-independent or in some other sense neutral, however, cannot simply be assumed *a priori.* One of the standard weapons employed against an unrestrictive sociology of knowledge that does not from the beginning exempt an important segment of human thought from its domain of inquiry (for example, scientific knowledge) is to insist that such a program is self-negating. While Mannheim was generally prepared to restrict his approach to the sociological analysis of ideas, he was nevertheless vulnerable to the criticism that his conception involved a *logical* flaw or even a contradiction. The argument that every unrestrictive sociology of knowledge becomes entangled in a self-defeating vicious circle is, no doubt, a kind of Archimedian standard of any critique of the sociology of knowledge. It is, however, worth asking whether the critique that the sociology of knowledge is self-negating—a critique that claims to point to a serious *logical* flaw in the design of the sociology of knowledge—can be dealt with only by logical discourse. It is quite doubtful that 'logical' discourse is as autonomous, sovereign, and distinct from other forms of discourse as adherents of the critique of self-contradiction tend to assume. The very assertion of the distinctiveness and authority of an argument is itself an indication that such convictions are by no means exclusively grounded in logical considerations. If, however, one begins with the assumption that the established boundaries among forms of discourse are in fact in flux, then even this argument appears to be open to challenge.

Recent theoretical reflections in the philosophy of science as well as empirical work in the sociology and history of science have seriously questioned the received wisdom concerning the possibility of objective scientific knowledge. The claim has been challenged that scientific knowledge can be generated, sustained and criticized in a social vacuum unmediated and independent of 'extraneous' cognitive and social processes.[13] The still widely held belief that there is a clear-cut distinction between cognitive and normative statements, that a neutral observations language is possible, and that there is a distinct demarcation between scientific and non-scientific knowledge now meets with considerable skepticism. An important outcome of these developments is the transcendence of the view that scientific knowledge is distinguished from other forms of knowledge by virtue of a range of unique qualities linked to the conditions of its production. Scientific knowledge is increasingly analyzed as contingent knowledge that is neither fully nor definitively determined by the world of facts (cf. Barnes 1977). Mulkay (1979: 53) aptly summarizes these far-reaching changes in the sociology of science: "[T]he criteria used in evaluating theoretical claims, like those applied to particular observational reports, seem to vary in meaning in accordance with the context in which they are used. They cannot be regarded, therefore, as providing a means of assessing knowledge-claims which are independent of specific analytical commitments." The assumption that *all* forms of knowledge, including scientific knowledge, are socially mediated has become the new dominant research hypothesis in science studies.

Applied to the issue at hand, these changes in our understanding of the nature and development of science mean that critics can no longer count upon the old secure and unassailable platform from which the sociology of knowledge could be examined objectively. What has been lost is the conventional conception of science, with its vantage point and knowledge claims believed to be immune to criticism because they were seen as uniquely capable of adjudicating cognitive conflicts across contexts and traditions. As a result, neither the critics of the classical sociology of knowledge nor the sociology of knowledge itself can avoid involvement in ethical or even political discourse.

If this argument is accepted, the convergence of Mannheim's position (especially his self-conception as an opponent of relativism) with that of his critics becomes plausible: both acknowledge the possibility of context-independent knowledge claims. The intellectual affinity of what initially appear to be opposed perspectives is the result of shared beliefs and similar convictions about the course of social developments. These assumptions in turn are linked to another set of shared beliefs, namely, that knowledge *is* capable of transcending its genesis and, moreover, that it is desirable to strive for this kind of valid knowledge. Mannheim and his critics agree that the goal of science ought to be the overcoming of intellectual dissent in science. This alone is sufficient reason to reject the claim that Mannheim's sociology of knowledge lacks standards of critical judgment, that it is forced to accept every cognitive development and every manifestation of cognitive activity as being of equal value. On the contrary, Mannheim was as convinced as his critics are that scientific work can greatly reduce intellectual and social strife.

[13] The typical response of critics unimpressed by the results of the research effort and unsympathetic to its philosophical and sociological underpinnings has been to assign the label of relativistic to such a research program (for example, Popper 1970: 56; Phillips 1977: 71).

The ethical and socio-political views of Mannheim and those of the opponents differ in some details in regard to their hopes for the future organization of society. What is most striking, however, is the basic convergence of their views, which also extends to their shared lack of specificity about future moral and social relations. Proponents of critical theory have been similarly unspecific in their description of the thrust of social evolution. Scholars opposed to theories of society which are more than mere descriptions of past or present social relations (e.g. Rule 1978) have objected to the ambiguity built into theories of society that include among their functions the instrumental function of helping to overcome prevailing social, political, and economic conditions. Such lack of specificity is seen as a way of immunizing theoretical discourse against falsification; the absence of concrete predictions regarding future developments makes it impossible, it is argued, to show that a theory failed to correctly anticipate a particular development.[14]

From this vantage point, Mannheim's sociology of knowledge and its intended function of reconciling reason and politics, as well as his repeated denial that his own approach was relativistic, are entirely consistent. Yet the apparent convergence of Mannheim's world view, as it is manifested in his sociology of knowledge, and the at least implicit theories of society advocated by his opponents requires explanation. Similarly, we should ask what implications the 'deep' linkage between a theoretical program and a world view has for the future development of the sociology of knowledge.

The largely successful restriction of Mannheim's design for a sociology of knowledge to a sociology of knowledge more narrowly concerned with the sociological analysis of the genesis of ideas[15] must be seen in connection with the growing institutional separation of sociology from neighboring disciplines.[16] Trespassing of boundaries can have serious professional repercussions since career decisions are made with reference to disciplinary traditions. The emerging cognitive identity of a discipline—the 'scientization' of theoretical programs—in turn legitimates institutional differentiation. Together with the growing professionalization of social science disciplines, the entrenched intellectual division of labor ensures that research programs which cross established boundaries, as is clearly the case with

[14] Whether one regards such lack of precision as a sign of a deficient theory depends on one's conception of an adequate theory. It is equally possible to argue that it is precisely the lack of precision in this regard that is a hallmark of good theorizing in that it avoids the fallacy of 'overgeneralization' so typical in social science.

[15] Mannheim (1952: 226–27) describes this process and its function for the delineation of researchable questions well when he stresses that "it is out of the question that a certain analysis should be stopped short once and for all at the most crucial point merely because the recognized domain of a different scientific department allegedly begins there (a mode of procedure typical of the bureaucratized organization of science)".

[16] The nature of the specialized cognitive identity that emerges is determined to a large extent by the dominant intellectual traditions within the discipline as well as by the nature of the relations of disciplinary perspectives to theoretical traditions in neighboring disciplines (cf. Coleman 1972; Offe 1985).

Mannheim's sociology of knowledge, are difficult to sustain and easily subjected to criticism.

The often rigid intellectual boundaries between sociology and its neighboring disciplines as well as general intellectual impulses derived from political developments (cf. Offe 1985) may well account for the frequent criticism that Mannheim's sociology of knowledge fails to properly delimit its cognitive aims. These criticisms cannot explain, however, why in the final analysis Mannheim and his critics share quite similar positions when they advocate a non-relativistic sociology of knowledge. This similarity should perhaps be understood as a manifestation of a shared belief in the value of *rational discourse* in science.[17]

The emphasis of intellectuals and scientists on rational discourse in science appears to be particularly strong in periods of intense social and institutional conflict. Rational discourse and a social organization that helps to reduce and perhaps ultimately even eliminate intellectual disputes are seen as a precondition for the resolution of conflict not only by the scientific community but by most other societal groupings.[18]

Modern science and a number of its normative principles that are widely viewed as the basic foundation of scientific work and of social relations in the scientific

---

[17] According to Gouldner (1976: 39), rational discourse implies that the justification of knowledge claims does not invoke 'authorities'; from an historical point of view, the emergence of rational discourse therefore requires that "the coercive power and the public credit of societal authorities has been undermined, restricted, or declared irrelevant, and that the use of manipulative rhetoric is limited either by institutional and moral restraints or by the prevailing technology of mass communication." However, this does not mean that reference or deference to authorities has been overcome completely in contemporary scientific discourse. This may not be possible in the first place. In the same spirit of skepticism, Bourdieu (1975: 24) has argued that "in the scientific field as in the field of class relations, no arbitrating authority exists to legitimate legitimacy-giving authority from the relative strength of the groups whose interests they express: inasmuch as the definition of the criteria of judgement and the principles of hierarchies is itself an issue in a struggle, there are no good judges, because there is no judge who is not also party to the dispute." In other words, there are good sociological grounds for a dismissal of the possibility of rational discourse that is pure rational discourse.

[18] The assumption that rational discourse in science, whatever the social structure, is exemplary in a political sense and a model for other social institutions is a central thesis of Popper's ([1945] 1962) political philosophy. As Lubasz (1981) has shown, Popper transposes his ideal-typical description of the 'rational' social and intellectual mode of scientific organization onto political institutions and decision-making processes. Popper assumes further that rational politics is identical with a politics committed to a trial-and-error approach. However, as Lubasz also points out, the "scientific attitude to politics *can* be a realistic one—so long as it is not applied as Professor Popper applies it, to the whole community, but to a political elite. A political elite such as a junta, or a bureaucracy, or a cabinet of government ministers *could* resemble the scientific community in its essential features. Such an elite might be in complete agreement as to what the world is like, what the problems are, how they are to be solved, and what is to count as a good solution. A political elite could resemble a community of scientific experts—so long as it was in a position to treat the rest of the political community the way physicists treat the physical universe: as something 'out there'. At that point political problems *would* resemble technical problems—but only at the price of abandoning the whole notion of the Open Society."

community first emerged as a response to the *proliferation* of systems of ideas in the seventeenth century.[19] Horton (1973: 286) observes that there

> can be little doubt that many features of the scientific outlook first developed *as responses to the proliferation of competing idea-systems* in seventeenth-century Europe. One effect of this proliferation was the growth of skepticism, of a readiness to challenge any and every idea regardless of its source. Another effect was the search for explicit criteria of choice in the realm of ideas. The most important outcome of this search was the gradual formulation of the ideal of objectivity.

Until now, in other words, the widely supported organizing principle of science has been the elimination of disagreement.

The socio-historical conditions of Weimar Germany, which helped to give rise to the sociology of knowledge (see Mannheim 1953: 211–13; Meja/Stehr 1990; Barth 1945; Hughes 1958; Remmling 1967, 1973, 1975; Ringer 1969), unquestionably represent a similar historical period in which doubt, skepticism, and intellectual experimentation were also typical responses to profound societal conflicts and changes. In this period, too, the ethic of rational discourse and its universalizing political implications was radically challenged; yet the hope arose as well that it might be affirmed on a new basis. Mannheim's diagnosis of these conditions in the Weimar Republic, the intense competition among world views and the thorough-going skepticism toward the possibility of rational and objective arguments, was hardly unique. His affirmation of the fundamental moral underpinnings of scientific discourse and his hope that it might serve as a political model are quite close to traditional convictions and expectations about the ultimate efficacy of discourse in science. The search for objectivity in science is closely linked to the perhaps utopian hope, as Gouldner (1973: 66–67) has observed, "of dissolving the differences that divide and the distances that separate men by uniting them into a single, peace-bringing vision of the world". Mannheim saw the sociology of knowledge as an important vehicle for transcending the opposition among contending world views. Such a moral-political conception is incompatible with an uncritical acceptance of the *status quo* in society or with an indiscriminate acceptance of every interpretation of the world and of society. Mannheim and most of his critics reject relativism because they all adhere to the predominant ethic of modern science which regards the sciences as a tool for the progressive transformation of the world.[20]

Mannheim's (frequently merely implicit) expectation that the model of rational discourse of science may point toward more consensual social and intellectual conditions represents a widely supported moral-political perspective among

[19] Merton's widely discussed analysis of the ethos of science (1957: 550–61)—that is, of the norms of universalism, communism, organized skepticism, and disinterestedness—is first and foremost an attempt to consider the *immanent* functions of adherence to these norms for the development of scientific knowledge (cf. Stehr 1978).

[20] The perhaps even naïve confidence with which scientists and non-scientists alike have subscribed to the ethic of modern science has experienced a perhaps decisive historical challenge during the past decade.

intellectuals and scientists. However, it is questionable whether the ethic of rational discourse remains a satisfactory model for a sociology of knowledge. A renewed sociology of knowledge must be based on a different moral-political foundation. Rather than being wedded, like the model of rational discourse, to the elimination of discourse, it must be supportive of its continuation and of the promotion of discussion rather than the creation of silence as the exemplary aim of science.

References

Aron, Raymond. [1935] 1953. *Die deutsche Soziologie der Gegenwart.* Stuttgart: Kröner.

Baldamus, Wilhelm. 1976. *The Structure of Sociological Inference.* London: Robertson.

Barber, Bernard. 1975. "Toward a New View of the Sociology of Knowledge", in: Lewis Coser (ed.), *The Idea of Social Structure.* New York: Harcourt Brace Jovanovich: 103–116.

Barnes, Barry. 1974. *Scientific Knowledge and Sociological Theory.* London: Routledge & Kegan Paul.

Barnes, Barry. 1977. *Interests and the Growth of Knowledge.* London: Routledge & Kegan Paul.

Barth, Hans. 1945. *Wahrheit und Ideologie.* Zurich: Manesse.

Berger, Peter and Thomas Luckmann. 1967. *The Social Construction of Reality.* Garden City, NY: Doubleday.

Bergmann, Gustav. [1951] 1968. "Ideology", in: May Brodbeck (ed.), *Readings in the Philosophy of the Social Sciences.* London: Macmillan. 123–138.

Boudon, Raymond. 1971. *La crise de la sociologie.* Geneva: Droz.

Bourdieu, Pierre. 1975. "The Specificity of the Scientific Field and the Social Conditions of the Progress of Reason", *Social Science Information,* 14: 19–47.

Coleman, James. 1972. *Policy Research in the Social Sciences.* Morristown, NJ: General Learning Press.

Coombs, Robert H. 1966. "Karl Mannheim, Epistemology, and the Sociology of Knowledge", *Sociological Quarterly,* 229–233.

Friedrichs, Robert. 1970. *A Sociology of Sociology.* New York: Free Press.

Gellner, Ernest. 1974. *The Legitimation of Belief.* London: Cambridge University Press.

Gouldner, Alvin. 1973. *For Sociology: Renewal and Critique in Sociology Today.* London: Allen Lane.

Gouldner, Alvin. 1976. *The Dialectic of Ideology and Technology: The Origins, Grammar, and Future of Ideology.* New York: Seabury.

Grünwald, Ernst. 1934. *Das Problem einer Soziologie des Wissens.* Vienna: Braumüller.

Hamilton, Peter. 1974. *Knowledge and Social Structure.* London: Routledge & Kegan Paul.

Horkheimer, Max. 1930. "Ein neuer Ideologiebegriff?", *Archiv für die Geschichte des Sozialismus und der Arbeiterbewegung,* 15: 33–56.

Horkheimer, Max. [1931] 1937. „Die gegenwärtige Lage der Socialphilosophie und die Aufgaben eines Instituts für Sozialforschung", *Frankfurter Universitätsreden,* 37: 3–16.

Horton, Robin. 1973. "Lévy-Bruhl, Durkheim, and the Scientific Revolution", in: Robin Horton and Ruth Finnegan (eds.), *Modes of Thought: Essays on Thinking in Western and non-Western Societies.* London: Faber & Faber: 249–305.

Hughes, H. Stuart. 1958. *Consciousness and Society.* Glencoe, IL: Free Press.

Kettler, David. 1967. "Sociology of Knowledge and Moral Philosophy: The Place of Traditional Problems in the Formation of Mannheim's Thought", *Political Science Quarterly,* 82: 399–426.

Kettler, David, Volker Meja and Nico Stehr. 1980. "Karl Mannheim frühe kultursoziologische Arbeiten", in: Karl Mannheim (ed.), *Strukturen des Denkens*. Frankfurt am Main: Suhrkamp: 9–31.

Kettler, David, Volker Meja and Nico Stehr. 1984. *Karl Mannheim*. London: Tavistock; New York: Methuen.

Köhnke, Klaus Christian. 1985. *Entstehung und Aufstieg des Neokantianismus: Die deutsche Universitätsphilosophie zwischen Idealismus und Positivismus*. Frankfurt: Suhrkamp.

Lenk, Kurt. 1961. *Ideologie, Ideologiekritik, und Wissenssoziologie*. Neuwied: Luchterhand.

Lieber, Hans-Joachim. 1965. *Philosophie, Soziologie, Gesellschaft*. Berlin: de Gruyter.

Louch, A. R. 1966. *Explanation and Human Action*. Oxford: Blackwell.

Lubasz, Heinz. 1981. "Popper in Utopia", *Times Higher Education Supplement*, 477: 10.

Luhmann, Niklas. 1980. *Gesellschaftsstruktur und Semantik*. Frankfurt: Suhrkamp.

Lukács, Georg. [1954] 1974. *Die Zerstörung der Vernunft: Vol. 3. Irrationalismus und Soziologie*. Neuwied: Luchterhand.

Lukes, Steven. 1977. *Essays in Social Theory*. New York: Columbia University Press.

Mandelbaum, Maurice. 1938. *The Problem of Historical Knowledge*. New York: Harper & Row.

Mannheim, Karl. 1931. "Wissenssoziologie", in: Alfred Vierkandt (ed.), *Handwörterbuch der Soziologie*. Stuttgart: Enke: 237–230.

Mannheim, Karl. [1929] 1936. *Ideology and Utopia*. New York: Harcourt, Brace & World.

Mannheim, Karl. 1952. *Essays on the Sociology of Knowledge*. London: Routledge & Kegan Paul.

Mannheim, Karl. 1953. *Essays on Sociology and Social Psychology*. London: Routledge & Kegan Paul.

Mannheim, Karl. 1964. *Wissenssoziologie*. Neuwied: Luchterhand.

Mannheim, Karl. 1980. *Strukturen des Denkens*. Frankfurt: Suhrkamp.

Maquet, Jacques J. 1951. *The Sociology of Knowledge*. Boston: Beacon.

Meja, Volker and Nico Stehr. 1988. "Social Science, Epistemology, and the Problem of Relativism", *Social Epistemology*, 2: 263–71.

Meja, Volker and Nico Stehr (eds.). (1990). *Knowledge and Politics: The Sociology of Knowledge Dispute*. London and New York: Routledge.

Merton, Robert K. 1957. *Social Theory and Social Structure*. New York: Free Press.

Merton, Robert K. 1971. "The Precarious Foundation of Detachment in Sociology", in: Edward A. Tiryakian (ed.), *The Phenomenon of Sociology*. New York: Appleton-Century-Crofts: 188–199.

Merton, Robert K. 1973. *The Sociology of Science*. Chicago: The University of Chicago Press.

Merton, Robert K. 1977. "The Sociology of Science: An Episodic Memoir", in: Robert K. Merton and Jerry Gaston (eds.), *The Sociology of Science in Europe*. Carbondale: Southern Illinois University Press: 3–141.

Merton, Robert K. 1978. "Strukturelle Analyse in der Soziologie", in: Peter M. Blau (ed.), *Theorien sozialer Strukturen*. Opladen: Westdeutscher Verlag.

Mulkay, Michael. 1979. *Science and the Sociology of Knowledge*. London: Allen & Unwin.

Nagel, Ernst. 1961. *The Structure of Science*. New York: Harcourt, Brace & World.

Neusüss, Arnhelm. 1968. *Utopisches Bewusstsein und freischwebende lntelligenz*. Meisenheim: Hain.

Offe, Claus. 1985. "Sozialwissenschaften zwischen Auftragsforschung und sozialer Bewegung", in: Bede von Greiff (ed.), *Das Orwellische Jahrzehnt und die Zukunft der Wissenschaft*. Opladen: Westdeutscher Verlag.

Parsons, Talcott. 1936. "Review of Alexander von Schelting, Max Webers Wissenschaftslehre", *American Sociological Review*, 1: 675–681.

Parsons, Talcott. [1937] 1949. *The Structure of Social Action*. Glencoe: Free Press.

Parsons, Talcott. 1951a. *The Social System*. New York: Free Press.

Parsons, Talcott. 1951b. *Toward a General Theory of Action*. Cambridge, MA: Harvard University Press.

Phillips, Derek, 1977. *Wittgenstein and Scientific Knowledge*. London: Macmillan.

Popper, Karl. [1945] 1962. *The Open Society and Its Enemies*. London: Routledge & Kegan Paul.

Remmling, Gunter W. 1967. *Road to Suspicion: A Study of Modern Mentality and the Sociology of Knowledge*. New York: Appleton-Century-Crofts.

Remmling, Gunter W. 1973. *Towards the Sociology of Knowledge: Origin and Development of a Sociological Thought Style*. London: Routledge & Kegan Paul.

Remmling, Gunter W. 1975. *The Sociology of Karl Mannheim*. London: Routledge & Kegan Paul.

Ringer, Fritz. 1969. *The Decline of the German Mandarins: The German Academic Community, 1890–1933*. Cambridge: Harvard University Press.

Ritzer, George. 1975. *Sociology: A Multiple Paradigm Science*. Boston: Allyn & Bacon.

Rüschemeyer, Dietrich. 1958. *Probleme der Wissenssoziologie*. Dissertation, University of Cologne.

Rule, James B. 1978. *Insight and Betterment*. New York: Oxford University Press.

Salomon, Albert. 1947. "Karl Mannheim, 1893–1947", *Social Research*, 14: 350–364.

Schelting, Alexander von. 1934. *Max Webers Wissenschaftslehre*. Tübingen: Mohr.

Schelting, Alexander von. 1936. "Review of Karl Mannheim, Ideology and Utopia", *American Sociological Review*, 1: 664–674.

Speier, Hans. 1936. "Review of Ernst Grünwald", *Das Problem einer Soziologie des Wissens. American Sociological Review*, 1: 681–682.

Stark, Werner. 1958. *The Sociology of Knowledge*. London: Routledge & Kegan Paul.

Stehr, Nico. 1978. "The Ethos of Science Revisited: Social and Cognitive Norms", in: Jerry Gaston (ed.), *The Sociology of Science*. San Francisco: Jossey-Bass: 172–196.

Stehr, Nico. 1982. "Sociological Languages", *Philosophy of the Social Sciences*, 12: 47–57.

Stehr, Nico, and Volker Meja. 1982. "The Classical Sociology of Knowledge Revisited", *Knowledge*, 4: 33–50.

Stehr, Nico, and Volker Meja. 1984. "The Development of the Sociology of Knowledge", in: Nico Stehr and Volker Meja (eds.), *Society and Knowledge: Contemporary Perspectives on the Sociology of Knowledge*. New Brunswick, NJ and London: Transaction.

Storer, Norman W. 1966. *The Social System of Science*. New York: Holt, Rinehart & Winston.

Sztompka, Piotr. 1979. *Sociological Dilemmas*. New York: Academic.

Toulmin, Stephen. 1972a. *Human Understanding*. Princeton, NJ: Princeton University Press.

Toulmin, Stephen. 1972b. "Conceptual Change and the Problem of Relativity", in: Michael Kransz (ed.), *Critical Essays on the Philosophy of Collingwood*. Oxford: Oxford University Press: 201–221.

Ultee, Wouter Cornelis. 1980. *Fortschritt und Stagnation in der Soziologie*. Neuwied: Luchterhand.

Walter, Benjamin. 1967. "The Sociology of Knowledge and the Problem of Objectivity", in: Llewellyn Gross (ed.), *Sociological Theory: Inquiries and Paradigms*. New York: Harper & Row: 335–357.

Wolff, Kurt H. 1978. "Karl Mannheim", in: Dirk Käsler (ed.), *Klassiker des soziologischen Denkens* (Vol. 2). München: Beck. 286–387.

Wolff, Kurt H. 1983. *Beyond the Sociology of Knowledge*. New York: University Press of America.

Chapter 14
The Power of Scientific Knowledge and its Limits

Nico Stehr

With few exceptions classical and contemporary sociological discourses have assumed that scientific knowledge will inevitably replace traditional forms of knowledge.[1] These latter forms of knowledge are portrayed as incapable of resisting the progressive dissemination of scientific knowledge. Traditional forms of knowledge are defined as institutionally based while scientific knowledge knows no time and place. It is further assumed that the carriers of conventional types of knowledge are not sufficiently powerful to slow or prevent their demise. Despite the broad consensus, thriving traditional forms of knowledge in modern society suggest distinct limits of scientific knowledge. This essay outlines the case made for the force of scientific knowledge in much of sociological theory but also examines the limits of the power of scientific knowledge, limits which are constitutive of scientific forms of knowledge rather than merely the outcome of resistance to the spread and dominance of such knowledge in contemporary society.

For decades it was the contagious optimism and, more rarely, the skeptical fascination with the apparently relentless success of modern science and technology that prompted social scientists to ignore the systematic practical limits of scientific knowledge as a theoretical issue. The present period, however, is seeing an apparently growing conviction that the widespread criticism of the practice of modern science and technology is both meaningful and pertinent. But whatever the intellectual grounds for the belief that practical limits of scientific knowledge do *not* represent a serious theoretical issue, the outcome is a rather elementary

This text was first published as: Stehr, Nico. 1991. "The power of scientific knowledge—and its limits", *Canadian Review of Sociology & Anthropology*, 28(4): 460–482. The permission was granted on 17 July 2017 on behalf of Wiley by Ms. Kelly Hoff, Permissions Coordinator.

[1] I am grateful for helpful critical readings of an earlier draft and suggestions to Zygmunt Bauman, David Bloor, Wolfgang Krohn, Volker Meja, Robert K. Merton, Trevor Pinch and a reviewer of this journal. All translations of previously untranslated quotations are mine.

M. T. Adolf (ed.), *Nico Stehr: Pioneer in the Theory of Society and Knowledge*, Pioneers in Arts, Humanities, Science, Engineering, Practice 16,
https://doi.org/10.1007/978-3-319-76995-0_14

misunderstanding of the nature of and constraints to the social power of scientific knowledge.

An intensive examination of the interplay of scientific knowledge and social relations clearly ought to have priority. As Hübner ([1978] 1983: 214), reflecting the widespread agreement among social scientists now and in prior decades, recently observed: "the way in which present-day human society, as an industrialized society, understands itself rests, to a very great extent, on genuine technological-scientific forms and ideas". In spite of some exceptions,[2] earlier generations of social scientists shared this conviction[3] and anticipated either with fear or in disillusionment, an "age of science and technology", and an increasing rationalization of irrational forces.[4]

In a sense, this essay seeks to complicate a matter which has often been rendered increasingly simpler in the history of *sociological* thought. Proponents and opponents of modern science alike agree that scientific knowledge transcends traditional knowledge and, in modern society, tends to take its place (for example, Marcuse 1964; Schelsky [1961] 1964; Roszak 1972; Richta 1969; Bell 1972). Social scientists favour this assertion, despite the growing strength and acceptance of the *philosophical* thesis that common-sense concepts about things, persons, agency or

---

[2] The paradoxical condition, formulated less radically, that social action is necessary in contexts (still) unaffected by scientific knowledge, was already noted by Kant. The antirationalist then converts such a 'dilemma' into necessity: "Zu den Dingen, welche einen Denker zur Verzweiflung bringen können, gehört die Erkenntnis, dass das Unlogische für den Menschen nötig ist." (Nietzsche, *Menschliches, Allzumenschliches* 1,1)

[3] Max Weber's views about the vital omnipotence of modern science for instance are well-know. They can be found in the *Zwischenbetrachtung* in his *Gesammelte Aufsätze zur Religionssoziologie* (Weber 1978: 564). For Parsons (1937: 752), Weber's work culminates precisely in his "conception of a law of increasing rationality as a fundamental generalization about systems of action". This law constitutes the most fundamental generalization that emerges from Weber's work. Scheler ([1926] 1960: 207), for example, asserts essentially the same fateful development. Emile Durkheim's discussion, in *Elementary Forms of Religious Life,* of the conflict or reciprocal relations between science and religion also proceeds from the premise that science will displace religion, although Durkheim (1965: 574–7) is prepared to grant a continual though limited role to religious knowledge in modern society. Of course, not all sociological classics assign culture of modern society. Vilfredo Pareto celebrates and justifies the societal function of illogicality (as a capacity for action), though he attempts to do so on the basis of strictly logical reasoning.

[4] Compare, for example, Jaspers' ([1932] 1979) description in his *Spiritual Condition of the Age*, of the effect of technology on the nature of everyday life, on work, politics, the family, leisure, sports, language, education, the psychology of the modern individual etc. and the corresponding crisis of traditional forms of life. The crisis which affects science is not, for Jaspers, linked somehow to its ability to radically transform and alter the entire spectrum of modern life. He takes it for granted that this is in fact the case. But it rather lies in the extent to which science continues to be able to provide meaning to and for these profound changes (cf. Jaspers [1932] 1979: 125). The ability of science to offer meaning is not threatened, according to Jaspers, because science, in the broad sense of the term, is incapable of providing meaning in principle, but because the practice of science and scientists are subject to the same forces which make science more and more a reflection of the culture and of the structure of social relations it produces in modern society.

intention may have an immovable character as well as evidence of thriving forms of non-scientific knowledge in modern society. In general, I agree with the methodological and theoretical strategy for science which Geertz (1973: 33) succinctly states as follows: "Scientific advancement commonly consists in a progressive complication of what once seemed a beautifully simple set of notions but now seems an unbearably simplistic one."

The question about the limits of the power of scientific knowledge does not have any narrow boundaries. On the contrary, it raises a multitude of pertinent concerns which may be examined, moreover, from a variety of legitimate perspectives. Thus, some care, patience and space has to be devoted—more at any rate than is usual in essays of this sort—to *situate* the specific issue chosen for examination.

The literature relevant to my topic employs a great variety of notions to establish meaning for the concept of 'knowledge', for example. For the purpose of this discussion, knowledge can best be defined as constituting a *capacity for social action.*[5] And, in this broad sense, different forms of knowledge may indeed be seen as functional equivalents. In fact, much of the literature on the role of scientific knowledge in modern society clearly assumes that *it* can be a substitute for other forms of 'knowledge'. But the question then becomes, why and how one form of knowledge, for example, traditional knowledge, is replaced by or fails to yield to knowledge generated by science. The ability of scientific knowledge to dislodge other forms of knowledge is almost taken-for-granted and is seen to be linked primarily to superior cognitive attributes of scientific knowledge. I will inquire into this assertion. The power or the limits of scientific knowledge at issue here are consequently primarily intellectual or 'ideological' limits, that is, the limits built into scientific knowledge itself.[6]

Almost all social science studies of science are governed by the conviction, at least until very recently, that modern scientific knowledge is increasingly replacing traditional collective beliefs—as well as labor and property—as the dominant media of social organization (cf. Weingart 1981). The strategic theoretical resource which produces such agreement in social science discourse is an *institutionally-based definition of non-scientific forms of knowledge*—for example, of religious knowledge—and the concurrent classification of scientific knowledge as unaffected by time and place. As the functioned differentiation of society increases, it is argued, the net loss in the influence of institutionally-based knowledge grows while

[5] According to this conception, knowledge is a condition for the possibility of action. Obviously, scientific and technical knowledge are such capacities for action. And, in this regard, social and natural scientific knowledge do not differ. The definition of knowledge as a capacity for action does not prejudice scientific knowledge as a 'superior' form of knowledge. The term capacity signals also that knowledge may be left unused or may be employed for 'irrational' ends. Moreover, it indicates that the *use* of knowledge is related to the *local* conditions of action. For a much more extensive discussion of these notions cf. Stehr (1991).

[6] The term 'built into' intends to convey the sense to which a non-essentialist, contingent thesis will be pursued in the essay. The limits at issue are not somehow 'essential' to scientific knowledge as scientific knowledge.

scientific knowledge, not bound by any institutional limits, fills the void and deals with emergent problems.

The dissemination and social implementation of the knowledge generated in science, and in particular the transformation of discoveries and inventions into technical forms, are indeed undoubtedly the most important elements of modern societal development. However, this does not necessarily imply that displacement and elimination of traditional forms of knowledge follow on the heels of the successful implementation of scientific knowledge; nor is it permissible to conclude with certainty that scientific knowledge represents a precise functional equivalent or even replacement for non-scientific knowledge; or that the dissemination of objectified science as technology is linked to a rationalization of world-views.

I will discuss the social and cognitive conditions which limit the dissemination of scientific knowledge and, conversely, the conditions which foster the persistence of 'traditional' or everyday knowledge[7] (where such knowledge has not been transformed by science)[8] in contemporary society insofar as such persistence is the outcome of specific attributes of scientific discourse and its cognitive products. I begin with a critical examination of a couple of exemplary macrosociological theories of social change which consider scientific knowledge one of the crucial principles of social organization. This examination focusses on how these theories deal with the 'nature' of scientific knowledge and its unique position in modern society.

Theories pertinent to the question of the power and authority of scientific knowledge in modern society, although most in fact limit their analysis to the broad assertion of the salience of science in and for modern society, are varied. One of the purposes of the following section is to show the extent to which past and present theories in fact converge. The theoretical conceptions to which these remarks pertain, for example, are the Weberian thesis of the increasing rationalization of society; Georg Simmel's notion of a cultural tragedy; the dichotomy of civilization and culture, first explicated in Alfred Weber's cultural sociology; Ogburn's theory of cultural lag; the assertion of a professionalization of culture; and, finally, the theory of postindustrial society. However, I will limit my discussion to a brief description of Max Weber's and Daniel Bell's influential theories of modern and postmodern society.

---

[7] This means that knowledge which is explicitly constructed to operate as a substitute (as competing knowledge) for scientific knowledge is not at issue here. The account of the origin of the universe offered by creationists as direct challenge to the account accepted in the scientific community would be such a form of knowledge (La Follette 1983; Montagu 1984).

[8] However, I do not intend to argue that non-scientific knowledge is somehow static; of course, everyday knowledge is affected by attempts to popularize scientific knowledge. A most incisive study by the historian Burnham (1987) demonstrates such effects on everyday knowledge: he shows how everyday 'superstitions' were transformed by efforts to decisively popularize science in the last one hundred years. The result is the emergence of something akin to a functional equivalent of traditional superstition, in direct response to attempts to popularize science.

14.1 Calculability and Power

Among the theoretical accounts in sociology of the overwhelming role of modern scientific knowledge, none has been more influential than Max Weber's world historical perspective of the practical impact of science on a rationalization and demystification process affecting ultimately all social relations. Weber ([1922] 1989: 13) describes this process and the special role of knowledge in his lecture "Science as a Vocation" in the following terms: "The knowledge or the belief that, *if one only wanted to*, one *could* find out at any time; that there are in principle no *mysterious, incalculable powers at work*, but rather that one could in principle master everything through *calculation*." The profound transformation in ways of living implies a destruction of traditional forms of life and principles of organization but also of conventional beliefs as obsolete. Weber's views are representative for the prevailing theoretical reflections in sociology concerned with, in this instance, methodical approaches to life and social relations as well as bureaucratic forms of organization. The destruction of traditional social structures and ideologies is the other side of the coin of the irresistible advance of modernity. Shortly after Weber, Mannheim ([1929] 1940: 101) formulated the same idea more skeptically: "The chief characteristic of modern culture [perhaps][9] is the tendency to include as much as possible in the realm of the rational and to bring it under administrative control—and, on the other hand, to reduce the 'irrational' element to the vanishing point." The chances of a survival of non-scientific knowledge in modern society are evidently minimal.[10] The reliability of the kind of everyday knowledge dominant up to this juncture in time disintegrates as a result of the increasing complexity of social and physical systems (Boulding 1967: 690).

[9] The qualifying term '*womöglich*' here translated as 'perhaps' (or, alternatively, as 'possibly'), is missing from the translation of *Ideologie und Utopie* into English. This term represents a 'skeptical' qualification by Mannheim of the Weberian thesis about irresistible rationalization of modern social life. For further discussion of the differences in the English and German editions of Mannheim's *Ideology and Utopia,* see Kettler et al. (1983).

[10] As a result, Seyfarth (1972: 354), for example, demands "a greater emphasis of de-structuralized processes which attached themselves to emerging structures" as an important addition and qualification of the Weberian thesis of the one-sided emphasis of the broad structural transformation of modern life resulting from the increasing application of the calculability of social action. However, such a qualification does not necessarily revoke the thesis of the irreversible and irresistible success of a methodic form of life or the importance of examining the limits of rationalization beyond those Weber himself was prepared to recognize.

14.2 Post-Industrial Society

Among the present-day macro-sociological theories, Bell's (1972) theory of post-industrial society stands out. It is of particular interest in our context because Bell explicitly designates scientific knowledge as the ultimate source and engine of societal transformations. Bell's theory of society, which deals primarily with changes in the social structures of advanced (Western) society, has won more critical acclaim and attention than similar designs (e.g., Touraine 1969; Richta 1969). According to Bell, the spheres of social structure, the polity, and culture form the order of society. These three realms of society are not assumed to be strongly linked. Bell does not proclaim that changes in the social structure determine all other spheres (cf. Bell 1987: 1). His theory lacks, as a result, a deterministic center,[11] his conceptual framework, especially the differentiation of social structure, on the one hand, and polity and culture, on the other hand, are strongly reminiscent of the dichotomies between civilization and culture or material and adaptive culture (see also Bell 1975). The social structure can be differentiated into the economic, technological and occupational systems. The developmental logic which Bell considers to be the "axial principle" of his theory of society, is the centrality of *theoretical knowledge* as the source of innovation and the basis of policy formations. The importance of theoretical knowledge can perhaps be measured by indicating that Bell (1972: 344) is convinced that many if not all decisions in production and business, as well as in politics, will be increasingly based on the fruits of research and development. The decision-making process becomes ever more a technical process, although the role of science and technology is not given quite the same prominent role as in the context of the "end of ideology" debate (cf. Bell 1969; Lipset 1962). It was thought (Lane 1966: 659) that political criteria were about to vanish completely and give way to universalistic conventions produced by science (see also Vidich/Lyman 1985: 289–94).[12]

---

[11] Bell (1972: 119) elaborates on the consequences of his decision to examine primarily axial changes in the social structure by indicating that the analysis of such changes does not mean a "specific determinism between a 'base' and a 'superstructure'; on the contrary, the initiative in organizing a society these days comes largely from the political system. Just as various industrial societies—the United States, Great Britain, Nazi Germany, the Soviet Union, post-World War II Japan—have distinctively different political and cultural features, so it is likely that the various societies that are entering a post-industrial phase will have different political and cultural configurations".

[12] In contrast to Alfred Weber and William Fielding Ogburg, Bell (1972: 20) attempts to specify certain attributes of modern theoretical (scientific) knowledge and indicates that knowledge has always played a role in the functioning of society. But what is "distinctive about post-industrial society is the change in the character of knowledge itself. What has become decisive for the organization of decisions and the direction of change is the centrality of *theoretical* knowledge—the primacy of theory over empiricism and the codification of knowledge into abstract systems of symbols that, as any axiomatic system, can be used to illustrate many different and varied areas of experience. Every modern society now lives by innovation and the social control of change, and tries to anticipate the future in order to plan ahead. This commitment to social control introduces the need for planning and forecasting into society. It is the altered awareness of the nature of innovation that makes theoretical knowledge so crucial".

According to Bell, the post-industrial society is no longer organized around the axis of man and machine but around scientific knowledge. The transformation is reflected in a shift away from the importance of manufacturing to the service sector of the economy. This change is accompanied by a reduction in workers employed in manufacturing. The occupational structure witnesses a noticeable growth in the number of "professional and technical" employees. The new kind of work increasingly requires theoretical knowledge. Scientific workers are the most significant resource of post-industrial society. The privileged position of this "new class' or stratum,[13] implies that the ethos of society will increasingly derive from the ethos of science (cf. Bell 1972: 386; see also Lane 1966; Gouldner 1979). On the basis of what Bell optimistically calls new possibilities and modes of technological forecasting, post-industrial society will be capable of planning and controlling its technological growth, thus reducing the indeterminacies of economic development. An increasingly complex society develops a new intellectual technology, which emerges as a result of the necessity to replace intuitive problem-solving and judgments with algorithms and formalized decision-making rules. The greater dependence of social institutions on theoretical knowledge increases the societal and political significance of universities, research institutes and other intellectual institutions as producers and codifiers of such knowledge.

In his theory of post-industrial society, Bell draws a rather optimistic portrait of Western society (see also Dahrendorf 1977: 79–82). Some of the oldest dreams of mankind are close to reality. It will be possible, for example, to control social change. Progress and change in virtually all spheres of society, as well as different forms of knowledge, are dependent on the centrality and primacy of theoretical knowledge. While Bell does not explicitly take up the issue of the displacement of traditional forms of knowledge, it appears that he simply assumes that the dilution and replacement of traditional knowledge may be taken for granted (cf. Bell 1976: 4).

A much more explicit commitment in this respect may be found in the analysis of the distribution of forms of knowledge in post-industrial society by Holzner/Marx (1977). They assert that the structure of everyday knowledge in post-industrial society will change qualitatively and take on characteristics of scientific knowledge: "Common sense in post-modern, knowledge-based societies is much more flexible and open since it requires acceptance of varied, differentiated frames of reference and confidence that rules exist or can be discovered for their translation into each other. It also requires a fairly complex understanding of, and faith in, rationality as an encompassing code that admits of differential specifications and modes of expressions, but also provides for their ultimate mutual intel-

[13] However, one ought to keep in mind that Bell (1979) expresses severe doubts that the notion of a "new class" itself is of great theoretical value.

ligibility" (Holzner/Marx 1977: 20).[14] Significant aspects of everyday knowledge function as "an orientation to sources of needed information and procedures for gaining access to specialized bodies of knowledge and their expert practitioners" (Holzner/Marx 1977: 25).

14.3 Knowledge as a Natural Force

I have limited my discussion to two representative theories of macrosocial change which assign particular importance to scientific knowledge, in order to demonstrate that the notion of knowledge, despite its apparent theoretical centrality, is treated as a black box, as a kind of mysterious natural force beyond the control of individual and corporate actors. Scientific knowledge does not constitute, according to these theories, a specific form of life nor a principle of social organization. Scientific knowledge is treated as a universalistic category, not as a varied cultural resource. As a means of organizing activities within the economy and the state it becomes the dominant resource, yet, in the final analysis, it is viewed as a medium beyond the control of actors. Equally clear, at least as an implicit assertion, is the thesis that these means of organizing social life must penetrate and affect traditional ways of life and displace these obsolete expressions and processes. However, in the context of these theories, neither the origins of modern science and modern scientific knowledge, the peculiar qualities of scientific knowledge claims, nor the ways in which and the reasons why such knowledge travels well, are examined sociologically in any serious sense. And in this neglect, virtually all theories of societies have many surprising communalities. For example, when Habermas ([1981] 1983: 159) observes that Max Weber considers the history of science and technology a central aspect of the development of Western culture "but in his *sociological* attempt to *explain* the origins of modern society", Weber treats it rather as a boundary condition, then this observation is actually self-exemplifying because much of critical theory is similarly based on an equally simplified version of scientific knowledge.[15]

The scientistic conception which underlies the theories of Max Weber and Daniel Bell reflects the still dominant view of the development of science and scientific knowledge as a semi-automatic process. The theory of science adopted by these authors asserts that neither the expansion of scientific knowledge nor its wide application is, in any serious sense, affected by local socio-historical circumstances. Scientific labor, scientific discourse and the practical use of scientific knowledge

---

[14] But Holzner/Marx (1977: 23) observe, at the same time, that the *contents* of everyday knowledge will be affected much less by scientific-technical knowledge because "the consequence of modern science and technology has not been to overwhelm common sense views and render them insignificant, but to modify them somewhat and increase their cultural significance to the point where they represent the unique and distinguishing characteristic of post-modern society". But his assertion is clearly quite abstract and, perhaps by design, ambivalent.

[15] Cf. Habermas (1982: 274).

are, according to this conception of the nature of science, generally homogeneous and ahistorical processes. The technical successes of science develop in unison with the expansion of scientific rationality into social conduct. In particular, the penetration of social relations by scientific rationality in the form of increasing planning, control, systematization of decisions and even the "formation of drives or impulses", as Mannheim (1935: 41) speculated, is judged to be a realistic outcome of an extensive employment of science in society. It is entirely possible and part of the logic of these theories that saturation with scientific rationality, and, with technology, is a gradual and uneven process. In principle, however, it is not felt necessary to distinguish between them. On the contrary, the two processes melt into each other and their consequences become a unity. As a result, the notion of 'scientificity' (*Wissenschaftlichkeit*) is treated within the theoretical frame of reference, on the one hand, as a natural component and, on the other hand, as immune to any sociological analysis. A sociological analysis of the development of scientific knowledge and technology is considered to be impossible in principle because its evolution is subject to entirely separate regularities.[16] The separation between scientific and non-scientific knowledge is, moreover, viewed as an impenetrable boundary and not as the expression of an historical process (Böhme 1978: 44–47; van den Daele 1975). Error and ignorance are considered, in so far as these theories of macrochange address the question of everyday knowledge and non-scientific knowledge at all, as static elements. Once scientific knowledge expands, its practical realization is virtually unproblematic. No loss of identity occurs in the process. Finally, a frequent and implicit assumption is that scientific knowledge is the most authoritative form of knowing (and doing) and is obligatory.

These scientistic conceptions which inform most of our understanding of the role of scientific knowledge in society have, however, been questioned and replaced by a more socio-historical conception of the development of science and scientific discourse.[17] This change enables me to consider the continued presence of non-scientific convictions not merely as an expression of delayed backwardness but to justify the question, to what extent are sociologically relevant features of scientific knowledge itself responsible for the apparent limits of the power of such knowledge in contemporary society?

[16] It is not surprising, therefore, to encounter many theories of society which advance and defend an understanding of science and technology which is linked to a prohibition of a sociology of scientific knowledge, in as much as such a position is reinforced by the classical sociology of knowledge and science which is animated by and faithful to the same conviction about its own limits (see Mulkay 1979; Stehr/Meja 1981, 1984).

[17] The literature which deals with the origins and the nature of this cognitive change in the philosophy, history and sociology of science has, in the meantime, grown considerably; an overview may be found in Brown (1984).

14.4 The Limits of the Power of Scientific Knowledge

There are a number of theoretical positions worth brief examination, though they do not directly address the issue of the possible limits of scientific knowledge from the point of view of such knowledge. However, these positions are, for the most part, the outcome of theoretical or empirical work concerned with other matters; moreover, despite suggestions about possible limits of scientific knowledge, such restrictions to its power are considered to be marginal at best, especially in the long run. It is rather difficult to separate, as will be seen, an examination of the nature of scientific knowledge from certain important features of social practice to which such knowledge is seen to apply and where it is widely expected to command considerable assent. In the final analysis, however, I would like to highlight those limits to scientific knowledge which are constitutive of such forms of knowledge.

14.4.1 Thinking Against Knowledge

Scientific knowledge has always had its opponents. Social strata and groups which have resisted scientific knowledge, may have developed a style of thought which has articulated opposition quite effectively and has contributed to a slowing in the pace of the dissemination of scientific knowledge.

As the capitalistic process of rationalization and the spread of rational conduct to many spheres of life go hand in hand with the development of modern science and technology, so does a critique of capitalism depend on criticism and opposition to modern technology and a rationalist style of thought. The critique of the capitalist economic system is inaugurated by an opposition to it from the conservative position of the political spectrum (cf. Mannheim [1925] 1986: 63–71).

Some of the basic features of the emerging conservative style of thinking and experiencing, in response to the Enlightenment and the capitalist economy, include, according to Mannheim ([1925] 1986: 100),

> [...] its experience of the qualitative, its concrete rather than abstract ways of experiencing, its experiencing on the grounds of what is and not what ought to be, its experience of imaginary spatial relationships in contrast with the linear experience of historical development, its substitution of landed property for the individual as the substratum of history, its preference for 'organic' associations over '- classes', and others.

Modern science strives to generate experiences which are generally demonstrable and universally valid. As Mannheim ([1925] 1986: 60) explicates further, science constitutes a project which strives for "socializable knowledge" in stark contrast to the kind of cognitive insights emphasized by conservative thinking which are possible and accessible only to particular, more circumscribed "experiential communities" (*Erfahrungsgemeinschaften*). The conservative argues, therefore, against scientific knowing to the extent that it involves an indifference towards all specific and concrete elements in the object or can only be demonstrated to a particular experiential community. For the conservative, scientific knowledge eliminates, to its own detriment and that of all human beings, "all *particular essential references* to man, nature, and things in which every piece of knowledge comes embedded" (Mannheim [1925] 1986: 61).

During the era of the Weimar Republic and the Nazi period, a rapprochement can be observed in the work of influential conservative intellectuals, especially Hans Freyer, Werner Sombart, Ernst Jünger, Carl Schmitt and Oswald Spengler, of 'culture' and technology. Herf (1983) has called the sympathetic connection between what used to be contradictory categories in conservative-romantic thought, an expression of "reactionary modernity". The genesis of this rapprochement may be traced to experiences of this generation of intellectuals during the First World War and the pragmatic political realization that certain national goals could only be reached in conjunction with an affirmative attitude toward modern technology. At the same time, intellectual resistance against the Enlightenment and its aims no longer required opposition to the manifestation of modern technology. The ideology of National Socialism in Germany represents the most radical manifestation of such a connection between *Innerlichkeit* and technology (Benjamin 1961). The rapprochement between conservative thinking and technology affirms the fact that technical-industrial modernization does not need to go hand in hand with the modernization and rationalization of intellectual life.

The present-day critique of science and technology resembles, to some extent, the conservative assault on science in the 1920s, although conceptual similarities cannot disguise essential differences. Among the resemblances are the shared conviction about the potency of scientific knowledge and technical artifacts and the belief that scientific rationality may somehow be monopolized in the hands of a few, forming the basis for the exercise of power in social relations. How else is one to interpret the persistent warnings about an impending "imperialism of instrumental rationality" (Weizenbaum), the danger of an aggressive "colonialisation of the life world" (Habermas), or the unavoidable "Taylorisation of the world of work" (Volpert)?

14.4.2 The Compartmentalization of the Life-World

Not only explicit resistance against a particular style of thought may limit the success or channel the dissemination of scientific knowledge, but the differentiation or compartmentalization of the life-world into more or less separate spheres may impede the process of the dominance of scientific knowledge. Moreover, individuals and groups can hold scientific and non-scientific beliefs concurrently.

Mannheim ([1925] 1986) has pointed out that the capitalist-rationalist process may have limits or be capable only of partially supplanting existing views and social processes. While the realm of public social relations is rationalized, the sphere of private relations and convictions might be more immune to transformation. Even strata, Mannheim ([1925] 1986: 64) suggests, at the forefront of the rationalizing process "did not entirely lose their original bearing towards life. It merely disappeared from what we may call the foreground of their *public* and *official* life. Their *intimate* relationships, insofar as they remained untouched by the capitalist process, proceeded in a non-calculating, non-rationalized manner. The relationship to life did not become abstract in these spheres". In fact, the phenomenon of gradual disappearance and recession into intimate spheres of certain public spheres, that is, into spheres of life in which personal and religious feelings prevail, complements the rationalization of work and exchange, among other realms (see also Elias [1939] 1978). But as Mannheim himself indicates, the spheres of life which retreat into intimacy are those left untouched by the process of rationalization. They do not remain unaffected, however: these intimate spheres evidently live a fragile and precarious existence.

Simmel's analysis of the limits of the 'intellect' are specific. He speaks about its averaging or levelling character and close proximity to the principles of individualism and egoism. In particular, in the *Philosophy of Money*, Simmel ([1907] 1978: 437) asserts that the nature of the (objective) contents of the intellect can be communicated universally and that, assuming its validity, everyone with a sufficiently predisposed mind will be persuaded by it. In this respect, there is no analogous function in the sphere of the will and the emotions. The basic antagonism of (individual) reason and inner feelings becomes evident further by virtue of the fact that the "contents of the intellect [...] do not possess the jealous exclusiveness that is common in the practical contents of life". The distinction between intellect and emotions finds its analogy in the dichotomy between means and ends; both fulfill distinct functions and cannot be substituted for one another:

> Certain emotions, for example, involved in intimate personal relationships, would completely lose their significance and value if others were entitled to share them. It is also essential for certain objectives of the will that other people are excluded from both pursuing them as well as gaining them. It has been rightly suggested that theoretical notions, on the other hand, are like a torch whose light does not become dimmer by igniting innumerable others from it. In as much as their potential boundless dissemination has no influence whatsoever upon their importance, they elude private ownership more than any other contents of life. (Simmel [1907] 1978: 437–8)

14.4.3 Mythological and Scientific Truths

In Durkheim's lectures on "Pragmatism and Sociology", delivered shortly before the First World War, a prominent place is reserved for consideration of the dichotomy of mythological and scientific knowledge. Mythological truths are accepted without further inquiry and testing, while scientific truths are always subject to verification and proof. The persuasive character of mythological knowledge, its constraining and objective nature, is linked to the fact that such knowledge is collective knowledge. But mythological knowledge is by no means unconnected to reality. The reality represented by mythological truths is the reality of society. Mythological knowledge is erroneous with respect to things or objects but it is true in relation to the thinking subject (cf. Durkheim [1955] 1983: 87). Scientific knowledge is, for Durkheim, also a form of collective consciousness. However, scientific knowledge becomes possible only in a society which is differentiated. Mythological and scientific knowledge correspond in many ways to organic and mechanical solidarity. The function of scientific truth is therefore to mediate and strengthen the collective consciousness. The difference between mythological and scientific knowledge may be found, above all, in their respective relationship to collective beliefs and individual consciousness. In other words, scientific thought like mythological knowledge, can form the basis for social solidarity. But what are the preconditions for social communication? According to Durkheim ([1955] 1983: 88), "either by uniting to form a single collective mind, or by communicating in one object which is the same for all, with each however retaining his own personality; like Leibnitz's monads, each expressing the entirety of the universe while keeping its individuality. The first way is that of mythological thought, the second that of scientific thought". Scientific knowledge represents and expresses societal formations and their state of development, in which individual differences and individualism are the foundation for collective solidarity rather than a threat to it. Durkheim ([1955] 1983: 92) therefore concludes that "intellectual individualism, far from making for anarchy, as might be expected during the period of the domination of mythological truth, becomes a necessary factor in the establishment of scientific truth, so that the diversity of intellectual temperaments can serve the case of impersonal truth".[18]

However, Durkheim is not as certain as Comte that scientific truths will displace mythological claims rapidly and completely. Durkheim suggests that it must be assumed that the power of scientific knowledge is, for the time being, limited to the world of physical objects. Sociology is faced with a most complex domain and therefore at best able to produce frail and limited hypotheses. Moreover, these knowledge claims have not affected public consciousness to any great degree. Social action is subject to constraints, especially to the pressure to act, and cannot be postponed until a scientific solution of social issues is at hand. Society is forced to operate with certain images of itself. The relative backwardness of sociology and the uneven development of science assures the survival of mythological claims. Mythological knowledge does not lose its social relevance fully, even in societies in which natural scientific knowledge appears to dominate. Actors are constrained to orient themselves in specific ways and the kinds of orientations provided by mythological truths are capable of serving as orientation for action. More generally, Durkheim ([1955] 1983: 91) insists that scientific truths cannot govern or dominate orientations by themselves because "there is, and there always will be, room in social life for a form of truth which will perhaps be expressed in a very secular way, but will nevertheless have a mythological and religious basis". What Durkheim has concretely in mind, are ideas or taken-for-granted 'dogmas' such as 'democracy',

---

[18] I have changed the translation slightly to capture the original meaning.

'progress', "the class struggle" which continue to contain, since they are rarely questioned, mythological components. Although Durkheim by no means suggests that the co-existence of scientific and mythological truth is inevitable, or even a fate we must bear and cannot transcend, the elimination of mythological claims will require a considerable period of time.[19]

14.4.4 Organized Knowledge and Traditional Knowledge

In contrast to these conceptions, there are accounts within sociological discourse about knowledge which imply, in the final analysis, that scientific knowledge is limited in its social effectiveness, although these limits are fixed a priori. The most important perspective[20] in this respect is derived from linguistic rather than sociological theoretical themes. A main example holds that the language of science, as differentiated from everyday language, cannot be formalized fully. Related is the sociological thesis that scientific knowledge is ultimately based on other forms of knowledge and cannot entirely relinquish ties to them, in particular everyday knowledge, and that specialized scientific knowledge cannot replace everyday knowledge (Luckmann 1981). Cicourel (1986) maintains, for example, that medical knowledge is dependent on everyday knowledge (*declarative knowledge*). The sociological variant of this argument points, therefore, to structural differences among forms of knowledge which perform different functions. Related to this conception is the thesis that, under more or less stable social conditions, the demand for traditional knowledge does not decline decisively. Finally, to this context also belongs the idea of competition among carriers of knowledge (cf. Znaniecki 1940; Böhme 1981). I would like to pursue, in greater detail, the notion of how structural features of knowledge depend on the conditions and the context of their production.

A critical analysis of the limits of the social power of scientific knowledge must incorporate some understanding about the special nature *as well as* the similarities of scientific and non-scientific knowledge and action. It can be shown that, from a *sociological* perspective, the dominant basis for the classification of scientific knowledge as a unique form of human knowledge is of limited use for the purposes at hand. That is, the dominant classification and attributes of scientific knowledge are linked too closely to now obsolete epistemological conceptions of science—notions and ideals of scientificity such as universality, experience, rationality,

[19] For Durkheim, the co-existence of mythological and scientific truths constitutes also a significant hurdle for the progress of sociological knowledge. On the other hand, one expects sociology to serve a crucial function in the establishment of any future dominance of scientific truths in the collective consciousness of society.

[20] The linguistic conception about the limits to scientific knowledge are of particular importance because it tends to be taken seriously within science itself, especially among theoreticians of knowledge and science. Its attractiveness to epistemology could well be linked to the fact that it allows the invocation of *logical* or formal considerations (cf. Goedel's Theorem).

necessity, practicality. Conceptions of scientific knowledge which are based on these attributes deny that scientific knowledge is socially based and a collective enterprise. They also ignore the fact that science is an historical enterprise. Merton ([1942] 1973) has suggested that everyday knowledge has a greater measure of plausibility and comprehension than scientific knowledge for most people as well as considerable substantive affinity to existing cultural 'prejudices' and so constitutes a possible source of competition for scientific knowledge claims, at least in everyday practical circumstances. Merton's conception is useful and indicative of an early theoretical conception of the limits of scientific knowledge which has the merit of transcending considerations primarily driven by epistemological concerns. Toulmin's (1972: 378) characterization of specific organized human activities may be seen as an initial way of delineating the conditions for the possibility of (organized) scientific knowledge, a superior basis of classification: "[...] human activities and enterprises [...] in which decisions are made, procedures followed, considerations taken into account, conclusions arrived at, new possibilities entertained, and 'reasons' given for the resulting conclusions or actions". But, as Toulmin indicates, not all human activities and decisions can be subjected to corresponding organized forms of scientific discourse.[21] Such limits are designed to stress, on the one hand, the special circumstances which prevail in the production of scientific knowledge, affecting the structure and the possibilities of reproducing such knowledge; and, on the other hand, that the special organizational form of scientific discourse does not extend to and incorporate the entire range of human problems, themes and issues. Toulmin's delineation of scientific discourse may serve, nonetheless, as a first approximation of the practical limits of scientific knowledge.

Toulmin's reference to scientific discourse as differentiated spheres of social action (see also Lepsius 1983) with special attributes, especially in the case of scientific laboratories, isolated from other social contexts, makes it evident that the structure of knowledge produced under these circumstances affects its reproduction in other social contexts. The specific attributes assumed by scientific knowledge under these conditions may be thought of as 'material' and cognitive attributes, but both sets of features impinge upon the conditions under which scientific knowledge may be reproduced.

First, I will consider what may be called the material attributes assumed by scientific knowledge as a result of the conditions of its production. Knowledge claims or knowledge effects produced under special conditions in scientific laboratories can undoubtedly only be reproduced outside the laboratory if the special conditions which allowed such outcomes are also reproduced outside the laboratory. That is, the special circumstances which allowed for the original observation

[21] Toulmin (1972: 405) stresses that "the boundary between disciplinable and non-disciplinable activities runs where it does because, in the course of their practical experience, men have discovered that it is both functionally possible and humanly desirable to isolate certain classes of issues, and make them the concern of specialized bodies of inquiries; while with issues of other kinds this turns out to be either impossible, or undesirable, or both at once".

of the effect must be extended to the context in which a successful transfer is to be made (Rouse 1987: 227). Thus, the notion that scientific knowledge, unlike other forms of knowledge, is not bound or limited institutionally, has to be questioned in light of the conditions necessary for the reproduction of scientific knowledge claims outside the circumstances of their initial discovery. It is by no means certain, in other words, that it is only the influence of conventional forms of knowledge that contracts as the functional differentiation of society progresses and once powerful institutions such as religion as well as institutionally-based knowledge in general diminishes in importance.

Secondly, knowledge claims not only take on features derived from the material conditions of their production but also reflect institutionally bound cognitive attributes. These attributes include a *suspension* of the pressure to act as constitutive of scientific discourse. Knowledge produced within the scientific community is released from the tasks it must perform outside of science. One of the most salient attributes of everyday life situations is, in contrast, the persistent constraint to act, the pressure to reach a decision, to observe a specific rule, to follow a particular course of action by discarding alternative possibilities or to provide an account of completed action ex post facto. This suspension of the constraint to act within scientific discourse may be described, on the one hand, as a virtue of intellectual activity taking place under privileged conditions which moderates the effect that the pressing interests, rapidly passing opportunities and ambiguous dependencies of everyday contexts can have on the production of scientific knowledge claims. On the other hand, the result of this suspension of the pressure to act is that scientific knowledge takes on qualities of incompleteness, provisionality, fragmentariness or expansiveness, which reduce its effectiveness as knowledge in circumstances in which action is the foremost requirement. For as Durkheim ([1912] 1965: 479) observed so well: "Life cannot wait" (cf. also Gehlen [1950] 1988: 296–7).[22] In most social contexts the need to act takes precedent over the need to know. Perhaps there is, as Simmel (1890: 1) surmises, a kind of anthropological constant in the form of a general and widespread preference among humans, namely to 'do' something rather than to know about something and knowing, in turn, may flow or require prior doing.[23] In his lectures on "Pragmatism and Sociology", Durkheim ([1955] 1983), discussing the scientific status of the discipline of sociology, refers to the same set of issues when he attempts to enumerate some of the reasons for the relative scientific backwardness of sociological knowledge. He underlines, for example, that the fragmentary and uncertain knowledge of sociology cannot but produce skepticism or doubt about the contingencies of practical action in the social

[22] Incompleteness or the lack of any impetus to action is constitutive for scientific knowledge: "Faith is before all else an impetus to action, while science, no matter how far it may be pushed, always remains at a distance from this. Science is fragmentary, and incomplete; it advances but slowly and is never finished." (Durkheim [1912] 1965: 479, compare also Luhmann 1983: 154–9).

[23] Though there is no direct textual evidence, Simmel's early observations may have been prompted by Goethe's dictum *Im Anfang war die Tat* ("in beginning was the deed") since Goethe's sentiments about the priority of doing over knowing were no doubt well-known to Simmel.

world. This might be appropriate with respect to natural science knowledge, but we have to live in the social world, and to live means to act. "[S]ociety cannot wait for its problems to be solved scientifically. It has to make decisions about what action to take, and in order to make these decisions it has to have an idea of what it is" (Durkheim [1955] 1983: 90). As a rule, scientific knowledge is, however, produced under conditions which consider 'waiting', distancing, careful reflection, the elimination of time-bound constraints to reach a decision or even the deliberate abstention from a judgment until the 'evidence' is in, a distinct attribute of the validity and the virtue of such knowledge claims. By reducing, and at times even eliminating, urgency as a part of the production process of scientific knowledge, gains for the point of view of epistemological ideals contrast with deficiencies from the perspective of everyday life in which urgency to act becomes a constitutive characteristic of action impossible to neutralize.

Laboratory studies of the production of scientific knowledge (e.g., Latour/Woolgar 1979; Knorr-Cetina 1981) show as well that the knowledge claims produced in science, which are neither based on nor an expression of an unique or special form of rationality or logic, result primarily in claims to non-local knowledge. Indeed, the site for the production of scientific knowledge does not differ much from the sites for the production of conventional or everyday knowledge. From this it follows that scientific rationality, once it appears outside the boundaries of the scientific community—for instance, in the form of expert knowledge in the determination of curricula, the allocation of public funds for research, or as expert witness or counsel—often is followed by severe disappointment in the eyes of the public because the scientific knowledge fails to display the expected reliability and consensus (cf. Barnes 1972).

In addition to these considerations following Bourdieu ([1980] 1987), for example, it is possible to assign to practical contexts a logic which is less stringent than the logic of logic. The social scientific analysis of everyday contexts reduces the urgency to act in such situations. The effect which is achieved might be called a depragmatization of everyday contexts or the elevation of practical circumstances to the level of theoretical contingencies. At the same time, the depragmatization of everyday contexts through social scientific discourse makes visible features of the former which offer resistance against theoretical transformation. Among them are aspects of a *practical* logic such as the ease of operation and control, subjective adequacy, economy and its practical persuasiveness represented in the union of a totality of judgments and their ambivalence. This opposition of practical and theoretical logic prompts Bourdieu to draw the radical conclusion that any theoretical reconstruction of practical situations amounts necessarily to a distortion of the 'truth' of praxis. The peculiar character of practical circumstances happens to be that it resists theoretical reconstruction because the truth of praxis resides in its blindness to its own truth. Scientific discourse and praxis have different purposes and attempt to realize different functions.

14.5 Science and Ignorance

I have already indicated that the assumption of the irreversible and forceful progress of scientific knowledge and the concomitant demise of traditional forms of knowledge, which cannot measure up, in its practical efficacy, to knowledge produced by science, includes the assertion, at least implicitly, that only scientific knowledge advances while non-scientific forms of knowledge are void of any transformational capacity. The same thesis of the inefficacy of conventional knowledge asserts, finally, that such knowledge is not really capable of defending itself. The conviction that traditional forms of knowledge are essentially helpless finds its parallel in the belief that scientific knowledge reduces the volume of conventional knowledge consistently rather than adding to it.

Is, however, the pool of conventional knowledge merely static knowledge? As a matter of fact, in an age of somewhat reduced enthusiasm for scientific knowledge, the idea that science might well be one of the important sources of the growth and the transformation of conventional knowledge becomes more plausible (cf. Brzezinski 1970: xii). When Ravetz (1986: 100) suggests, for instance, that "while our knowledge continues to increase exponentially, our relevant ignorance does so even more rapidly. And this is ignorance generated by science", he alerts us to the possibility that advances in scientific knowledge and their practical application are accompanied by a wide range of unsolved problems, often in the forms of risks, and by unanticipated consequences. From the point of view of scientific discourse, "scientific progress" produces immediately and incessantly non-knowledge or even ignorance (i.e. "certified ignorance"). In an indirect manner, therefore, the production of scientific knowledge and its realization—for example, in the responses of affected groups and individuals to unintended and unexplicated consequences, multiple risks and costs—contributes to the persistence and transformation, perhaps even expansion of conventional knowledge. Of course, these patterns occur in response to developments in science and their dissemination to and translation into practical contexts. But as a result science itself can be seen to represent a source for the perpetuation and the dynamics of traditional forms of knowledge in society.

14.6 Conclusions

The point of departure of my considerations was the widely-held view among sociologists, past and present, that, as Weingart (1981: 228) perceptively points out, our primary means of action orientation, are replaced "in more and more spheres of life, by the production *and* application of systematic knowledge" and that this process of replacement occurs as different aspects of life become successively the object of scientific scrutiny. Concretely, the displacement in dominant forms of knowledge implies that, on the basis of scientific knowledge, "different frames of reference and modes of attribution are established for social action and/or existing orientations are proven to be irrational or erroneous with respect to accepted purposes". The result is, according to Weingart, that "reflective reasoning, in light of competing components of systematic knowledge, takes the place of the internalization—on which its taken-for-granted status depends in the first place—of norms and values".

I have tried to show that this assumption about the powerlessness of non-scientific knowledge represents one of the more important themes on which both classical sociologists and theorists of present-day society concur. These ideas are injected into sociological discussion, sometimes in the form of prophecies and dire warnings about the impending destruction or colonization of traditional ways of life, but mainly in the form of a black box thesis. The thesis about the power of scientific knowledge becomes, in an unreflective manner, a premise of theoretical work and assumes, for the most part, a taken-for-granted status.

I have critically examined the premise of the unlimited power of scientific knowledge by suggesting constitutive features of scientific knowledge itself which could be seen as co-determinants for the social and intellectual limits of such knowledge. The perspective adapted implies, of course, that I could also pursue the question of the unique 'functions' of conventional forms of knowledge in modern and highly differentiated societies. Undoubtedly, traditional knowledge too has specific constitutive features which assist in its resistance to competing knowledge claims and helps its survival even in societies which have achieved a high degree of scientification. While scientific knowledge becomes itself more and more differentiated and represents, without question, one of the motors of many social transformations, to the extent that contemporary society is a society built by and on scientific and technical knowledge, the same dynamism does not apply to traditional knowledge. Conventional knowledge is somehow an island of tranquility within the storm of functional differentiation. Such observations and the repeated complaint about the consequences of 'dilettantism' and extreme 'specialism' may in fact be taken as a point of departure for a more systematic inquiry into the peculiar role conventional forms of knowledge may play as media of social interaction and as social cement. Or one may assert with Mannheim ([1929] 1940: 170) that politics as politics remains possible only as long as the sphere of the irrational exists; where it tends to disappear, "administration takes its place". The theoretical and empirical challenge remaining is to bring into union the analysis of the interaction of those forms of knowledge which are characterized by changeability and recalcitrance.

References

Barnes, Barry. 1972. "On the Reception of Scientific Beliefs", in: Barry Barnes (ed.), *Sociology of Science*. Harmondsworth: Penguin: 269–291.

Bell, Daniel. 1969. *The End of Ideology*. New York: Free Press.

Bell, Daniel. 1972. *The Coming of Post-Industrial Society. A Venture in Social Forecasting*. New York: Basic Books.

Bell, Daniel. 1975. "The Disjunction of Culture and Social Structure: Some Notes on the Meaning of Social Reality", *Daedalus*, 94(1): 208–222.
Bell, Daniel. 1976. *The Cultural Contradictions of Capitalism.* New York: Basic Books.
Bell, Daniel. 1979. "The New Class: A Muddled Concept", in: Barry Bruce-Biggs (ed.), *The New Class?* New Brunswick, NJ: Transaction Books: 169–190.
Bell, Daniel. 1987. "*The World and the United States in 2013*", *Daedalus*, 16(3): 1–31.
Benjamin, Walter. 1961. *Illuminationen.* Frankfurt am Main: Suhrkamp.
Böhme, Gernot. 1978. "Alternativen in der Wissenschaft – Alternativen zur Wissenschaft", in: Christoph Hubig and Wolfert von Rahden (eds.), *Konsequenzen kritischer Wissenschaftstheorie.* Berlin: De Gruyter: 40–57.
Böhme, Gernot. 1981. "Wissenschaftliches und lebensweltliches Wissen am Beispiel der Verwissenschaftlichung der Geburtshilfe", in: Nico Stehr and Volker Meja (eds.), *Wissenssoziologie. Sonderheft 22 der Kölner Zeitschrift für Soziologie und Sozialpsychologie.* Opladen: Westdeutscher Verlag: 445–463.
Boulding, Kenneth. 1967. "The 'Two Cultures'", in: Melvin Kranzberg and Carroll W. Pursell Jr. (eds.), *Technology in Western Civilization.* Volume II. New York: Oxford University Press: 686–695.
Bourdieu, Pierre. [1980] 1987. *Sozialer Sinn. Kritik der theoretischen Vernunft.* Frankfurt am Main: Suhrkamp.
Brown, James Robert (ed.). 1984. *Scientific Rationality: The Sociological Turn.* Dordrecht: D. Reidel.
Brzezinski, Zbignew. 1970. *Between the Ages. America's Role in the Technotronic Age.* New York: Viking Press.
Burnham, John C. 1987. *How Superstition Won and Science Lost: Popularizing Science and Health in the United States.* New Brunswick, NJ: Rutgers University Press.
Cicourel, Aaron. 1986. "The Reproduction of Objective Knowledge: The Common Sense Reasoning in Medical Decision-Making", in: Gernot Böhme and Nico Stehr (eds.), *The Knowledge Society.* Dordrecht: D. Reidel: 87–122.
Daele, Wolfgang van den. 1975. "Die soziale Konstruktion der Wissenschaft", in: Gernot Böhme et al. (eds.), *Die gesellschaftliche Orientierung wissenschaftlichen Fortschritts.* Frankfurt: Suhrkamp: 129–182.
Dahrendorf, Ralf. 1977. "Observations on Science and Technology in a Changing Socio-economic Climate", in: Ralf Dahrendorf et al. (eds.), *Scientific-Technological Revolution.* London: Sage: 73–82.
Durkheim, Emile. [1912] 1965. *The Elementary Forms of Religious Life.* New York: Free Press.
Durkheim, Emile. [1955] 1983. *Pragmatism and Sociology.* Cambridge: Cambridge University Press.
Elias, Norbert. [1939] 1977. *The Civilizing Process. Volume 1: The History of Manners.* New York: Urizen Books.
Elias, Norbert. 1983. "Knowledge and Power", in: Nico Stehr and Volker Meja (eds.), *Society and Knowledge. Contemporary Perspectives on the Sociology of Knowledge.* New Brunswick, NJ: Transaction Books: 251–291.
Geertz, Clifford. 1973. *The Interpretation of Cultures.* New York: Basic Books.
Gehlen, Arnold. [1950] 1987. Man. *His Nature and Place in the World.* New York: Columbia University Press.
Gouldner, Alvin W. 1976. *The Dialectic of Ideology and Technology. The Origins, Grammar, and Future of Ideology.* New York: Seabury Press.
Gouldner, Alvin W. 1979. *The Future of Intellectuals and the Rise of the New Class.* New York: Seabury Press.
Habermas, Jürgen. 1982. "A Reply to My Critics", in: John B. Thompson and David Held (eds.), *Habermas: Critical Debates.* Cambridge, MA: MIT Press: 219–283.
Habermas, Jürgen. [1981] 1983. *The Theory of Communicative Action. Volume One: Reason and the Rationalization of Society.* Boston: Beacon Press.

Herf, Jeffrey, 1983. *Reactionary Modernism. Technology, Culture, and Politics in Weimar and the Third Reich.* Cambridge: Cambridge University Press.

Holzner, Burkart, and John H. Marx. 1977. *Knowledge Application: The Knowledge System in Society.* Boston: Allyn and Bacon.

Hübner, Kurt. [1978] 1983. *Critique of Scientific Reason.* Chicago: University of Chicago Press.

Jaspers, Karl. [1932] 1979. *Die geistige Situation der Zeit.* Berlin: de Gruyter.

Kettler, David, Volker Meja, and Nico Stehr. 1983. *Karl Mannheim.* London: Tavistock.

Knorr-Cetina, Karin. 1981. *The Manufacture of Knowledge: An Essay on the Constructivist and Contextual Nature of Science.* Oxford: Pergamon Press.

La Follette, Marcel C. (ed.). 1983. *Creationism, Science and the Law. The Arkansas Case.* Cambridge, MA: MIT Press.

Lane, Robert E. 1966. "The Decline of Politics and Ideology in a Knowledgeable Society", *American Sociological Review*, 31: 649–662.

Latour, Bruno, and Steve Woolgar. 1979. *Laboratory Life: The Social Construction of Scientific Facts.* London: Sage.

Lepsius, M. Rainer. 1983. "Über die Institutionalisierung von Kriterien der Rationalitat und die Rolle der Intellektuellen", *Jahrbuch des Wissenschaftskollegs.* Berlin.

Lipset, Seymour M. [1960] 1962. *Soziologie der Demokratie.* Neuwied: Luchterhand.

Luckmann, Thomas. 1981. "Vorüberlegungen zum Verhältnis von Alltagswissen und Wissenschaft", in: Peter Janich (ed.), *Wissenschaftstheorie und Wissenschaftsforschung.* München: C.H. Beck: 39–51.

Luhmann, Niklas. 1983. *Ökologische Kommunikation.* Opladen: Westdeutscher Verlag.

Mannheim, Karl. 1935. *Mensch und Gesellschaft im Zeitalter des Umbruch.* Leiden: A.W. Sijthoffs Uitgeversmaatschappij.

Mannheim, Karl. [1929] 1940. *Ideology and Utopia. An Introduction to the Sociology of Knowledge. Second Impression.* London: Kegan Paul, Trench, Trubner & Co.

Mannheim, Karl. [1925] 1986. *Conservatism. A Contribution to the Sociology of Knowledge.* Edited and introduced by David Kettler, Volker Meja and Nico Stehr. London: Routledge and Kegan Paul.

Marcuse, Herbert. 1964. *One-Dimensional Man. Studies in the Ideology of Advanced Industrial Society.* Boston: Beacon Press.

Merton, Robert K. [1942] 1973. "The Normative Structure of Science", in: Robert K. Merton (ed.), *The Sociology of Science. Theoretical and Empirical Investigations.* Chicago: University of Chicago Press: 267–278.

Montagu, Ashley (ed.). 1984. *Science and Creationism.* Oxford: Oxford University Press.

Mulkay, Michael. 1979. *Science and the Sociology of Knowledge.* London: Allen & Unwin.

Parsons, Talcott. 1937. *The Structure of Social Action.* New York: McGraw-Hill.

Ravetz, Jerome R. 1986. "Usable knowledge, Usable Ignorance", *Knowledge*, 9: 87–116.

Richta, Radovan. 1969. *Civilization at the Crossroads. Social and Human Implications of the Scientific and Technological Revolution.* White Plains, NY: International Arts and Sciences Press.

Rouse, Joseph. 1987. *Knowledge and Power. Toward a Political Philosophy of Science.* Ithaca: Cornell University Press.

Roszak, Theodore. 1972. *The Making of a Counter Culture.* New York: Doubleday.

Scheler, Max. [1926] 1960. "Probleme einer Soziologie des Wissens", in: Max Scheler (ed.), *Versuche zu einer Soziologie des Wissens.* München and Leipzig: Duncker & Humblot: 1–146.

Schelsky, Helmut. [1961] 1964. "Der Mensch in der wissenschaftlichen Zivilization", in: Helmut Schelsky (ed.), *Auf der Suche nach der Wirklichkeit. Gesammelte Aufsätze.* Düsseldorf: Diederichs: 439–480.

Seyfarth, Constans. 1972. "Protestantismus und gesellschaftliche Entwicklung: Zur Reformulierung eines Problems", in: Constans Seyfarth and Walter M. Spronde (eds.), *Seminar: Religion und gesellschaftliche Entwicklung.* Frankfurt am Main: Suhrkamp: 338–366.

Simmel, Georg. 1890. *Über sociale Differenzierung. Sociologische und psychologische Untersuchungen*. Leipzig: Duncker & Humblot.

Simmel, Georg. [1907] 1978. *Philosophy of Money*. London: Routledge and Kegan Paul.

Stehr, Nico. 1984. "The Development of the Sociology of Knowledge", in: Nico Stehr and Volker Meja (eds.), *Society and Knowledge. Contemporary Perspectives on the Sociology of Knowledge*. New Brunswick, NJ: Transaction Books: 1–18.

Stehr, Nico. 1991. *Praktische Erkenntnis*. Frankfurt am Main: Suhrkamp.

Stehr, Nico and Volker Meja. 1981. "Zur gegenwärtigen Lage wissenssoziologischer Konzeptionen", in: Volker Meja and Nico Stehr (eds.), *Der Streit um die Wissenssoziologie*. Vol. 2. Frankfurt: Suhrkamp: 893–946.

Toulmin, Stephen. 1972. *Human Understanding*. Princeton: Princeton University Press.

Touraine, Alain. 1969. *La sociéte post-industrielle*. Paris: Denol.

Vidich, Arthur J. and Stanford M. Lyman. 1985. *American Sociology: Worldly Rejections of Religion and Their Directions*. New Haven: Yale University Press.

Touraine, Alain. 1984. *American Sociology. Worldly Rejections of Religion and Their Directions*. New Haven: Yale University Press.

Weber, Max. [1920] 1978. *Gesammelte Aufsätze zur Religionssoziologie*. Vol. 1. Tübingen: J.C.B. Mohr (Paul Siebeck).

Weber, Max. [1922] 1987. *Science as a Vocation*. Edited by Peter Lassman and Irving Velody. London: Unwin Hyman.

Weingart, Peter. 1981. "Verwissenschaftlichung der Gesellschaft – Politisierung der Wissenschaft", *Zeitschrift für Soziologie*, 12: 225–241.

Znaniecki, Florian. 1940. *The Social Role of the Man of Knowledge*. New York: Columbia University Press.

Chapter 15
The Culture and Structure of Social Inequality

Nico Stehr

> *For just as the substance of education – in spite of, or because of its general availability – can ultimately be acquired only through individual activity, so it gives rise to the most tangible and thus the most unassailable aristocracy, to a distinction between high and low which can be abolished neither (as can socio-economic differences) by decree or a revolution, nor by the good will of those concerned.*
>
> Georg Simmel ([1907] 1978: 439–440)[1]

Almost a century ago, Georg Simmel speculates, in his *Philosophy of Money*, about a most invidious and insidious form of social inequality in modern society. Though it remains widely invisible, at least to the professional eye,[2] its importance may well have increased manifold since the turn of the century. But one does not have to share the anticipatory enthusiasm of Bell (1973: 213) and his reference, in analogy to the system of inequality in the Republic of Plato—to the Scientific City of the future and its three classes of elite scientists and top professional administrators, the middle class of engineers and the professoriate, and the proletariat of technicians, junior faculty, and teaching assistants—to realize that most theories of social inequality lack correspondence to the social conditions to which they are intended to apply. Class analysis appears to have fallen into disrepute. Its demise perhaps accelerated by the disappearance of many states which attempted to abolish class in practice. And what self-consciously remains of class analysis in social science and desires to be apart from the examination of class in strictly Marxist terms, can be

This text was first published as: Stehr, Nico. 1994. "The Culture and Structure of Social Inequality", *International Journal of Group Tensions*, 24(4): 361–382. The permission was granted by Springer.

[1] I am grateful to Herman Strasser for his comments and to the Social Sciences and Humanities Research Council, Ottawa, Canada for research support.

[2] But as Simmel ([1907] 1978: 440) also observes, "there is no advantage that appears to those in inferior positions to be so despised, and before which they feel so deprived and helpless, as the advantage of education".

M. T. Adolf (ed.), *Nico Stehr: Pioneer in the Theory of Society and Knowledge*, Pioneers in Arts, Humanities, Science, Engineering, Practice 16,
https://doi.org/10.1007/978-3-319-76995-0_15

reduced to an investigation of social inequality in industrial society driven by conventional classifications and considerations pertaining to employment relations in labor markets and production units. As a result, the traditional, Marxian definition of social class becomes but a special case of sweeping concepts of class, advanced for example by Goldthorpe/Marshall (1992) or Dahrendorf ([1957] 1959: 247).[3]

But the central issue emerging with some urgency in inequality relations, it seems to me, is not anymore the compactness, even a persistent concentration of forms of inequality but an enduring increase in the level of affluence and wealth generally in modern society.[4] The possible emancipation from the worst forms of economic vulnerability and subjugation of large segments of the population—which neither occurs to the same degree and at the same pace in all industrialized countries nor is an entirely irreversible phenomenon—provide, I would like to suggest, for the *material foundation of new forms of inequality.* And, contrary to Marcuse (1964: 2), for these new forms of inequality it does in fact make a difference, whether the "increasing satisfaction of needs is accomplished by an authoritarian or a non-authoritarian system."

This essay therefore has two aims: I would like to briefly show that most of our theorizing and much of our research about social stratification and social mobility is intimately linked to a particular form of society, namely *industrial society.* The choice of "focal variables" or spaces (as Sen 1992, calls them) of inequality theory, is as much else a child of its times. However, as industrial society as we have come to know it begins to undergo a profound transformation, the social features which have served as the backdrop for inequality theory begin to recede in significance as well. The result is that these developments, that is, the change from industrial society to *knowledge society,* require a fresh examination of social inequality and factors constitutive for inequality generally.

---

[3] For Dahrendorf ([1957] 1959: 247) social classes and therefore class conflict are "present wherever authority is distributed unequally over social positions." Perhaps gradients in authority or what Comte had called "involuntary authority" relations are a kind of anthropological constant and essential to human conduct. However, in opposition to theses defended by Goldthorpe and Marshall, Dahrendorf ([1957] 1959: 269–270) observes or deduces from his theory of the institutional isolation of industrial conflict, a *delimitation* of spheres and roles in advanced industrial society linked to the economy with the result that industry in post-capitalist society lost its all-embracing social significance and has to a certain extent been capsuled off from society." Social roles other than *employment position* (of industrial workers and likely other occupational groups) gain in significance, thus, the "sector of social behavior which is not immediately determined by occupation is extending steadily" (Dahrendorf [1957] 1959: 273).

[4] Compare Gellner's (1987: 102–104) discussion of the role of wealth in agrarian and early industrial society and in modern society.

15.1 Industrial Societies and Social Inequalities

The varied contemporary theoretical efforts concerned with social inequality (various versions of Marxian theory, approaches in the Weberian tradition, functionalist conceptions or more eclectic approaches) are first and foremost reflections of forms of social inequality in industrial societies. The historical specificity is quite appropriate. Since the theories in question (e.g. Lipset 1968) are well known, I will only sketch its outlines, stressing however the linkages and dependence of these perspectives, despite its diversity, to socio-economic and socio-cultural processes constitutive of industrial societies (cf. Kreckel 1983).

Classical theories of social inequality as well as its more recent derivatives explicate themes which all display an interest in the *vertical* nature of social class or strata[5] and the relation of individuals or groups to differences in property and its price, including the wages derived from the property of labor power. The result of the basic premise of theories of inequality in industrial society is that social hierarchies in the final analysis always are generated and legitimized with reference to the productive process and its organization. This implies of course that inequality becomes directly but more often indirectly a function of the relation of the individual to work or capital and its benefits in the form of monetary income, interest, rent and profit. The identity of individuals is mediated if not determined by their relation to the work process. Social strata and classes form in the same manner.[6] Both Marxist and non-Marxist approaches are convinced that industrial society[7] is still primarily a society of labor (*Arbeitsgesellschaft*). And this in turn strongly suggests to social scientists that social hierarchies continue to be "shaped predominantly by class rather than other forms of social cleavage, no matter whether the phenomena under scrutiny are structural or cultural in nature" (Marshall et al. 1988: 183; also Adorno 1969; Giddens [1973] 1980: 264; Maryanski/Turner 1992: 159) and that the class based distributions of occupations tends to be closely reproduced intergenerationally whereby social classes become the product of social class. In other words, the performance imperatives and attitudes oriented toward

[5] Differences of lower and higher are then seen as essential to the very phenomenon of social inequality (cf. Kreckel 1983: 6–12). The result at times is a metaphorical overemphasis on matters vertical.

[6] Perhaps the most obvious distinction between Marxist and non-Marxist theories of social inequality in industrial society is related to the conceptions of what ought to constitute the central unit of analysis in research and theory concerned with inequalities. Non-Marxist theories of *social stratification* tend to generalize about inequality based on *individual* characteristics while Marxist theories prefer social *collectivities* of the basic unit of social inequality. The individual dimension is then seen as essentially 'subjective' by its critics while 'objective' units such as social class draw the objection of lacking precisely such a subjective dimension (e.g. Adorno 1969: 13).

[7] Another, though less pertinent, commonality of contemporary theories of social inequality is that they share as a boundary individual nation states constitutive of the political limits of industrial society.

and necessary for production remain at the social center of modern society (cf. Narr [1979] 1985: 35–39).

However, these assertions beg the question whether there is a form and world of work typical of industrial society (cf. Offe 1983) that, even as it changes, continues to exercise the same control over social inequality or, whether work, its social organization and outcome does not undergo significant transformations in the course of the development of the forces of production in industrial society affecting forms of inequality and whether economic development itself does not basically alter the structure of inequality in modern society. One should inquire, whether changes in the productive process, for example, the extent to which 'direct' labor gives way to another form of work, do not become as important a yardstick for concomitant changes in inequality regimes as comparable changes did in the course of the emergence of industrial society. Moreover, in industrial society, the importance of labor is said to insure that ascription, at least for those who work, gives way to achievement. But the percentage of individuals active in the work process in modern industrial society is the minority of the population[8] and likely will continue to decline even further. Nevertheless, the social status of those not active in the labor force continues to be seen, in an almost ascriptive fashion and contrary to the trend toward greater individualization in modern society, as dependent on the productive process. And, finally, it certainly would appear to be reasonable to expect that economic development reduces the sense of economic deprivation among significant segments of the public (e.g. Inglehart/Rabier 1986).

As a matter of fact, one does encounter observations about changes in the basis of inequality in contemporary society but the repercussions of these changes for the most part are considered to be virtually immaterial. Vertical social hierarchies for example are effectively retained as are many of the same economic, social and cultural consequences of inequality. Descriptions of new forms of social inequality therefore amount, as critics have argued with respect to Bell's theory of post-industrial society, to a further elaboration and evolution of the logic of the industrial social structure and a perpetuation of its inherent contradictions (cf. Steams 1974: 17).

In so-called multidimensional theories of stratification (cf. Barber 1968) of modern industrial societies, the dimensions usually identified as stratifying individuals in a vertical regime, such as occupation, income, *occupational* prestige and education, are for the most part viewed as derivatives of the primary attribute, namely the relation of the person to his or her work, but also as attributes which tend to converge in their relative merit, that is, toward status consistency (cf. Goldthorpe 1966; Stehr 1971).

The expectation was, at least a decade or two ago, that the industrial society and even more generally, the modernization process is bound to produce a system of

[8] In 1989, in Canada the proportion of individuals actively employed was 51.8 percent of the total population, in the U.S.A. it was 47.2 percent (117 million of 248 million), in England 47.36 percent (27 million of 57 million) and in (West) Germany 45.16 percent (28 million of 62 million) to mention but a few examples.

social inequality which in the end will be less hierarchical, more standardized and leveled, reflect individual abilities more closely, be more open and, therefore, much less rigid than inequality in the initial phases of industrialization with its often impenetrable class boundaries (cf. Schelsky 1955: 218–242; Dahrendorf 1959: 274; 1967: 68; Goldthorpe 1966: 650).[9] But a general lack of interest in questions of social inequality just a decade later was by no means the result of a triumph of equality in industrial society.[10]

However, consistently growing results of the productive process would appear to have compensated for the lack of any significant redistribution of social rewards and costs (cf. Galbraith 1957: 95) and account for the widespread comfort and content of the day. But in this respect, too, the optimism of the heroic days, of the late fifties and early sixties has given way to a more differentiated picture, in particular to the realization that the expected modification or even decline in economic inequality has not been achieved (e.g. Williamson/Lindert 1980; Hradil 1983a, b), although a noticeable leveling in some respects has, of course, taken place, e.g. in the area of life expectancy, health care and certain provisions of social security.

However, changes in the power structure or in authority relations of industrial society are much too difficult to discern or, for that matter, to agree on. Some of the anticipated consequences of industrial society for social inequality may well be realized yet such changes may well have occurred for entirely different reasons; for example, they might be the result of structural changes in society rather than individual mobility and achievement.

The emphasis on the dimensions of class and status has generally restricted the sensitivity and the vision of researchers to possible changes in the nature of industrial society which in the final analysis do not merely represent reiterations of the same processes of stratification but significant transformations of the entire

[9] Schelsky's (1955: 222) notion of a "nivellierte Mittelstandsgesellschaft" or a leveled middle-class society, to which I intend to refer in this context, is, however, primarily based on the assumption of a process of leveling which resembles a regression toward the mean of inequality although the leveling of inequality Schelsky anticipates is one based mainly on downward rather upward mobility. Parsons (1954: 431, 434), on the other hand, feels compelled to comment, in addition to stressing the paramount importance of occupational status and of occupational earnings, in his description of the system of social inequality in American society using his "Revised analytical approach to the theory of social stratification" that there is "the amount of 'compression' of the scale, so far as the income is concerned, which has occurred in about the last generation" and that the traditional bottom of the occupational pyramid is in the course of "almost disappearing". The American class structure and society will become even more predominantly middle-class than it already is.

[10] Galbraith (1957: 85) is of the opinion that the lack of interest in phenomena of social inequality at the time can be attributed, on the one hand, to the fact that the degree of inequality in capitalistic societies, contrary to Marxist expectation, did not increase and, on the other hand, that the societal influence of the wealthy, at least in the United States was less pronounced.

structure of inequality resulting from linear transformations of societal inequality.[11] Some recent analyses of social inequality have stressed novel and quite powerful attributes of inequality, for example, differences in access to social security, working conditions, leisure opportunities, living arrangements, access to social services, or unequal treatment based on ascriptive features which signal shifts away from the centrality of occupation as the primary stratifying principle (cf. Hondrich 1984). More recent theories of social inequality in industrial society emphasize therefore a much larger and wider range of factors which effect the life chances and the social standing of individuals.

The merit of classical sociological theories in general and theories of social inequality in particular was to stress certain then operative dimensions constitutive of social inequality. More comprehensive theories of social inequality which add more and more dimensions which may affect status relations in society not always result in a stringent or more adequate theory of inequality in modern society. The important question now is whether new, governing dimensions of inequality are emerging in contemporary society which have to be taken into account and perhaps replace those typical for industrial societies. Chief among the new dimensions indeed may be the rise of 'knowledge' as a stratifying principle or, the extent to which knowledge may be employed as a capacity to acquire social standing.

15.2 Knowledge Societies and Social Inequalities

To suggest that knowledge plays an increasingly important role in shaping the nature and the structure of social inequality of modern society does not mean that knowledge as a resource for action is a *novel* phenomenon in the production and the analysis of social inequality. On the contrary, knowledge representing a variety of cultural competencies and abilities has, of course, always played a significant role throughout history in determining aspects of inequality and its evaluation in society; for example, the ability to read and write the dominant language in a society has had an important place in inequality systems as have other abilities, for example, knowledge of the laws and procedures governing transactions in society or religious knowledge affected the position of a person in past societies (cf. Stinchcombe 1968: 326–327).

The important question that therefore needs to be addressed, in this context, not only is, how the growing importance of knowledge affects patterns of social inequality but also why knowledge, in the first instance, is 'capable' of eroding and

[11] Beck (1986: 121–160) has focused on the linear transformation and *elevation* of material inequality since the decade of the fifties sustaining established relations and concentrations of inequality. His primary focus then is on the extent to which the elevation in the general standard of living has allowed for the dissolution of class-based social conduct or for a further individualization. But his discussion remains transfixed by what might well be reversible material pre-conditions of changes in the life world of individuals.

ultimately perhaps replacing what have been for centuries, and continue to be seen by many observers, the solid foundations for patterns of social inequality in industrial society.

It would appear that Bell (1973: 374–375) follows a somewhat analogous perspective since he describes the possible emergence of a "knowledge class" in post-industrial society (see Table 15.1). Bell maintains that in terms of social status (that is, esteem and recognition, and possibly income), the knowledge class may be the highest class in the new society, but it also is in the nature of post-industrial structure that there is no intrinsic reason for this class, on the basis of some coherent or corporate identity, to become a new *economic* interest class or a new *political* class which could bid for power in the sense in which the organized working class is seen to be poised to assume power by Marxists. The scientific and professional strata which constitute the knowledge class thus stand apart from and outside of the resident classes of capitalism, namely those of labor and property. And since the knowledge basis cannot be easily exchanged or translated into either one of the other dimensions of social inequality, Bell suggests that the notion of *situs* may be a more germane concept to capture the changes underway in the structure of social inequality in post-industrial society. In short, Bell is convinced, in concert with the views of some of his critics, that the *foundations* of social inequality in post-industrial society, in comparison to industrial society (and contrary perhaps to some images his summary table of sources of social inequalities, reproduced here, conveys) will not really change all that fundamentally.

If one desires to advance a more radical image of changes in the structure of inequality to modern society, the first question one perhaps should pose is what evidence—with respect to structural changes of society—is there that knowledge acquires special and therefore more than routine significance in modern societies for patterns of social inequality?

Table 15.1 Condition for and sources of social inequality

	Pre-industrial	Industrial	Post-industrial
Resource	Land	Machinery	Knowledge
Social locus	Farm plantation	Business firm	University Research Institute
Dominant	Landowner	Businessmen	Scientists
Means of power	Direct control	Indirect influence on politics	Balance of technical-political forces Franchise and rights
Class base	Property Military force	Property Political Organization Technical skill	Technical skill Political Organization
Access	Inheritance Seizure by armies	Inheritance Patronage Education	Education Mobilization Co-optation

Source Bell (1973: 359)

There are at least three sweeping sets of societal changes that may serve as evidence for the emergence of knowledge as a stratifying principle.[12] Modern socio-structural—as well as socio-economic, socio-legal and socio-political—conditions rather than socio-cultural attributes[13] that should be seen as providing the foundation for the emergence of knowledge as a stratifying principle of course are (1) the relative decline in the immediate and unmediated importance of the economy for individuals and households. I mean a decline in the direct material subordination of individuals and households on activities centered in the market economy, in particular, their occupational roles and therefore their singular dependence on what is still for many their basic role as economic actors.[14] What diminishes is the tightness of the linkage in the material dependence of many actors on their occupational status only[15] and what increases is the relative material emancipation from the labor market in the form of personal and household

[12] For a discussion of a number of additional relevant elements that are likely part of the same social figuration (which reduce the identity of work and life), for example, the reduction in the hours per year and years per lifetime worked or the substantial rise in occupational income that increase freedom to find and afford opportunities outside of work and decrease the marginal benefits derived from further incremental additions to earned income, see Kern/Schumann (1983).

[13] Kalberg (1992) has noted that the de-coupling of work from social status in modern society or of debates on the place of work in post-industrial society is not necessarily universal but strongly mediated by national, cultural, political and historical *milieus.* In contemporary Germany, occupational values do not enjoy an elevated status above other social domains (e.g. leisure or the private sphere) while in the United States the hegemony of work remains unchallenged. The result is, Kalberg argues, that the place of work in the United States and Germany differ considerably. It is noteworthy that Kalberg does not refer to the work of Dubin (1992: 123–124), particularly his notion of work as a "central life interest" and the empirical research this has spawned in the last decades showing that the belief in the centrality of work varies considerably by type of occupation (e.g. industrial workers, professionals or managers) and country (e.g. Japan and the United States). Dubin's (1956) original research in the United States, among industrial workers, found only 24 percent reporting that work was their central life interest (see also Lane 1978; Offe 1983: 50–57 for a discussion of the changing subjective value of work in modern society).

[14] This assertion undoubtedly has at least some resemblance with the diminishing marginal utility thesis of economic returns found in Ronald Inglehart's (e.g. 1987; also Lane 1978) recent discussion of the reasons for the emergence of postmaterialistic values in advanced societies, that is, the significant rise in *per capita GDP* in advanced societies brings about a decline in any further marginal gains that can be achieved with additional income units. Inglehart's general observations mainly refer to growing levels of per capita income and the idea that a smaller and smaller proportion of the income is needed to defray the costs of absolutely necessary expenditures. As a result, the priority individuals attach to economic matters should decline as well. Rising income levels are important; however, they may well only constitute *one* of the conditions for the possibility of the liberation of individuals and households from the labor market and the economy.

[15] The broad change in the total wealth or assets of households in industrialized countries in the last three decades in particular can be illustrated in a number of ways. For example, one point to the immense growth, although inflation is not taken into consideration, in aggregate savings on deposit with commercial banks in different countries (this information is not available for all industrialized countries). See Table 15.2.

Table 15.2 Aggregate savings deposits (commercial banks) in billions of local currency 1960–1990 for selected industrialized countries

	1960	1965	1970	1975	1980	1985
Canada	7.1	9.4	16.5	32.9	74.5	117.9
Germany	–	–	189.9	356.3	464.0	599.6
Switzerland	–	–	40.7	76.7	107.4	145.9
United Kongdom	–	–	16.2	32.8	61.2	133.0

Source OECD, Main Economic Indicators. Historical Statistics, 1969–1988

Table 15.3 Net equity of households on life insurance and pension fund reserves (in millions of local currency), 1978–1990

	1978	1981	1984	1987	1990
Canada	9070	14074	16214	18191	26886
USA	95022	117911	172922	113598	207860
Japan	4042	5899	9672	17217	18905
Germany	27050	36050	44140	49480	53500
United Kingdom	8283	14863	18523	20574	–

Source OECD, *National Accounts. Detailed Tables*, Volume 2: 1978–1990
In the case of the United States, the net equity is fluctuating considerably. In 1986 net equity reaches 231 billion, only to fall to 113 billion in 1987

wealth.[16] Since the savings rates vary considerably from country to country, the populations in some countries have of course amassed a much greater fortune in savings (see Table 15.2). The considerable *aggregate* growth of net equity of households on life insurance and pension illustrates the same point (see Table 15.3). The decreasing material subordination to one's occupational position of course not only affects those who work but applies with even greater force,

[16] For any audience but particularly for a North American audience and during recessionary times, this is not an easy contention to advance. Not too long ago, a prominent essay in the *New York Times* diagnosed a decline in middle-class prosperity (Kevin Phillips, "Down and out: Can the middle class rise again?" *New York Times Magazine*, January 10, 1993). This is an observation anticipated by many months by the voting middle class and politicians in the United States. Rebuilding a sense of prosperity in the middle class is crucial not only to the political fortunes of those who promised greater shares of wealth to the middle class but perhaps the key to the future of American culture. At the same time, the stagnation or even decline in middle-class prosperity in the United States in the 1980s and early 1990s is a useful reminder that developments in the United States do not necessarily signal patterns to be repeated, after a certain lag, in other modern economies. The North American political systems and their economies have not produced widespread levels of prosperity which lead the world. In other words, the thesis to be expounded here does not necessarily apply with particular force to the case of Canada and the United States.

paradoxically perhaps, to the rising segment of the population who is out of work and who therefore involuntarily is cut off from the labor market.[17]

Compelling evidence illustrating the extent and the relative significance of at least a linear transformation of prosperity and wealth are most difficult to obtain because considerations of the distribution of personal wealth, household assets, various entitlements etc. still are singularly driven, for obvious ideological reasons, at one end of the spectrum, by an interest in the concentration of wealth (cf. Atkinson 1980; Field 1983; Wolff 1991), especially the proportion of the wealth controlled by the upper percentiles of the wealthy or, at the other end of the spectrum, the focus is exclusively on attempts to measure poverty.[18] However, both enduring wealth inequalities[19] that indeed at times defy comprehension or the real prospect of an increasingly divided society[20] should not lead one to simply ignore the substantial rise in the *general level of wealth and prosperity* and of course ask what consequences this may have for manifestations of social inequality in highly developed nations.[21] In Canada for example—which has no estate or wealth tax, Statistics Canada occasionally has undertaken household sample surveys to estimate the changing wealth in the country. An analysis of the information obtained in

[17] One should include here the growing though uncounted number of individuals and households who have made, for a variety of reasons, the self-copious derision to reduce their dependence on the labor market by cutting or drastically changing their consumption patterns and expectations. This group includes the growing number of individuals who retire early: As reported in an article in the *Globe and Mail* (February 27, 1993, B22), in 1988, "18 percent of Canadians aged 55–64 were pensioners—a considerable jump from the 3 percent in 1969 and even the 8 percent in 1979."

[18] Often much of the wealth data made available is limited to the upper percentiles of the wealth distribution. In addition, information about wealth inequality frequently is plagued by methodological difficulties, for example, with respect to the definition of wealth and/or the unit of analysis, systematic under reporting of assets, sapling problems, and the statistical treatment of wealth data.

[19] Despite patterns of high income concentration of household wealth, there has been a gradual decline in this century, at least until the mid-1970s, in wealth inequality and a convergence in the level of wealth differences among industrialized societies for which longitudinal information is available; in those countries, Sweden, the United Kingdom and the United States, "the share of total household wealth held by the top 1 percent of wealth holders declined from 50 in 1920 to 21 percent in 1975 in Sweden. The share of the top 1 percent in Great Britain fell from 61 percent in 1923 to 23 percent in 1974. The decline in wealth inequality in the United States was less dramatic. The share of the top 1 percent of individuals fell from a peak of 38 percent in 1922 to 27 percent in 1956" (Wolff 1991: 128). Although these three countries differ considerably in regard to economic performance, social and political institutions, policy, entitlements etc. the data show a convergence in the level of household wealth inequality. Since the mid-1970s the decline in wealth inequality has been arrested.

[20] These have become much more since Gorz (1982) for example initially sketched the emergence of a divided society in post-industrial societies. Lyotard's (1989: 10) observation about the division of mankind into two parts, namely "one part confronted with the challenge of complexity; the other with the terrible ancient task of survival", resonates with the same sentiments.

[21] The limited visibility of wealth in social science during the seventies and the eighties is at least partly related to the critique of and deep seated resentment by social critics, in same countries, toward wealth and their bad conscience to live off well of what is being criticized.

the years 1970, 1977 and 1984—that is prior to the significant asset inflation which took place latter in the decade of the 1980s—reveals that the most remarkable observed change has been in the *growth* of the general level of wealth. The *average* wealth for "the combined universe of families and unattached individuals was $85,344 in current dollars (1984)—4.7 times the $18,189 average for 1970" (Oja 1987: 7).[22] During the same period, the consumer price index rose by a factor of 2.98. Increases in the median wealth during the same time period were even more significant, implying a simultaneous decrease in wealth inequality in Canada.[23]

A further, (2) major reason for a transformation in the basis of social inequality in modern society is the establishment and guarantee of a bundle of *social citizenship rights,* especially a floor of existential welfare below which no one is allowed to slip (cf. Dahrendorf 1987). The establishment of such social entitlements restricts and diminishes the immediate and unmediated dependence of individuals and households on the dynamics of the economy in general and the labor market in particular. The collective or aggregate entitlement structure in modern societies of course is dependent on the availability of paid employment and the future aggregate amount of work at hand is crucial to a sustainable future—and any growth potential in—entitlements. The implication of these developments at the same time is that the effect of lifestyles on the economic structure (cf. Gershuny 1986) multiplies in significance. The uncoupling from such immediate and direct subordination, the decline of intimidation and a diminished vulnerability opens up and sustains a new sphere of action.[24] The decentering and uncoupling process is even more general. The last major structural change which accompanies and feeds on the diminishing importance of the economy therefore is (3) a decline in the degree to which modern society, with respect to many activities, as I will enumerate, looses previous authoritative centers and therefore exemplary or tightly constraining patterns of conduct. Modern societies no longer—despite what may be said about the

---

[22] The wealth concept used by Statistics Canada excludes important personal assets such as equity in pension funds, insurance policies, the value of collectibles and consumer goods and entitlements. What is counted in the survey as a component of wealth includes the equity in the home, cash, saving deposits, certificates, bonds, stocks, registered savings plans, equity in a business, farm or profession and these figures then reduce the indebtedness to arrive at total wealth. As the report states, "there is reason to believe that large wealth holdings may have been underestimated in the survey" because of the characteristics of the survey method alone (Statistics Canada 1986: 83); for example, refusal rates are higher among higher income families and individuals. The number of respondents (family units) in the 1984 survey of Statistics Canada (1986) is approximately 14.000. Of all the family units sampled approximately 72 percent supplied data on which the wealth estimates are biased.

[23] Although the author of the comparative study of the development of wealth in Canada observes that one is struck by how little change there has been over the period 1970–1984 in the relative distribution and composition of wealth and how dramatic the growth in overall wealth was at the same time (Oja 1987: 7), virtually no further attention is paid to the latter development in her analysis.

[24] What these developments do not imply is that identifiable and systematically (economically) disadvantaged groups or a distinct underclass simply vanish. As a matter of fact, the handicaps experienced by such an underclass multiply.

increasing globalization or homogenization for example—have a few dominant (at least in those societies in which the electoral laws do not discourage a multiplication of political parties) and/or coherent political parties, family patterns, labor unions, gender structures, religions, scientific disciplines, ethnic groups, social strata, communities, cities, corporate structures etc. In each instance a process of *decentering* is underway. For example, in most modern societies we do not find that the traditional family continues to be the dominant family. On the contrary many patterns not only in the area of family life prevail. The decentering provides malleable structures which can be reconstructed in many ways enhancing the very process underway only further. The reconstruction of the rules which govern the structural patterns in enable one to employ one's 'knowledge' in productive ways.[25]

15.3 Knowledge as Bundle of Competencies

If one wants to acquire an idea of how knowledge as a 'multi-faceted' *resource and capacity of action* becomes implicated in social inequality and does so not merely as another set of *means* among traditional, especially material instruments and devices which produce and nurture along a chain of social hierarchies, it is necessary to refer to these structural changes as a background and envision that knowledge under these circumstances more *immediately* and directly yields social advantages and disadvantages in the form of power and authority. It seems to me that knowledge in this context has to be conceptualized as a *bundle of competencies,* for example, in the sense of a particular agent's socially recognized capacity to speak and act legitimately (cf. Bourdieu 1975: 19), yielding different social benefits (or costs) for those who command and are able to mobilize its capacities in appropriate situations.[26] The relation between material and cognitive factors of

[25] It might be suggested that the views about the role of knowledge in social inequality elaborated so far resonate strongly with observations made in the sixties and seventies about the growing affluence, even a saturation of material needs, the sectoral shift in employment and resulting change in attitudes and values, what Inglehart (1971, 1977) called the 'silent' revolution, toward concerns away from the bare, material necessities of life toward 'postmaterialist' values and goals. The political agenda and the basis for political conflict changes, as does politics. But the demand for material goods really has not diminished with increasing affluence and declining scarcity and, as I already observed, the development of the service sector is best examined once disaggregated. Finally, Inglehart's (1987: 1289) approach to the "diminishing marginal utility of economic determinism" is, it seems to me, primarily functionalist, that is, directed toward an examination of these consequences of the declining returns from economic development such as life expectancy, economic equality or political preferences.

[26] In a somewhat different context, Lyotard ([1979] 1984: 18–19) describes the enabling social features of knowledge in a rather similar sense. Lyotard explicates the function of knowledge in postmodern society and emphasizes that knowledge is by no means limited to scientific, that is, instrumental knowledge. On the contrary, knowledge refers to a much more general and "extensive array of competence-building measures and it is the only form embodied in a subject constituted by the various areas of competence composing it."

social inequality is reversed. Knowledge commands material well-being, its components and extension. Even if knowledge is considered a mere mean or basis of social inequality in modern society then at least it has to be conceptualized as a 'meta-level' instrument capable of affecting the acquisition, protection and disposition of more traditional means of stratification. From a material point of view, but not only from such a perspective, knowledge should be seen as a resource insulating and protecting individuals and households from the immediate impact of the vagaries of the market and coercion. In knowledge societies, inequality becomes a much less obvious, concrete and visible social phenomenon than in industrial society. The choice and definition of knowledge as a "focal variable" of inequality also signals that its prerequisites and consequences in practice tend to be, for the time being, less definitive and consensual than alternative dimensions of inequality such as income, education and occupation and tend to lead to less firmly established structures of inequality. Knowledge as a resource for generating and maintaining patterns of inequality is more tolerant of ambiguity, dissensus, unresolved disputes and essentially contested pattern of inequalities. My focus therefore is less on outcomes or the results of allocative processes (and the extent to which they are linked to what are almost ascriptive attributes of individuals) but on capacities to assert oneself or to achieve different desired ends in widely varying social contexts.[27]

The question then is how such a viewpoint on knowledge either complements or differs from recent theoretical conceptions of social inequality which view inequality as a game of cognitive representations of social differences and as a struggle to dominate discourse and strategies about classifications of stratification. The concern of Bourdieu (1984) for example is precisely such an analysis on social inequality as maps of cognitive representations. It is argued that classificatory contests in contemporary society have moved beyond a smuggle for advantages which revolve around income. New types of conflicts emerge and center on struggles for certification, matters of and ethical superiority (cf. Eder 1989). While a perspective adds considerably to our understanding of the now multi-faceted re-construction of categories of inequality, it can easily remain a (functionalist) view concerned *not* with the enabling (disabling) or constitutive features of inequality but merely with outcomes or products of classificatory struggles.

---

[27] My focus resonates with that of Amartya Sen who concentrates on an examination of spaces of functioning or on the capability to achieve functioning as his focal variable for a normative analysis of inequality structures. The notion of capabilities of course is very broad indeed, as is the idea of knowledge as a general resource in the formation of inequality structures, and it therefore may suffer in the eyes of critics who prefer a less inclusive concept. Nonetheless, the capabilities approach differs, as Sen (1992: 6) observes, quite substantially "from the more traditional approaches to equality, involving a concentration on such variables as income, wealth, or happiness [...]. In particular, judging equality and efficiency in terms of capability to achieve differs from the standard utilitarian approaches as well as from other welfarist formulations". The utilitarian and the welfarist perspectives employ a much more restrictive and narrow calculus of inequality.

The focus on knowledge implies a focus on the new bases for inequality. What cognitive resources of action allow for the possibility of widely processing new types of classificatory struggles, in the first place? And how broadly do these resources have to be socially distributed to make such an argument about inequality not merely restricted to an elite of members of society? More concretely, how does knowledge function as a resource in struggles for status or dominance in cognitive representations in society?

As briefly indicated, in the context of focusing on social inequality formation and consciousness,[28] knowledge should be conceptualized as a bundle of social competencies that drive the process of forming and maintaining social prestige and status. Knowledge as a broad and heterogeneous bundle of competencies which have noticeable effects on the *process* of inequality formation, tough the *outcomes* continue to involve hierarchies along traditional and not so traditional dimensions. Given this notion, knowledge is unlike earlier, much more singular mechanisms of social inequality. Unpacking the bundle of competencies can only mean to enumerate some of the important specific capacities knowledge confers which are mobilized depending on the demands of the specific contexts. The concrete distribution, the substitutive possibilities and multiple interdependencies among competencies result in a less and less 'coherent' and transparent, at times at the surface even invisible, system of social inequality in knowledge societies. Social inequality becomes a heterogeneous and context dependent figuration.

Now I will list what I consider some of the most important social competencies which drive modern social inequality:

The *capacity to exploit discretion*: Since social rules and legal norms and regulations governing ordinary and extra-ordinary social conduct are never constituted and enforced in ways that they do not provide for discretionary interpretation and execution or leeway for reconstruction, the competence to mobilize discretion refers to the capacity of individuals to gain comparative advantages, for example in the area of taxation, investment, schooling, income etc.

The *facility to organize protection*: The capacity to put protective devices and measures in place is a matter of specialized competence which enables actors to mobilize access to differential knowledge in order to ensure, for example, that one's assets and entitlements are protected against structural or inordinate depreciation. The symbolic or material opportunity costs of being incapable to organize protection can be considerable.

[28] If one proposes to retain, especially against those perspectives arguing that class location has declined in significance in recent decades as a source of ideological consciousness, the notion of a (subjective) "class consciousness" as relevant for modern society, it would follow that the formation and retention of such class consciousness should perhaps be more effected by factors other than solely the material status of individuals and households (cf. Evans 1992 for an empirical critique of a narrow conception of factors linked to class consciousness in Great Britain).

The *authority to speak* (cf. Bourdieu 1975) is increasingly based on differential knowledge and immediately implies a parallel social division in opposition to those not authorized to speak. The authority to speak applies to many features and situations but also extends to the ability of a lay audience or a lay person to enter a discursive field of expertise as "speakers and confront the alleged truth of the discourse that justifies those practices" (Larson 1990: 37). By the same token, the inability to master knowledge, independent of the modes of exclusions/inclusions always associated with differential education, increasingly is interpreted as a sign of personal failure.

The *ability to mobilize defiance* constitutes another crucial component of the stratifying mode of knowledge. The ability to challenge the practices of experts, the state or corporations becomes an important asset of the inequality-folding capacity of knowledge. In the same sense, the ability of evade surveillance and obtain spaces of self-regulated autonomy acquires considerable significance and is based on the capacity to mobilize tools that often are seen as instruments which exclusively enhance scrutiny.

The *capacity of avoidance* and exclusion is a further stratifying trait which can be enlisted on the basis of different knowledge bases. I have in mind strategies which ensure that some of the risks of modern society tend to be distributed differentially, for example in the general areas of safety concerns, exposure to conflict or violence, incidents of health risks or environmental degradation. By the same token, the tremendous growth of the "informal economy" in most advanced societies, that is, of all kinds and forms of economic transactions uncontrolled by the state and legal system, irrespective of their legality, should be seen among the socio-economic consequences of the rise of knowledge as a stratifying principle effecting the material basis of inequality.

More generally, the range of social competencies counting to stratified facilities to immediately *master one's life,* for example, one's health (life expectancy), financial well-being, personal life, aspirations, career, long term security etc. or the ability to locate and gain assistance toward mastering these tasks, are generalized effects of a differential command of relevant knowledge bases.[29] The ability to mobilize defiance, exploit discretion, develop strategies for coping and organize protection of course are a significant part of such strategies and therefore the conviction that one somehow is in charge and not merely the object of fortuitous circumstances.

[29] In the United States, the social location of many of these activities can be found in what Drucker (1989: 187) calls the 'third' sector of non-profit, non-governmental, "human change" institutions. The third sector is actually the "country's largest employer, though neither its workforce nor the output it produces show up in the statistics. One out of every two adult Americans—total of 90 million people—are estimated to work as volunteers in the third sector" (Drucker 1989: 197).

15.4 Conclusion

In the context of proposing that advanced contemporary society is increasingly a knowledge society I have tried to show that the standard account as well as more recent, so-called multidimensional perspectives on social inequality fail to adequately capture the new social and economic realities. Theories of social inequality for the most part continue to be linked in a rather close fashion to the production process, its organization and outcomes. Moreover, the lingua franca of inequalities theories remains tied to an imagery which displays agents as *inflexible creatures* entangled in single-purpose, dedicated structures of inequality which impose their rhythm, often over generations, on passive individuals and groups. As long as inequality regimes are reconstructed as inflexible and largely one-dimensional, stratification is steep and the effects of material inequalities are long lasting, discourse on inequality naturally has to be with the coercion, that is, the constraints, the vulnerability and effective impotence of individuals and groups in the face of power exercised by the powerful. But new realities require a new language of inequality. The new language should, in contrast, stress agency, malleability, flexibility, multi-purpose resources, volatility, heterogeneity etc. of social structures and the extent to which individuals and groups have the capacity to employ and transform these structures, once the generalized vulnerability to the forces of inequality has been reduced to a significant degree. The condition for the possibility of greater and more broadly based agency is knowledgeability or a bundle of more widely accessible social competencies and their impact on social structures of inequality accelerating chances for actors to re-fashion social constructions. The consequences these developments have for social inequality are both immediate and difficult to trace. However, as knowledge replaces property and labor as the constitutive mechanisms for social inequality, the relation of individuals and groups to knowledge acquires fundamental significance for patterns of social inequality in society.

References

Adorno, Theodor W. 1969. "Spätkapitalismus oder Industriegesellschaft", in: Theodor W. Adorno (ed.), *Spätkapitalismus oder Industriegesellschaft. Verhandlungen des 16. Deutschen Soziologentages*. Stuttgart: Ferdinand Enke: 12–26.

Atkinson, Anthony B. (ed.). 1980. *Wealth, Income and Inequality*. Second Edition. Oxford: Oxford University Press.

Barber, Bernhard. 1968. "Social Stratification: Introduction", in: David Sills (ed.), *Encyclopedia of the Social Sciences*. Volume 15. New York: Crowell Collier and Macmillan: 288–295.

Beck, Ulrich. 1986. *Risikogesellschaft. Auf dem Weg in eine andere Moderne*. Frankfurt am Main: Suhrkamp.

Bell, Daniel. 1973. *The Coming of Post-Industrial Society. A Venture in Social Forecasting*. New York: Basic Books.

Böhme, Gernot and Nico Stehr (eds.). 1986. *The Knowledge Society*. Dordrecht: D. Reidel.

Bourdieu, Pierre. 1984. *The Distinction: A Social Critique of the Judgement of Taste*. Cambridge: Harvard University Press.

Dahrendorf, Ralf. [1957] 1959. *Class and Class Conflict in Industrialized Society*. Stanford: Stanford University Press.

Dahrendorf, Ralf. [1967] 1974. "Soziologie und industrielle Gesellschaft", in: Ralf Dahrendorf (ed.), *Pfade aus Utopia. Arbeiten zur Theorie und Methode der Soziologie*. München: Piper: 64–73.

Dahrendorf, Ralf. 1987. "Changing Perceptions of the Role of Government", in: Organisation for Economic Co-Operation and Development, *Interdependence and Co-Operation in Tomorrow's World*. Paris: OECD: 110–122.

Drucker, Peter F. 1989. *The New Realities in Government and Politics/In Economics and Business/In Society and World View*. New York: Harper & Row.

Dubin, Robert 1992. *Central Life Interests. Creative Individualism in a Complex World*. New Brunswick, NJ: Tractions Books.

Dubin, Robert. 1956. "Industrial Workers' Worlds: A Study of the 'Central Life Interests' of Industrial Workers", *Social Problems*, 3: 131–142.

Eder, Klaus. 1989. "The Cognitive Representation of Social Inequality: A Sociological Account of the Cultural Basis of Modern Class Society", in: Hans Haferkamp (ed.), *Social Structure and Culture*. Berlin and New York: de Gruyter: 125–146.

Evans, Geoff. 1992. "Is Britain a Class-Divided Society? A Re-analysis and Extension of Marshall et al.'s Study of Class Consciousness", *Sociology*, 26: 233–258.

Field, Frank (ed.). 1983. *The Wealth Report-2*. London: Routledge and Kegan Paul.

Gellner, Ernest. 1987. *Culture, Identity and Politics*. Cambridge: Cambridge University Press.

Gershuny, Jonathan I. 1986. "Lifestyle, Economic Structure and Time Use", in: Ralf Dahrendorf, Eberhard Kohler and Franocise Piotet (eds.), *New Forms of Work and Activity*. Dublin: European Foundation for the Improvement of Living and Working Conditions: 248–282.

Giddens, Anthony. [1973] 1980. *The Class Structure of the Advanced Societies*. London: Hutchinson.

Goldthorpe, John H.1966. "Social Stratification in Industrial Society", in: Reinhard Bendix and Seymour M. Lipset (eds.), *Class, Status and Power*. New York: Free Press: 648–659.

Goldthorpe, John H. and Gordon Marshall. 1992. "The Promising Future of Class Analysis: A Response to Recent Critiques", *Sociology*, 26: 381–400.

Gorz, André. [1980] 1982. *Farewell to the Working Class. An Essay on Post-Industrial Socialism*. London: Pluto Press.

Hondrich, Karl Otto. 1984. "Der Wert der Gleichheit und der Bedeutungswandel der Ungleichheit", *Soziale Welt*, 35: 267–293.

Hradil, Stefan. 1983a. "Entwicklungstendenzen der Schicht- und Klassenstruktur in der Bundesrepublik", in: Joachim Matthes (ed.), *Krise der Arbeitsgesellschaft?* Verhandlungen des 21. Deutschen Soziogentages 1982. Opladen: Westdeutscher Verlag.

Hradil, Stefan. 1983b. "Die Ungleichheit der 'sozialen Lage': Eine Alternative zu schichtungssoziologischen Modellen der Ungleichheit", in: Reinhard Kreckel (ed.), *Soziale Ungleichheiten. Soziale Welt*, Sonderband 2. Göttingen: Verlag Otto Schwartz & Co.: 101–118.

Inglehart, Ronald. 1987. "Value Change in Industrial Society", *American Political Science Review*, 81: 1289–1303.

Inglehart, Ronald. 1971. "The Silent Revolution in Europe: Intergenerational Change in Post-industrial Societies", *American Political Science Review*, 65: 991–1017.

Inglehart, Ronald. 1977. *The Silent Revolution*. Princeton: Princeton University Press.

Inglehart, Ronald and Jacques-René Rabier. 1986. "Political Realignment in Advanced Industrial Society: From Class-Based Politics to Quality-of-Life Politics", *Government and Opposition*, 21: 456–479.

Kalberg, Stephen. 1992. "Culture and the Locus of Work in Contemporary Western Germany. A Weberian Configurational Analysis", in: Richard Münch and Neil Smelser (eds.), *Theory of Culture*. Berkeley: University of California Press: 324–365.

Kern, Horst and Michael Schumann. 1983. "Arbeit und Sozialcharakter: Alte und neue Konturen", in: Joachim Mattges (ed.), *Krise der Arbeitsgesellschaft. Verhandlungen des 21. eutschen Soziologentages in Bamberg 1982*. Frankfurt am Main: Campus: 353–365.

Kreckel, Reinhard. 1983. "Theorien sozialer Ungleichheit im Übergang", in: Reinhard Kreckel (ed.), *Soziale Ungleichheiten. Soziale Welt*, Sonderband 2. Göttingen: Verlag Otto Schwartz & Co.: 3–12.

Lane, Robert E. 1978. "Markets and the Satisfaction of Human Wants", *Journal of Economic Issues*, 12: 799–827.

Larson, Magali Sarfatti. 1990. "In the Matter of Experts and Professionals, or How Impossible It Is to Leave Nothing Unsaid", in: Rolf Torstendahl and Michael Burrage (eds.), *The Formation of Professions. Knowledge, State Strategy*. London: Sage: 24–50.

Lipset, Seymour M. 1968. "Social Class", in: David Sills (ed.), *Encyclopedia of the Social Sciences*. Volume 15. New York: Crowell Collier and Macmillan: 296–315.

Lyotard, Jean-Francois. [1979] 1984. *The Postmodern Condition: A Report on Knowledge*. Minnesota: University of Minnesota Press.

Lyotard, Jean-Francois. 1989. "Defining the Postmodern", in Lisa Appignanesi (ed.), *Post-modern: ICA Documents*. London: Free Association Books: 7–10.

Marcuse, Herbert. 1964. *One-Dimensional Man. Studies in the Ideology of Advanced Industrial Society*. Boston: Beacon Press.

Marshall, Gordon, Howard Newby, David Rose, and Carolyn Vogler. 1988. *Social Class in Modern Britain*. London: Hutchinson.

Maryanski, Alexandra and Jonathan H. Turner. 1992. *The Social Cage. Human Nature and the Evolution of Society*. Stanford: Stanford University Press.

Narr, Wolf-Dieter. [1979] 1985. "Toward a Society of Conditioned Reflexes", in: Jürgen Habermas (ed.), *Observations on "The Spiritual Situation of the Age"*. Cambridge: MIT Press: 31–66.

Offe, Claus. 1983. "Arbeit als soziologisches Schlüsselkategorie", in: Joachim Mattges (ed.), *Krise der Arbeitsgesellschaft. Verhandlungen des 21. Deutschen Soziologentages in Bamberg 1982*. Frankfurt am Main: Campus: 38–65.

Oja, Gail. 1987. *Changes in the Distribution of Wealth in Canada, 1970–1984. Income Analytic Report No. 1. Statistics Canada, Labour and Household Surveys Division*. Ottawa: Statistics Canada.

Parsons, Talcott. 1954. "A Revised Analytical Approach to the Theory of Social Stratification", in: Talcott Parsons (ed.), *Essays in Sociological Theory*. New York: Free Press: 386–439.

Schelsky, Helmut. 1955. *Wandlungen der deutschen Familie in der Gegenwart*. Dritte, durch einen Anhang erweiterte Auflage. Stuttgart: Ferdinand Enke.

Sen, Amartya. 1992. *Inequality Reexamined*. Cambridge, MA: Harvard University Press.

Simmel, Georg. [1907] 1978. *The Philosophy of Money*. London: Routledge and Kegan Paul.

Statistics Canada. 1986. *The Distribution of Wealth in Canada, 1984*. Ottawa: Statistics Canada.

Steams, Peter N. 1974. "Comment on Daniel Bell's *The Coming of Post-Industrial Society*", *Transaction*, 11: 10–22.

Stehr, Nico. 1971. "Statuskonsistenz", *Kölner Zeitschrift für Soziologie und Sozialpsychologie*, 23: 34–53.

Stehr, Nico. 1994. *Knowledge Societies*. London: Sage.

Stehr, Nico and Richard V. Ericson (eds.). 1992. *The Culture and Power of Knowledge*. New York and Berlin: de Gruyter.

Stinchcombe, Arthur. 1968. "The Structure of Stratification Systems", in: David Sills (ed.), *Encyclopaedia of the Social Sciences*. Volume 15. New York: Crowell Collier and Macmillan: 325–332.

Williamson, Jeffrey G. and Peter H. Lindert. 1980. *American Inequality. A Macro-Economic History*. New York.

Wolff, Edward N. 1991. "The Distribution of Household Wealth: Methodological Issues, Time Trends, and Cross-Sectional Comparisons", in: Lars Osberg (ed.), *Economic Inequality and Poverty. International Perspectives*. Armonk, New York: M. E. Sharpe: 92–133.

Chapter 16
The Economic Structure of Knowledge Societies

Nico Stehr

There are, in the context of this study, several important reasons for addressing the nature of the changing economic structure of modern society. First, development of knowledge societies is connected to basic transformations in the structure of economic activity. The engine of much of the dynamics of economic activity and the source of much of the growth of added economic value can be attributed to knowledge. Paradoxically perhaps, the self-transformation of the economy diminishes the importance of the economy to individuals and society. Of course, it does not eliminate it. But from the point of view of the individual, for example, the economy of knowledge societies has the enabling quality of allowing central-life interests to progressively drift away from purely economic ones or, from a macro-perspective of social conflicts, for instance, a shift toward more generalized struggles not primarily driven by material clashes can be discerned. The conditions which allow for such displacements also render traditional economic discourse (and policy derived from such premises) less powerful. At the same time, natural, social and cultural conditions, namely scientific-technical, environmental and institutional change, often treated as exogenous factors by neo-classical economics, become increasingly important to economic activity.

Secondly, the emergence of a primarily knowledge-based labor force cannot be understood apart from a profound transformation in the economic system; and, thirdly, public debate and political discourse on modern society is still captivated and impoverished, perhaps dominated but certainly often limited, by reference to economic considerations. Prevalent popular orientation and a frequent basis for political judgments are references to economic imperatives and therefore how

This text was initially published as: Stehr, Nico. 1994. "The Economic Structure of Knowledge Societies", in: Nico Stehr, *Knowledge Societies*. London: Sage Publications: 121–159. The permission to republish this text was granted on behalf of the SAGE permissions team in London on 31 May 2017 by Mr. Craig Myles.

M. T. Adolf (ed.), *Nico Stehr: Pioneer in the Theory of Society and Knowledge*, Pioneers in Arts, Humanities, Science, Engineering, Practice 16,
https://doi.org/10.1007/978-3-319-76995-0_16

policies and conduct generally fit with 'free' market forms; that is, political realities are frequently defined in a most restrictive fashion.

A fourth reason is the quality of the link between the economic system and other societal sectors, for example the private world, which is changing and justifies a re-examination of the assumption of a self-propelled development of economic change, on the one hand, and the firm separation between the economy and other sectors of society, on the other. In addition, the extent to which situational and contingent rather than transcontextual and universal effects play a role in economic relations justifies a theoretical approach to economic relations informed by sociological considerations.

Fifthly, sociological discourse, in the past few decades, has been increasingly separated from economic discourse. It is possible to conceive of this distancing as a matter of the increasing differentiation of social science discourse. Economics lost interest in the analysis of social institutions, while sociology conceded the study of socio-economic phenomena to economics (cf. Swedberg 1987; Granovetter 1990).[1] Finally, both economic analysis and sociology lost interest in the study of the societal and socio-economic impact of science and technology. However, it is now, in the light of existing economic conditions, less certain that such a state of affairs represents a proper cognitive priority and intellectual division of labor. The sociological contribution to the analysis of economic relations should not merely be peripheral nor should the treatment of scientific and technical change be considered exogenous to economic analysis (cf. Dosi et al. 1988).

Central to my analysis is the thesis that the origin, social structure and development of knowledge societies is linked first and foremost to a radical transformation in the *structure of the economy,* including a set of novel and largely unintended consequences, for example in the area of terms of trade, inflation, productivity, competitiveness and employment. Moreover, the ways in which society is affected and co-ordinates its economic activities and agents are changing. Although the central thesis will be that we are witnessing the emergence of a new structure and organization of economic activity on the basis of a new combination of the forces of production, Emile Durkheim's specification of the general status of economic factors within a primarily sociological analysis, first introduced at the end of the nineteenth century for his own central theoretical category of the division of labor in society, remains valid. That is, Durkheim ([1893] 1864: 275) draws a distinction in the use of the category of division of labor by sociologists and economists; for the latter, "it essentially consists in greater production. For us, this greater productivity is only a necessary consequence, a repercussion of the phenomenon. If we specialize, it is not to produce more, but it is to enable us to live in new conditions of existence that have

[1] As Freeman and his colleagues (1982: ix), for example, resolutely stress: "The development of industrialized economies cannot be reduced to statistics of the growth of GNP, of industrial production, of capital stock, investment, employment etc., valuable though these statistics undoubtedly are. Underlying these statistical aggregates are the growth of entirely new industries and technologies and are the decline of old ones and many social and institutional changes in the structure of industry and government."

been made for us." Attention in this instance indeed centers only secondarily on outcomes. The primary focus is on how outcomes become possible, are sustained, organized and perhaps even continue to grow.

Productive processes in *industrial society* are governed by a number of factors all of which appear to be on the decline in their relative significance as conditions for the possibility of a changing, particularly *growing* economy: the dynamics of the supply and demand for primary products or raw materials; the dependence of employment on production; the importance of the manufacturing sector which processes primary products; the role of labor (in the sense of manual labor); the close relation between physical distance and cost and the social organization of work; the role of international trade in goods and services; and the nature of the limits to economic growth. The most common denominator of the changes in the structure of the economy seems to be a shift from an economy driven and governed, in large measure, by 'material' inputs into the productive process and its organization to an economy in which transformations in productive and distributive processes are determined much more by 'symbolic' or knowledge-based inputs and outputs. However, social science discourse and official data collection still tend to think of economic activity primarily in terms of the production of commodities.

The economy of industrial society is initially and primarily a material economy and then changes gradually to a monetary economy. Keynes' economic theory, particularly as outlined in his *General Theory* (1936), reflects this transformation; it becomes, as evident recently, a symbolic economy. The changes in the structure of the economy and its dynamics are increasingly a reflection of the fact that knowledge becomes the leading dimension in the productive process, the primary condition for its expansion and for a change in the limits to economic growth in the developed world. In short, the point is that for the production of goods and services, with the exception of the most standardized commodities and services, factors other than "the amount of labor time or the amount of physical capital become increasingly central" (Block 1985: 95) to the economy of advanced societies.

A close examination of the literature in economics indicates, however, that the function of knowledge and information in economic activity is, for the most part, ignored by economists. Either that, or they introduce knowledge as an exogenous variable, as an expense and generally treat it as a black box.[2] There are significant exceptions of course, and I will refer to them.[3] But the general and disparaging

---

[2] The economic concept of *capital* is narrowly defined and refers to fixed, physical capital equipment in plant and organizations. Such capital is recognized as *investment*, that is, as objects which must be purchased. However, the acquisition of knowledge, for example, in the sense of research and development, creating of organizational structures, educational programs or the development of skills are treated as *expenses* and not as contributing to the capital formation of organizations.

[3] One of the significant exceptions among economic theorists is Friedrich von Hayek (e.g. [1945] [1948]) for whom the central problem of economic theory is the problem of knowledge and who, despite his methodological individualism, views social institutions such as economic markets as knowledge-bearing phenomena. Markets are not allocative mechanisms but rather epistemological

observation by Stigler (1961: 213) is still close to the mark: "One should hardly have to tell academicians that information is a valuable resource: knowledge *is* power. And yet it occupies a slum dwelling in the town of economics." Knowledge is a residual, even invisible component of production and assets. Knowledge has many 'qualitative' components and quality has not yet prospered within economic discourse. Despite its apparent ascent as a source of added economic value, for example, knowledge remains elusive.

The specific changes in the economic *structure* may be described briefly as follows. The important changes in the relations of production, the nature of work and the composition of the labor force will be dealt with in greater detail below.

16.1 The Diminishing Role of Primary Materials

The striking change here is the 'uncoupling' of the raw material economy from the industrial economy.[4] The uncoupling has been accompanied in recent decades, perhaps slowed, by a secular decline in the price of commodities when compared to the price of manufactured goods. The decline of commodity prices has been uneven. It has been particularly strong in the case of metals (Fig. 16.1a; cf. Grilli/Yang 1988; for trends see IMF 1992: 86–87). In general these developments imply that the recent "collapse in the raw materials economy seems to have had almost no impact on the world of industrial economy" (Drucker 1986: 770). The traditional assumption of economists has of course been that changes in the price structure, most surely dramatic changes, ought to have a profound impact on the cycle of economies.

However, the significant decline in the price of most raw materials has not brought about an economic slump, except perhaps in those countries which rely to a large degree on trade with raw materials. On the contrary, production has grown (Fig. 16.1b). As the regression lines in Fig. 16.1c for example show, in the case of the (apparent) consumption of aluminum in OECD countries from 1962 to 1990, the increase in consumption departs, or slows more and more from the growth of the manufacturing sector in these countries. What accounts for the uncoupling is the decline in demand for primary commodities[5] and the simultaneous increase in the supply of, for example, food.

devices "in which knowledge that could not be collected by a single mind is yet rendered accessible and usable for human purposes" (Gray 1988: 55). Markets embody tacit knowledge.

[4] In the context of this section, attention will focus on the fact of the uncoupling process itself and on the possible reasons for such a development; however, I will not draw attention to and discuss national, regional or even global environmental consequences of the uncoupling process in some countries (cf. Simonis 1989), or the overall impact of a shift in material intensive production from some regions of the world to other groups of countries.

[5] Although these particular estimates should be taken with a grain of salt, Drucker (1986: 773) relates that "the amount of industrial raw materials needed for one unit of industrial production is now no more than two-fifths of what it was in 1900. And the decline is accelerating." These estimates need to be viewed with some skepticism simply because they are unusual ways of viewing production costs when compared to more conventional methods of accounting among economists. But this also ensures that such comparisons are challenging perspectives.

(a)

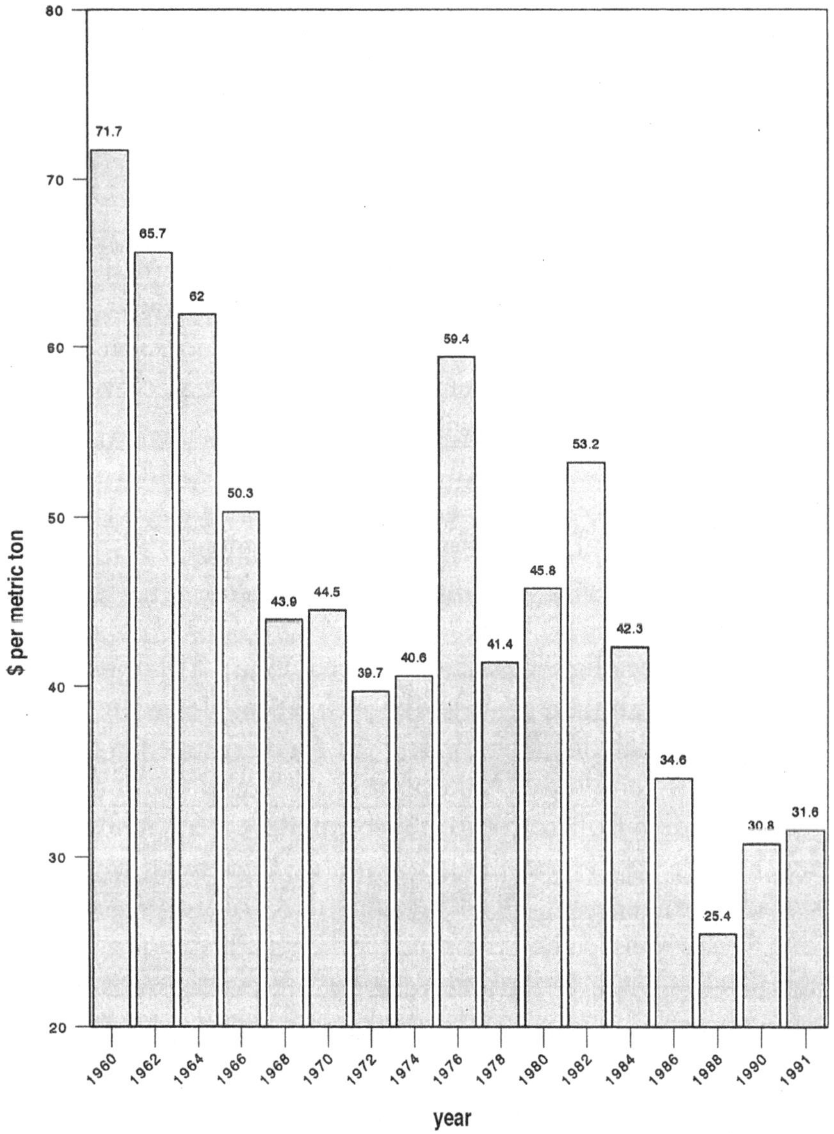

Fig. 16.1 **a** Iron ore prices per metric ton, 1960–1991 in 1990 constant US dollars (deflated by G-7 consumer price index). *Source* World Bank, Market Outlook for Major Primary Commodities, Volume 2. Washington, DC: World Bank, 1992: 139. **b** Annual real value added in manufacturing and annual growth in (apparent) consumption of primary aluminum in OECD countries, 1962–1990. *Sources* Real value added in manufacturing: OECD, *Historical Statistics,* 1992a: 48, 50; 1983: 44, 46; consumption of primary materials compiled by World Bank. **c** Annual real value added in manufacturing and annual growth in (apparent) consumption of primary aluminum in OECD countries, 1962–1990 (log regression). *Sources* Real value added in manufacturing: OECD, *Historial Statistics,* 1992a: 48, 50; 1983: 44, 46; consumption of primary materials compiled by World Bank.

(b)

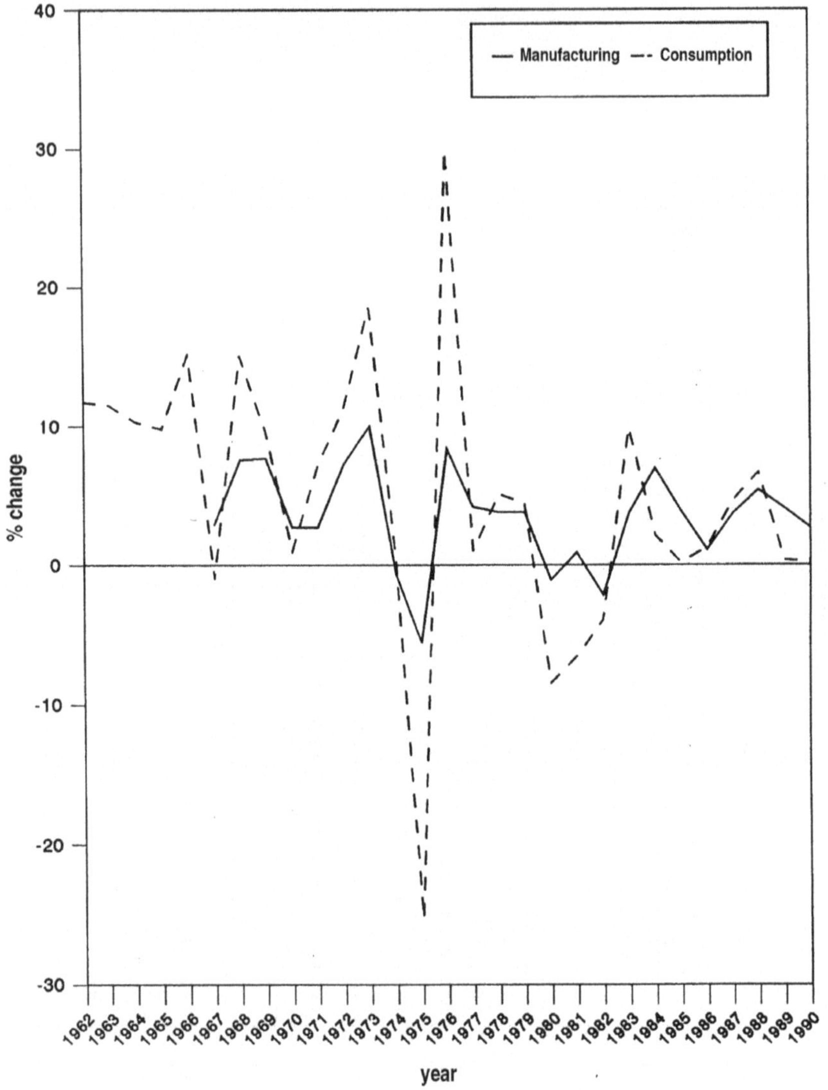

Fig. 16.1b (continued)

The demand for raw materials in manufacturing diminishes "not only because of miniaturization (e.g. chips) and the reduction of energy requirements, but also because of the revolution in material science. One asks less for specific materials [...] and more for the properties needed (e.g. tensility, conductivity) and the material combinations that can provide those properties" (Bell 1987: 9).

(c)

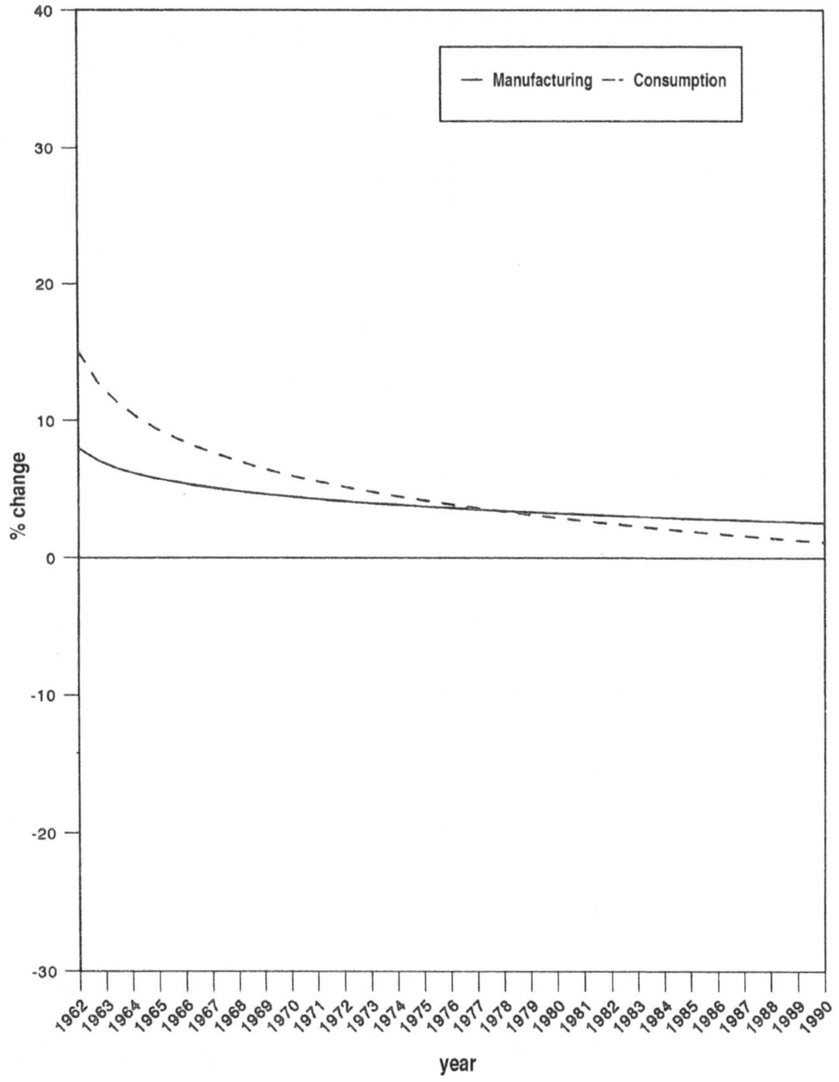

Fig. 16.1c (continued)

In terms of the sheer quantity of the consumption of primary materials, the amount of primary aluminum utilized in the OECD countries has not risen much since 1979. In the case of steel it has declined in the same period[6] as exemplified in

[6] According to World Bank data, in 1979, the total consumption of primary aluminum OECD countries was 11.422.000 tonnes and in 1990 it had reached 11.994.000 tons; even in low and

the United States motor industry; from the mid-1970s to the mid-1980s, the iron and steel content of cars manufactured was reduced by approximately one-third while the amount of plastics and composites increased by 33 percent (cf. OECD 1992e: 22). These shifts in the input of raw materials result from a combination of factors, including of course technological changes, but also from the relative price changes of factors, environmental regulations, market pressures and consumer preferences. It is, of course, very difficult to factor out the contribution of each of these influences and possibly others to the use of raw materials in production. However, it is evident that the thesis that growing production and consumption invariably entails increasing natural resource use has to be re-examined in light of these developments.

16.2 The Changing Manufacturing Sector

It is by no means unusual to encounter assertions even today which treat work in the manufacturing sector of the economy as relatively homogeneous.[7] That is, work found in the manufacturing sector continues to be understood as one of the few remaining examples of the kind of work which classic economic and sociological theory had in mind when it spoke of labor, namely work carried out by dependent and exploited laborers in industrialized settings. This type of work in the manufacturing sector is also often perceived as one of the last bastions against the increasing trend toward greater differentiation, new rationality, more extensive flexibility, self-determination and increased reflexivity of work in the service sector of the economy (cf. Offe 1984: 24–25). The upshot is, of course, that the worlds of work and economic sectors are seen as operating according to different rationalities, one technical, or perhaps better, 'functional' rationality and the other a kind of 'substantial' rationality (cf. Mannheim [1935] 1940: 51–56).[8] Consistent with some of the traditional general assumptions of contemporary theories of society, especially the notion of functional differentiation, the governing theme in understanding the modern economy becomes a reference to the normative and material *differentiation* among economic sectors rather than *linkage* and complementarity.

mid-income countries it increased by a small percentage. The steel consumption in OECD countries declined from 428 million tons in 1973, to 395 million tons in 1979 and to 358 million tons in 1990.

[7] Affirming this perspective, Offe (1984: 23), for example, contends that "by far the largest proportion of the work carried out in the 'secondary', that is, the industrial, commodity producing sector can be reduced to the common abstract denominator that such work is subject to the common technical-organizational productivity regime as well as the criteria of singular profitability while these standards of the work and evaluation process lose their (relative) definite character where labor becomes reflexive, namely within the largest segment of the 'tertiary' sector of service industry".

[8] The upshot according to Offe (1984: 27) is that we are faced with a significant "doubling of the term labor, with a proximity and an opposition of dual and contradictory standards of rationality corresponding to the images of the 'efficient producer' or the 'effective safeguarding of accomplishments' and therefore with an evident loss of once unequivocal conception of work".

Assuming that the differentiation of the modern economy into three sectors remains plausible, what is less contentious is that the character of labor in modern society is changing. The central questions for analysis become whether the changes in work are mainly confined to a specific sector of the economy or whether the transformation occurs in all sectors, perhaps driven by similar forces and constraints and whether the pattern of labor found in one sector increasingly extends into other sectors of the economy?[9]

By the same token, it is widely postulated that the overall economic importance of, or the capacity to add value to, the manufacturing sector is declining and that the primary significance of economic activity has shifted to the service sector. These conjunctures at times have been discussed under the heading of 'de-industrialization' (for example, Bluestone/Harrison 1982). It is worth noting that Daniel Bell does not, as far as I can tell, ever expressly advance, in the context of his extensive comments on the nature of the post-industrial economic structure of society, the thesis that the manufacturing sector actually *shrinks* in economic importance. Bell (1979a: 163) explicitly contends that post-industrial society involves a "change from a goods producing to a service society". As a result perhaps, the frequent reading of his theory that the aggregate importance of the manufacturing sector diminishes under post-industrial conditions acquires a certain credibility. As a matter of fact, the designation 'post-industrial' itself supports such an image. And the almost 'logical' inference of that kind of interpretation is that the industrial sector of the economy becomes largely dispensable in an affluent economy.

But exactly the opposite is the case. Contrary to many assumptions, the manufacturing sector and industrial production are not declining in importance in contemporary society.[10] Since several observers have seen fit to assert with some urgency that *manufacturing matters* (Cohen/Zysman 1987), and have done so quite convincingly, it is evident that the post-industrial economy thesis has been widely interpreted to mean that the secondary economic sector not only loses out as a source of employment but is also a loss leader in terms of its contribution to the overall

[9] I take it that the latter is the position argued by Cohen/Zysman (1987: xiii) in their book *Manufacturing Matters*. Summing up their position, Cohen and Zysman observe that the new technical division of labor has "extended the production processes outside the confines of the traditional manufacturing firm. We are experiencing a transition not from one kind of industrial economy to a service economy, but from one kind of industrial economy to another".

[10] This conclusion can be expressed quantitatively in a number of ways. One interesting set of figures (Forstner/Ballance 1990: 38–39) which confirms the consistent importance of manufacturing is its almost constant share between 1970 and 1986 among countries with "developed market economies (this is a group of twenty-five countries from Australia to the United States) of the value of total exports of these countries. In 1970, approximately three quarters of the value of total exports comes from manufacturing; in 1986, the figure has risen slightly. During the same period of time, in the case of developing countries, the share of manufacturing increases from 25.9 to 63.7 percent (exports excluding oil)". According to Forstner/Ballance (1990: 38), the rising share of "manufactures in the exports of developing countries cannot be attributed primarily to price effects, but was the result of more fundamental changes in the structure of production." The figures dispel the notion that these countries are dependent on the export of agricultural commodities.

material well-being of the modern economy. In other words, is the composition of the relative contribution to the gross national product across economic sectors really changing? When did the change commence and how significant is it? In addition, what exactly is the linkage between sectors of the economy? Do we have to assume that *economic* connections too are driven by the process of differentiation, or is it the case that the service economy, for example, perhaps represents more of a complement rather than a substitute or successor to manufacturing?

As the distribution of the contribution of the manufacturing sector (at *constant prices*)[11] to the GDP of selected economies in Table 16.1 indicates the share of the manufacturing sector between 1978 and 1990 has declined somewhat in some of the countries, remained stable in others and increased in the case of the Japanese economy. In other words, the repeated observation that the "shift to services" represents a change toward an increased consumption of services *at the expense* of manufactured (or agricultural) products is mistaken. The data indicate that there has not been a significant shift in the relative contribution of the different sectors of the economy to the total output (see also Baumol et al. 1985).

However, within the manufacturing sector rather significant transformations are taking place. First and foremost, production is switching away from commodities which are material intensive. Although Drucker (1986: 773) does not offer a precise source for his estimate, he relates that the raw materials in a semi-conductor microchip account for 1–3 percent of total production cost; in an automobile their share is 40 percent, and in pots and pans it is 60 percent. But also in older industries the same scaling down of raw material needs goes on, and with respect to old products as well.[12] The result is, for the time being, that two forms of manufacturing are emerging: one is material based, represented by the industries that provided economic growth in the first three quarters of this century. The other is information- and knowledge-based: pharmaceuticals, telecommunications, analytical instruments and information processing such as computers (Drucker 1986: 779).

[11] Using constant rather than current prices and therefore focusing on quantity rather than both volume changes and price changes generates a somewhat more accurate picture of the relative position of the manufacturing sector. When measured in current prices, the share of manufacturing in the US has constantly *fallen* since 1978; at constant prices, however, its contribution remains stable as Table 6.1 demonstrates. The difference in the two trends reflects the higher productivity in the manufacturing sector relative to other sectors and therefore a relative decline in its prices, for example, compared to prices in services. By the same token, the decline in the contribution of the manufacturing sector to the manufacturing sector to the GDP in the United Kingdom, measured in *current prices*, could be a consequence of the same factors. But in the case of the United Kingdom, constant prices were not available.

[12] Peter Drucker and Daniel Bell refer almost in unison, though without citing a source for their information, to the following relations for illustrative purposes: (a) Drucker (1986: 773) contends that "fifty to 100 lb of fiberglass cable transmits as many telephone messages as does one ton of copper wire"; and (b) Bell (1987: 8) recounts that "one hundred pounds of optical fibers in a cable can transmit as many messages as one ton of copper wire". I assume that both authors found their information in a newspaper or magazine.

Table 16.1 Percentage of gross domestic product (at 1985 prices)[a] generated by manufacturing activity,[b] 1978–1990

	Canada	US	Japan	Australia	Austria	France	Germany (FRG)	UK
1978	19.0	22.7	25.6	19.6	26.4	24.9	33.2	26.1
1979	19.0	22.7	26.3	19.81	26.8	24.8	33.5	25.0
1980	17.8	21.8	25.9	19.7	26.7	24.2	32.5	23.2
1981	17.8	21.7	27.3	19.4	26.4	23.7	32.1	21.4
1982	16.0	20.9	27.4	18.4	26.3	23.3	31.3	21.1
1983	16.6	21.3	27.9	17.9	26.1	23.3	31.2	20.4
1984	17.6	22.3	30.0	17.6	26.5	22.6	31.8	20.5
1985	17.7	22.4	29.5	17.3	26.9	22.1	31.4	20.7
1986	17.3	22.3	28.0	17.8	26.7	21.0	31.4	20.7
1987	17.4	22.4	28.9	17.5	26.0	20.8	30.4	19.8
1988	17.5	–	29.7	17.6	27.0	21.0	30.3	–
1989	17.5	–	30.6	17.4	27.4	21.1	30.4	–
1990	16.1	–	31.3	16.7	27.9	20.9	30.4	–

[a]Canada at 1986 prices; United States at 1980 prices; Australia at 1984–85 prices; France at 1980 prices and United Kingdom al current prices
[b]The manufacturing activities include: (1) food, beverages and tobacco; (2) textile. wearing apparel and leather industries; (3) wood, and wood products, including furniture; (4) paper and paper products, printing and publishing; (5) chemicals and chemical petroleum, coal, rubber and plastic products; (6) non-metallic mineral products, (7) bask metal industries; (8) fabricated metal products, machinery and equipment; and (9) other manufacturing industries
Source OECD, *National Accounts: Detailed Tables, 1978–1990,* volume 2 (1992c)

And, most of the economic growth in the manufacturing sector, in terms of value added, occurs in the knowledge-based industries.

Secondly, connected with these changes is a persistent modification in the kind of employment activity typical of the manufacturing sector. Most available official statistics on employment reflect only insufficiently continuing changes in predominant occupational skills and employment patterns in manufacturing. However, information on the percentage of "administrative, technical and clerical workers" in the manufacturing industries, for example in Great Britain between 1959 and 1982, indicate an appreciable shift in the balance of occupations within the manufacturing sector (cf. Cutler et al. 1986: 77). In the case of other countries for which the same type of information is available, consistent shifts toward a larger proportion of "administrative, technical and clerical" employees as a proportion of all employees in manufacturing can be observed (Table 16.2) (cf. also Roach 1991).[13] None the less, these figures should only be considered as representing a crude quantitative

[13] The OECD statistics on the proportion of "administrative, technical and clerical" employees are limited to a few of the member countries. Statistics for Germany and France as well as a number of other countries are not available or limited to just a few years of reporting.

approximation and likely significant underestimation of a trend toward knowledge-based manufacturing and therefore a demand for skills very much unlike those traditionally expected and practiced in industry.[14] Most importantly, these figures do not show the transformation of job skills and tasks faced by employees in manufacturing who are not, based on conventional typologies, classified as "white-collar workers" or as "administrative, technical and clerical workers". A disaggregation of the information for the manufacturing sector as a whole would, in addition, show that the scale of the shift toward a greater proportion of "administrative, technical and clerical" personnel and the extent of the substitution of knowledge-based occupations depends on the type of industry.[15] But what is needed even more urgently, independent of conventional occupational labels, is a detailed examination of the actual work tasks carried out by employees in industry.

Moreover, the proportion of the production cost in the industrial sector which accrues to knowledge grows and assumes remarkable proportions. In other words, the major source of revenue increasingly comes from software rather than the production of hardware. As a case in point, and as Drucker (1986: 778) points out: the manufacturing costs of the semiconductor microchip are about 70 percent knowledge —that is, research, development and testing—and no more than 12 percent labor. Similarly with prescription drugs, labor represents no more than 15 percent, with knowledge representing almost 50 percent. By contrast, in the most fully robotized automobile plant labor would still account for 20 or 25 percent of the costs.

A distinction which becomes quite central in this respect, therefore, is between technical means of production characteristic of industrial society which were, for the most part, experience-based or craft-based technologies and the means of production of the knowledge society which are based, also for the most part, on a 'scientification' of skills. New industries within the manufacturing sector will be derivative of knowledge rather than experience. The skills most valued in manufacturing will not be (practical) experience but systematic knowledge.

[14] By the same token, Wolff/Baumol (1989: Table 3) estimate that the total percentage of "*information workers*" in the "non-durable manufacturing" (and "durable manufacturing") sector of the US economy increased from 28.0 (32.0) per cent in 1960 to 35.8 (38.2) percent in 1980. Wolff/Baumol (1989: 33) indicate also that a substantial portion of the total increase in the number of information workers, namely 53 percent, is the result of the *substitution* of information workers for non-information employees in production.

[15] The classification of employees in industry into "administrative, technical and clerical" and "operatives" has been used for a considerable time in the British *Census of Production*. Census returns indicate that the ratio of administrative, technical and clerical employees to operatives in British industries already begins to rise in the 1920s and 1930s. In 1924 the ratio is 9.5 and in 1935 it has risen to 12.9 (Florence 1948: 142). Disaggregating these figures already indicates that the shift is least pronounced in coal mining, metal mining, jute, slate quarrying, tin-plate, non-metal mining, cotton spinning, linen, wool, brick making, coke, shoes, china, shipbuilding and building, while the highest ratios may be found in the 1920s and 1930s, in oil and tallow, ink, gum and sealing-wax, starch and polish, brush wholesale bottling, chemicals, electrical engineering, aerated waters, fur, butter, preserved foods, scientific instruments, musical instruments, aircraft, plate and jewelry and mechanical engineering (see Florence 1948: 142).

Table 16.2 Percentage of administrative, technical and clerical workers in manufacturing industries in selected industrialized countries, 1962–1989

	Canada	US	Japan	Denmark	UK
1962	–	25.9	25.1	21.7[a]	22.6
1965	25.6	25.6	26.8	22.9	23.4
1970	27.5	27.5	28.4	24.6	26.5
1975	29.4	28.8	33.1	27.2	27.8
1980	30.1	29.9	34.1	28.1	30.0
1985	–	32.0	34.3	29.4	–
1989	–	31.8	35.3	35.8	–

[a]1963

Source OECD, *Labour Force Statistics, 1963–1983* (1985); OECD, *Labour Force Statistics, 1969–1989* (1991b)

The analysis of employment trends exclusively based on the number of *jobs* or the aggregate of employment and unemployment has the disadvantage of failing to capture important transformations in the form, content and volume of work. Information about the number of jobs does not reflect changes in the type of work; for example, a growth in part-time work. If one is interested in the volume of work an economy requires or provides and how the volume of work in different sectors of the economy may have changed over time, a better measure to consider is total number of hours actually worked (cf. Block 1987: 130–134).

The total aggregate number of hours worked, based on the actual number of hours worked in different sectors, has steadily declined in Germany since 1960 (Table 16.3).[16] The volume of work has been reduced during the past thirty years by 20 percent. The only sector that has experienced an increase in the number of hours worked is the state sector. However, the initial rapid increase in the 1960s has slowed since 1980 and stagnates at the present. Aside from the agricultural sector of the economy in Germany, the number of hours worked in the manufacturing sector during the past three decades has declined by a third while it has remained at the same level in the (private) service sector. The data presented demonstrate how important it is to separate, within the service sector, what rarely is disaggregated, namely state employment and employment trends in private service industries.

The 'scientification' of the manufacturing (and the service) sector of the economy is reflected in a dramatic shift in the type of investment and therefore the kind of capital formation typical of industry. Such trends can be documented despite the fact that, as I have already indicated, the conventional accounting of capital formation is quite restrictive. Investments only count as capital formation if the investment involves tangible capital, that is, investments in either machinery or physical plant. Such a procedure underestimates, to a considerable extent, the degree and volume of likely structural shifts in investments in a knowledge society.

[16] A consideration which may play a role in the future, though it clearly did not in the past, is that further reductions in working hours may be negotiated or legislated as a means of securing places of work for the unemployed (cf. also Hinrichs et al. 1988).

Table 16.3 Aggregate hours of work by sector in Germany (FRG), 1960–1991 (millions of hours)

	Agriculture		Manufacturing		Service		State	
	Total	%	Total	%	Total	%	Total	%
1960	8478	(100)	26 101	(100)	17 032	(100)	4259	(100)
1965	7027	(84)	25 868	(99)	16 894	(99)	5172	(121)
1970	5141	(61)	24 632	(95)	16 343	(96)	4468	(105)
1975	3909	(46)	20 438	(79)	15 869	(93)	6105	(143)
1980	3084	(37)	19 922	(77)	16 355	(96)	6553	(154)
1985	2943	(34)	17 379	(67)	16 226	(95)	6735	(159)
1990	1973	(23)	18 186	(70)	17 148	(101)	6754	(159)
1991	1898	(22)	18 359	(70)	17 728	(104)	6711	(158)

Source Kalmbach (1988: 174) and private communication

Given the conventional approach, in the United States, between 1983 and 1988, the manufacturing sector capital formation was quite low; however, "this sector's spending on information technology has risen at about an 8 per cent average since 1982" (Roach 1991: 120). These gains were sufficient to account for the entire additional capital formation in the industrial sector in the United States. By the same token, within industry, the technology intensity declines significantly with commodity-based manufacturing.

These changes, characteristic of the change from industrial society to knowledge society, and the different systems of production in each, are differences in the wake of growing mutual dependence and integration during the twentieth century of scientific knowledge and technical practices and objects (Böhme et al. 1978). Production becomes increasingly an extension and specification of laboratory knowledge and leads to the construction of 'idealized' technical objects, which makes it possible to transcend, to some extent at least, more technical and craft-based knowledge in production (cf. Layton 1976; Channell 1982). The greater the comprehensiveness of scientific models and idealized technical objects, the more scope they offer for rationalizing technical expertise and practices since they provide a measure of efficiency for machines, and hence designs, which are not derived from current practices" (Whitley 1988: 394). Estimates about the impact of these developments on the production process, changes within sectors of the economy or employment patterns which utilize rather conventional statistical data, for example, the relative presence of "white-collar" occupations within the manufacturing or service sector of the economy, likely tend to seriously underestimate the transformations because these categories themselves, inherited from industrial society, become obsolete.

Moreover, and also contrary to many conventional assumptions, the normative and material differentiation of economic sectors in modern society is by no means the most significant or important observation one is forced to advance on the basis of the evidence about the pattern of interrelations among economic sectors. For in this instance too, linkage, complementarity or interdependence among sectors is the more incisive observation. For example, service-type inputs are needed at every stage of the production process in the manufacturing sector, or the delivery of most service-type jobs directly or indirectly requires products which originate in the industrial sector of the economy.

In many respects, therefore, the well-established conceptual distinction between the economic sectors, especially between the service and the manufacturing sectors is misleading, at least under contemporary conditions. More specifically, most *goods* purchased are intended to provide a service or a function and there are few 'pure' *services* unconnected to certain commodities. The distinction between goods and services becomes even more ambivalent when one takes into consideration the option of the consumer to purchase, lease or rent (and thereby enjoy the service of) a commodity. In short, "the output of economic activity may range from that of pure goods to pure services. However, most—and indeed an increasing proportion of—goods embody some non-factor intermediate services, and most services embody some intermediate goods" (Dunning 1989: 4).

16.3 Production Against Employment

The future of work has been a central concern of social theorists for centuries ever since employment in the sense of a commitment to paid work, in eighteenth-century Europe, became not only an existential imperative but an emblem of civilization. In the eighteenth and nineteenth centuries, the notion of the future of work for most social theorists meant an emancipation *from* the worst physical drudgery and toils associated with work at the time, namely a reduction in work time, greater autonomy and generally more challenging work tasks, whereby liberation from work was thought to be more likely than emancipation *within* the work context. Indeed, the emancipatory potential of industrialization has been fulfilled in the sense that the average work week has been reduced dramatically from more than 80 to less than 40 h.[17] But in this century too, skeptics remained convinced that workers would find self-fulfillment outside work (e.g. Friedmann [1956] 1992). That is to say, the standard account of the (inherent) constraints of production and the implicit coercive if not exploitative consequences of a concentration of ownership continued to provide the apparently persuasive background for the thesis that labor in the end is really only another *means of production.*

In the context of the theory of post-industrial society (Bell 1976: 148–149), however, new hopes seem to be raised about the chances of an emancipation of workers within the context of work because post-industrial society is a *communal* society in which the social unit is the community organization rather than the individual, a world in which the modalities are "cooperation and reciprocity rather than coordination and hierarchy". In the salient experience of work, "men live more and

[17] A consideration which may play a role in the future, though it clearly did not in the past, is that further reductions in working hours may be negotiated or legislated as a means of securing places of work for the unemployed (cf. also Hinrichs et al. 1988).

more outside nature, and less and less with machinery and things; they live with, and encounter only, one another". The question and the reality of any changes in the *quality* of the work context and the demands placed on workers, for example in terms of skill requirements, will be briefly discussed in the next section and at greater length once I analyze the fastest growing segment of the labor force in modern society, namely the group of knowledge-based occupations or "experts, counselors and advisers".

The issue which should be taken up first concerns the question of the future of work in a different sense, though not new to discussions about labor in this century, namely the relative *scarcity of work,* the threat of persistent secular unemployment and therefore the much strained relation between economic growth and full employment. The grim question becomes whether, in a knowledge-intensive economy, technology and knowledge not merely eliminate jobs but also work since not so long ago the much more positive conclusion reached after intensive study was that technology destroys jobs but not work (cf. United States National Commission on Technology, Automation and Economic Progress 1966).

The specific change I have in mind concerns the extent to which employment, especially but not only in the manufacturing sector of the economy, ceases to be a (positive) function of output in this sector; that is, "increased manufacturing production in developing countries has actually come to mean *decreasing* blue-collar employment. As a consequence, labor costs are becoming less and less important as a 'comparative cost' and as a factor in competition" (Drucker 1986: 775). In traditional terms, this development is of course a reflection of increased productivity in the manufacturing sector and therefore of a decrease in the labor/output ratio.[18] The output of the manufacturing sector in advanced economies increases and retains its relative economic importance while its contribution to employment declines. Drucker (1986: 776) predicts, therefore, that developed countries will in twenty-five years "employ no larger a proportion of the labor force in manufacturing than developed countries now employ in farming—at most, ten percent".

But the uncoupling of production from labor is more general since the overall increase in the number of unemployed persons has been soaring in the past two decades. The traditional close link between output and employment ceases to accompany shifts in the economy and creates the 'paradox' of growth and unemployment (cf. Therborn 1986). While the evidence in this regard for the manufacturing sector is clear and can draw on a heritage of information dating back a number of decades, there is no comparable experience for the service sector and its growing efforts to search for efficiencies. For the time being, productivity gains in

[18] A study prepared for the United States Department of Labor (1984) documents these developments for the four industries of hosiery, folding paperboard boxes, metal cans awl cleaning for the period 1962–1982. In each of the four industries, advances in technology ate responsible for a significant increase in productivity, a decline in employment and a rise in the overall output as well as the output per person employed in these industries. During this period, employment in hosiery manufacturing in the US declined by 41 percent, in metal cans 16 percent, and in laundry and cleaning the decline amounted to 34 percent in twenty-two years.

the area of service-type occupations appear to be confined to the manufacturing sector.[19] However, it is very likely that the search for efficiencies will be concentrated in the service sector in the future.

Between 1970 and 1989, the number of unemployed increased from 10 million to more than 25 million in the Organization for Economic Cooperation and Development (OECD) countries (Table 16.4).[20] Unemployment rates and patterns over time differ across countries (cf. Therborn 1986) because similar economic policies and constraints or straightforward patterns of persistent growth during this period cannot be found in all countries. According to OECD figures, in the United States, for example, the unemployment rate during the same period actually declined, while in Japan it continued to remain virtually unchanged (see Table 6.4). However, in absolute terms, the number of the unemployed in Japan doubled; it increased from 680,000 in 1973 to 1,420,000 in 1989 and 1,390,000 in 1991.

One might of course, in the light of these figures, be prompted to offer the entirely correct observation that, despite the increase in unemployment in the past three decades in OECD countries, both the *total labor force* (that is, the number of people registered as working or available for work) and the number of individuals employed or *total employment* has increased substantially. The dramatic increase in unemployment rates, it could be noted further, might therefore at least also be the result of a rapid growth in the supply of labor. However, a growing supply of labor must not invariably, as least as far as relevant historical precedents in the United States, Japan and Germany are concerned, go hand in hand with rising unemployment. A rapid growth in the labor force has been associated with sustained rates of low unemployment. Yet, given the now widely institutionalized expectation in advanced industrial societies that the state and the economy must find ways to guarantee their citizens acceptable and perhaps steadily improving standards of living (cf. Dahrendorf 1987: 110–111), growing and significant rates of unemployment constitute a serious political challenge to governments. Moreover, sustained and consequential unemployment represents a critical challenge to the maintenance of

[19] According to results reported by Roach (1991: 119), who defines service-type occupations as while-collar jobs, in the United States the "service industries could have doubled their white-collar productivity growth over the past five years if they had matched the employment efficiencies realized in manufacturing" during the same period.

[20] There is a persisting controversy among economists whether the rise in unemployment in the past three decades is also the result of a structural transformation of the economy and especially the outcome of 'technological' changes. Some recent econometric studies (e.g. Jackman/Roper 1987) report that structural changes do not account for the rise in unemployment in the 1970s and 1980s or that such changes represent but a small portion of the total variance (Layard/Nickell 1985). In the end, however, these studies do really manage to demystify the reasons for the dramatic rise in unemployment. According to the theory of long-term, Kondratiev economic cycles, technical innovations have long been associated with long-run economic waves. Each cycle lasts about fifty to sixty years. At the present time, we are in a 'downswing' of the wave during which "information technology" matures and cost-saving investments increasingly, as has been the case before in previous cycles, tend to displace labor (cf. Freeman 1979). However, these conceptions are much too crude to allow for sensible inferences about the impact of technology on economic activities, let alone the task of forecasting its effect on employment.

Table 16.4 Annual unemployment rates for selected industrial countries, 1960–1993

	1960	1965	1970	1975	1980	1985	1991	1993
Canada	7.0	3.9	5.6	6.9	7.4	10.4	10.2	10.0
United States	5.5	4.5	4.8	8.3	7.0	7.1	6.6	6.5
Japan	1.7	1.2	1.2	1.9	2.0	2.6	2.2	2.3
Australia	–	–	1.6	4.9	6.1	8.2	9.6	9.9
Germany (FRO)	1.0	0.4	0.8	3.6	2.9	7.2	4.5	4.8
France	0.7	1.5	2.5	4.0	6.3	10.2	9.4	9.8
Spain	2.4	1.5	–	3.0[a]	11.1	21.1	15.2	–
Belgium	3.1	1.6	2.1	5.0	8.8	11.3	9.3	9.6
United Kingdom	1.6	1.2	3.0	4.3	6.4	11.2	9.4	9.4

The rates are not strictly comparable among countries (cf. for example, Freeman et al. 1982: 4–5)
[a]1974
Sources United Nations, *Economic Survey of Europe in 1991–1992* (1992); OECD, *Main Economic Indicators: Historical Statistics, 1969–1988* (1990); OECD, *Employment Outlook* (1992b)

social citizenship rights. In short, it is possible that we are confronted with a new and sustained volume as well as a new structure of unemployment and novel political consequences.

The nature of unemployment has definitely changed in the past two decades. The number of persons who have been long-term unemployed (given the standard definition, those unemployed for more than one year) has tripled to 9.4 million. The increase in both the absolute number of long-term unemployed as well as the rapid increase in their share of total unemployment was one of the main transformations of the labor market in the 1980s. At the beginning of that decade, the proportion of long-term unemployed in OECD countries was on average around a quarter of the total unemployed. By the end of the decade, it had risen to about a third. The rise and the persistence of long-term unemployment is both related to, but also independent of, the overall unemployment rate. During the 1980s, long-term unemployment grew despite economic recovery and expansion during the later part of the decade. Obviously, the personal and social costs of long-term unemployment are enormous. In the case of many countries of the European Union, high levels of unemployment are now often associated with declining chances of finding a job at all (cf. Bean 1990). Thus, if the experience of the 1980s is any indication of future patterns of unemployment, the incidence of long-term unemployment will continue at a high level, or even increase in times of more significant economic downturn *and* growth. The rise and the persistence of long-term unemployment is likely due to a number of factors,[21] such as the increase in nonstandard work (part-time and short-term employment), and produces different patterns of exposure to long-term

[21] For some economists (e.g. Rahman/Gera 1990: 2), however, the matter is much more straightforward: "One major reason for the increase in the duration of unemployment is the increase in the proportion of individuals experiencing prolonged unemployment."

unemployment depending, for example, on the age, education, gender or regional residence of workers.[22] In the long run, however, the most significant factor may well be reduced requirement for labor as a result of fundamental transformations in the economy preventing entry into the workforce in the first place, changing skill requirements displacing workers for extensive periods of time and a mismatch between competencies in demand and those available among unemployed persons.

None the less, as the figures in Table 16.5 indicate there are some countries in which long-term unemployment is relatively low.[23] This is the case for some Scandinavian countries, Canada and the United States. In Japan, the long-term unemployment rate is about three times higher than it is in Canada. Whether these differences in long-term unemployment rates are primarily the result of distinct accounting (or state-labor market)[24] regimes, or whether they reflect, in North America for example, as the OECD (1991a: 41) maintains, differential rates of flows into and out of unemployment, is open to debate. Across OECD countries, as Kolberg/Kolstad (1992: 185) observe, the incidence of "long-term unemployment follows the general level of unemployment closely, and seems to be independent of the sex- and age-specific distributions of unemployment." This allows, it seems to me, the interpretation that long-term unemployment reflects a specific secular trend in the economy of these countries and will not disappear with the regular economic fluctuations (cf. Walsh 1987). The increase in unemployment and long-term unemployment, at this point referring to the management of the process of entering and exiting the labor market only, will likely hasten the decline in the autonomy of market forces and increase the *fusion* of welfare-state regimes and the labor market. This has consequences both for our understanding of the welfare state and the principles which govern economic conduct (cf. Kolberg/Esping-Andersen 1992). Part of the same structural change in employment patterns is the rapid increase in the proportion of part-time workers and temporary employees as a proportion of the total labor force, suggesting an increase in the 'precariousness' and 'flexibility' of employment patterns (cf. OECD 1991a: 53).

---

[22] According to OECD (1992b: 271) figures, during the 1980s, the proportion of relatively young men (25–44 years of age) among the long-term unemployed in most industrialized countries rose (the proportion of 'youths' under 25 years of age and older workers tended to decline) while the ratio of men to women among the long-term unemployed remained, for the most part, fairly stable.

[23] The standard measure for long-term unemployment is twelve months. The available information about longer durations of unemployment confirms the trend about a rising proportion of people who are out of work for extensive periods of time.

[24] For example, while Belgium reports one of the highest incidences of long-term unemployment, not only is it without a waiting period after which insurance benefits commence, but it also provides an unlimited unemployment benefit duration; in the United States, in contrast, the maximum duration for unemployment benefits is 26 weeks and in Sweden it is 60 weeks (cf. OECD 1991a: 200–201). Thus, it is not surprising that there is an overall close correlation between the duration of benefits and the duration of unemployment (United Nations 1991: 201) and that economists begin to refer to the existence of a "culture of unemployment".

The data on which most of these observations are based are aggregate unemployment rates. The structural composition of unemployed individuals according to different criteria and on reasons for unemployment are not easy to obtain; nor is the comparative analysis of unemployment rates without serious problems since unemployment is conceptualized and politically managed in very different ways across OECD countries. Aside from national unemployment regimes, for example, the *links* between the welfare state and the labor market, the result is that both the level and the development of unemployment rates will be structured differently. However, this does not mean that observers are completely immobilized in discerning certain fundamental employment changes.[25]

But it is not the welfare state which contrives, especially in the strong sense of producing, such a 'secular' or structural trend in unemployment rates, for example, as some social critics maintain, by offering disincentives to work; nor is the welfare state any more in a position today to achieve full employment with the aid of national aggregate fiscal and monetary policies, as envisioned say by Keynesian economic policies (cf. Stehr 1992). The decline in employment in both the agricultural and industrial sectors of the economy in developed economies is taken for granted by economists. Further, it is assumed that this pattern will continue. The question therefore becomes, given the premise of the three-sector differentiation, whether the service sector is able to compensate for the losses with employment growth and how significant its ability will be to create jobs. Baumol's (1967) answer is a negative one.[26] But the weakness of his argument may well rest with introducing the notion of the service sector as a black box (cf. Scharpf 1988; Esping-Andersen 1992).

In any event, the degree to which economists and other social scientists as well as policy-makers are seriously alarmed about rising and persistent unemployment in advanced industrial societies in recent decades depends on the degree of faith they display in the compensatory significance and the efficacy of market mechanisms (or, for that matter, intervening policies) to secure conditions approaching full employment and therefore an equilibrium in the demand and supply of labor. But even from a purely economic point of view, it is of course widely recognized that markets

[25] Among the most remarkable differences in unemployment rates is that between *skill levels*. In Great Britain (1985) and the USA (1987), for example, the unemployment rates of semi- and unskilled workers was approximately four times that of professional and managerial workers (see Layard et al. 1991: 286–287).

[26] Baumol (1967) distinguishes between economic activities in which productivity is relatively constant and technologically progressive activities in which a cumulative rise in productivity can be observed. The former class of activities includes, for the most part, service-type jobs in which labor is an end in itself. Such a differentiation between sector productivity is also part of Jean Fourastié's (1950) classification of economic sectors. In addition, Baumol assumes that labor costs of both types of activities increase in concert. It follows that unbalanced productivity growth threatens to drive many services, as ultimately too costly, from the market (including a growth in self-service activities), or the state has to finance (subsidize) these services.

Table 16.5 Long-term unemployment in industrialized countries, 1973–1991 (percentage of unemployed workers out of work for more than 12 months)

	1973	1979	1983	1986	1989	1991
Belgium	51.0	58.0	62.8	68.9	76.3	–
France	21.6	30.3	42.6	47.8	43.9	3.73
Germany (FRG)	8.5	19.9	28.5	32.0	49.0	–
Italy	–	35.8	41.9	56.4	70.4	–
Netherlands	12.8	27.1	43.7	56.3	49.9	–
United Kingdom	26.9	24.5	36.2	41.1	40.8	–
Sweden	–	–	10.3	–	6.5	–
Canada	–	3.5	9.5	10.9	6.8	7.2
United States	3.3	4.2	13.3	8.7	5.7	6.3
Japan	–	16.5	15.5	17.2	18.7	17.9
Australia	–	18.1	27.5	27.5	23.0	24.9

Source OECD, *Economic Outlook* (1992a) and *Employment Outlook* (1992b)

in general and labor markets in particular are far from perfect (and economic policies are distant from aiding the self-adjusting mechanism of the market). Thus, the 'lags' and disequilibrium in labor markets already observed for an extensive period may not only constitute a prolonged stalemate because the conditions responsible are not merely frictional, cyclical and short term, but the labor market dislocations could last for an unprecedented time and even grow considerably.[27]

It may be objected that the experience with profound labor market dislocations and shrinking employment due to technological changes is not really novel. The plight of the coal, steel and textile industries in many countries in recent decades provides relevant examples; similarly, technological inventions in the nineteenth century produced the first such severe shocks and labor market dislocations. But each time, discussions about the specter of technological unemployment abated.[28] Concern usually diminished as each of the dislocations was absorbed and *compensated* for by re-organizations, migration, a growing demand for labor in other industries and sectors, re-training and other measures.[29] The present problem,

[27] In one study, designed to estimate the economic effects of computer-based automation technology on the American economy, Leontief (1985: 39) forecasts, assuming considerable investment into the new technologies, that the labor force required in the year 2000 would be significantly smaller than today. The new technology would require approximately 20 million fewer workers to produce the same bill of goods. Similarly, the composition of the labor force would be different; the proportion of professionals would rise while the percentage of managers and clerical workers would decline.

[28] Early discussion of the impact of technological change on employment may be found in Hobson (1910), Lederer ([1931] 1938) and Salz (1932).

[29] Compensation theories suggest, as Marx for example already maintained in the first volume of *Das Kapital,* that technological progress, in the final analysis, adds to the aggregate quantity of work (cf. zur Nedden 1930). But chronic, persistent unemployment invariably undermines the intellectual and political status of compensation theories (e.g. Salz 1932: 1608).

therefore, boils down to the question, as Lowe (1986: 3) for example asks, "whether the employment effect of the new technology basically differs from those earlier impacts".

There is serious reason, I believe, to assume that restoring full employment—and I do not mean to refer to the more demanding traditional definition of such a state—is no longer feasible in knowledge societies. Economic processes which in the past may have compensated for severe dislocations of the labor market cannot be counted on anymore. Because the technological changes under way will ultimate that impact with special force in the private and public service sector, this sector is no longer capable or can no longer be expected to absorb displaced employees from the other economic sectors, but will itself contribute to a decline in the quantity of available work.[30] Governments and companies in almost every field of economic activity are forced and determined to do more with fewer employees. The repercussions of such a development are considerable. They are significant because full-time paid labor in industrial society was not merely a matter of existential necessity but basic to citizenship rights of individuals and because the volume of the compensatory fiscal activities of the welfare state is dependent on the employment performance of the economy.[31]

16.4 The Social Anatomy of Work

A discussion of the relevant changes of labor and of the workplace in the economy of knowledge societies may be separated into two more or less distinct considerations. First, there is the question of the *quantity* of labor which likely obtains in knowledge societies, and associated with specific structural trends in available work, changes in the meaning of the social construct of 'work'. I have already discussed this issue briefly. Secondly, there is the question of the *quality* of work activities, particularly the required qualifications, typical work activities and the social organization of production. But in each instance, the most significant common issue is whether labor in knowledge societies primarily consists of an extension of trends established in industrial societies, though some of the conditions which justify speaking of a continuation of entrenched patterns may not be the same any more. In the case of the quality of work activities, for example, one might conclude, although low-skilled manual labor may not be the typical work activity in

[30] The image of the dilemma of the laboring society increasingly losing work but also of being emancipated from the burden of work may be found in Arendt ([1960] 1981).

[31] Hinrichs and his colleagues (1988) offer a number of resolutions and options to this dilemma which do not simply either reduce the legal claims individuals are allowed to make on the income-maintenance programs of the state or increase the contributions of employers and those still employed to the welfare state. The option which assumes particular significance in the context of their considerations is the working time and the possibility of devising ways to reduce it across the board in order to generate additional employment.

knowledge societies anymore, that so-called "intellectual labor" is subject to the same processes of rationalization, coercion and control that affected manual labor in industrial society. In this section, I will concentrate on the second set of issues although a few observations of the social construct of labor are in order.

The meaning associated with the term 'work' today is a product of industrial society. What constitutes work in industrial society is much more narrowly defined than was the case in pre-industrial society. In a number of ways, work activities in industrial society became more clearly separated from non-work activities. The emergent boundaries between the economic and social spheres correspond to the distinction between work and non-work activities: the *spatial* division between the place of work and the location for other types of conduct is among the most important distinction in industrial society. Equally self-evident is the differentiation of work *time* and leisure. Finally, the use of the term 'work' is often restricted to work associated with *employment*, or self-employment.

One of the potentially contentious but also crucial questions of contemporary society is whether the primary meanings associated with the term 'work', especially its narrowness, will or can persist in a world in which work, in the traditional sense of the term, will likely become much more scarce.[32] In knowledge societies, individuals who never join the 'regular' workforce, who are forced out of work, or decide to be unemployed, do not simply drop out of society. They are integrated into society by knowledge as the new principle of sociality. However, they are likely to be mainly *objects* of knowledge and not subjects of knowledge.

It is by no means a novel observation that the social organization of work is changing and that the nature of the change has to do with what originally constituted, at least according to Marx and Engels, the condition for the possibility of the division of labor in society, namely, the separation of labor into manual and intellectual labor (cf. Marx/Engels [1932] 1960: 28). The shift is away from manual to intellectual labor, and therefore to a corresponding increase in the role knowledge and learning play in shaping work and the ability to work.

Within this set of issues, at least two distinct questions may be identified. First, the extent to which, in the course of these changes, the nature of work *activities*, of work organization and experience with work undergoes changes; secondly, the

[32] One social theorist who has reflected on the appropriateness of the narrow, employment-centered term of work and offered suggestions for a broader understanding of work is Enzo Mingione. For example, he proposes that work should include "all types of formal employment, but also a variety of irregular, temporary or occasional activities undertaken to raise cash and various activities that produce use values, goods and services for direct consumption either by the individual and his/her household or other individuals or households, which are more or less necessary for the survival" of the individuals and households (Mingione 1991: 73). One of the purposes of the new broader definition of work is to join conceptually more closely what already has been joined in practice, namely the social and economic spheres of activity. Another consequence of the broader conception, then, is to offer an analysis of the importance of the 'informal' economy. The notion of work, as Sabel (1991: 24) therefore agrees, refers to "such disparate and rapidly changing experiences that it is at least as reasonable to treat the word as a popular shorthand for survival as to regard it as a category of activity that gives similar contours to our different understandings of life".

changing conditions of production raise questions about the relation between work and other social arrangements, for example, social inequality, education, culture, leisure, the family, and their respective boundaries. These issues will be dealt with separately.

As Marx ([1939–1941] 1973: 705) outlines in his *Grundrisse,* with the advance and application of technology and science,[33] the worker is no longer the principal agent of production and what appears to be the mainstay of production and wealth is neither the immediate labor performed by the worker, nor the time that he works, but the appropriation by man of his own general productive force, his understanding of nature and the mastery of it; in a word, the development of the social individual. In contrast, the productive forces of industrial society are based, for the most part, "on the direct labor of workers, measured and exploited in terms of labor time" while the productive forces of advanced society are "based on the capacity of people to learn" (Block/Hirschhorn 1979: 367). The quality or, as Marx formulates it, the "power of the agencies", rather than the quantity of labor and the social organization of work, becomes crucial. For Marx, this transformation already signals the end of bourgeois society and the demise of an economy based on exploitation and of exchange value as the measure of the use value of commodities.

However, as most of the categories used by economists and others still indicate, the unit of production which is still invoked is primarily the *individual* employee and the time spent on the job by the individual; that is, the leading assumption that remains dominant to this day is that the crucial components in the determination of the "quality of labor", for example, are the individual characteristics that workers or employees generally bring to their place of work "rather than of the organizational environment in which the labor is employed" (Block 1985: 441, n. 62). The 'discovery' of the importance of the *social* organization of work dates at least to the 1930s and the Hawthorne studies of Roethlisberger and Dickson; but today the importance of institutional features for productivity and profitability have increased to the point that Block (1985: 95), for example, elevates it to the "determining role" of the factor of labor quality.[34]

One cross-national survey of attitudes to technological change in the workplace, carried out in the second half of 1982 in six industrialized countries, namely, the United States, Britain, West Germany, Sweden, Japan and Israel, indicates that respondents who have experienced technological changes on the job generally appear to rate the effect quite favorably. They report not only that their jobs have

[33] It should be noted that Karl Marx, in tracing the evolution of the capitalist mode of production, was always careful to trace, in contrast to many of his followers, both the 'negative' and the 'emancipating' effects of the forces of production. However, as long as the advance of science and technology occurs within the frame of capitalist relations of production, their development fosters, at the expense of the workers, the profits of the owners of the means of production. After all, for Marx, socialist modes of production do not involve scrapping modern technological means of production (cf. Sohn-Rethel 1978).

[34] As Block (1985: 95–96) therefore asserts, "two factories might be quite similar in the quality of their labor forces and the nature of their capital stock, but their output might differ greatly because of institutional differences that lead in one factory to greater downtime and poorer quality control".

become more interesting, responsible, cleaner, but also involve less physical strain. None the less, a large number of respondents sense that monotony or loneliness, dependence and difficulty have increased. In Japan, unfavorable reactions predominated (cf. Yuchtman-Yaar 1987).

The increased 'scientification' of work has already been emphasized. However, Whitley (1988) has cautioned against excessive estimates of the extent to which new scientific knowledge is transforming working procedures and practices in all sectors of the economy and at all levels of labor. The extent to which labor is affected by new scientific knowledge depends, as Whitley (1988: 415) for example observes, on a range of factors which either facilitate or present obstacles to the adoption of new knowledge. The diffusion of knowledge is, of course, not independent of the "organization and control of occupational practices and expertise, the type of the new knowledge and its relations to current practices and the degree to which the objects and systems being treated are separable from their environment and/or the latter are controllable" (Whitley 1988: 415). At the same time, Whitley's note of caution and his enumeration of factors which influence the 'scientification' of work does not extend to forces which may quite severely constrain an organization or corporation finding itself, not in splendid isolation, but in a competitive environment, as the result of market forces.

In a more general sense, in recent years the focus of the sociology of work, the sociology of organizations and industrial sociology has been on the extent to which the work environment and the social organization of work activities have been transformed enough to justify the conclusion that work in advanced industrial society justifies either the label of a perpetuation of trends already in place in industrial society (perhaps even a worsening of some attributes of work activities), or whether, as a result of the changing working conditions, work and its environment changes dramatically and constitutes a break with the kind of work typical of industrial society.

In the case of the dominant perspective of industrial society in which a certain kind of technological progress or regime is closely linked to mass production systems, intensive productivity gains and the capacity to produce an abundance of goods as well as hierarchical forms of work organization and control, the answer is almost self-evident. In the end, the (capitalist) logic at work always contributes, as emphasized at one time, to a massive alienation of workers, or as argued more recently, to an extensive de-skilling of the work force (e.g. Braverman 1974).[35] The ability of management to preserve and exercise domination is assured by virtue of

[35] Although the de-skilling thesis is often associated with the work of Harry Braverman, there are numerous predecessors, reviewing developments in production processes of industrial society, that conclude, as Schelsky (1954: 20), for example does: "The closer we approach automation, though without ever fully reaching it, the greater the degree to which work becomes spiritless and stressful and the lesser the extent to which it requires interest in technical matters and skills or, even initiative of any sort". However, Schelsky anticipates a further state in the evolution of work in which, after automation has been achieved, the worker will be required to perform highly skilled tasks, for example, in the supervising and controlling highly complex production equipment.

holding on to or monopolizing *knowledge* about the conceptions on which production is based. That is, the successful separation of execution and conception is the key to the control and persistent degradation of the worker. The new version of the oppression thesis also generalizes about the workplace without any credit to the imagination of the worker and specific conditions of work. The thesis minimizes, consistent with Marx's portrait of the labor process in capitalist society, the ability of the worker to affect his or her working conditions. Technological developments simply reproduce the domination of capital over labor, often on a more repressive scale[36] and contribute, as Merton (1947: 80) already expresses it, to an "enforced obsolescence of skills".[37] These observations tend to make inferences about processes from outcome. In general, therefore, the much discussed Braverman thesis, developed decades later, about the irreversible de-skilling process of labor treats the change in the nature of technological paradigms, the specificity of the workforce, the dynamics of work and local conditions as a black box. But one might also ask whether it is something inherent in the process of material production which requires the detachment of planning and execution, or whether the disconnection stems rather from the desire of management to control and exploit labor?[38]

---

[36] The authority for and classic example of Marx's view of the role of technology in the labor process in capitalist society, as outlined in volume 1 of *Capital,* is the alleged displacement of textile workers in England as the result of the introduction of the self-acting mule in cotton spinning. As Marx ([1867] 1967: 435–436) puts it, "machinery not only acts as a competitor who gets the better of the workman, and is constantly on the point of making him superfluous. It is also a power inimical to him, and as such capital proclaims it from the roof tops an as such makes use of it. It is the most powerful weapon for repressing strikes [...] it would be possible to write quite a history of the inventions, made since 1830, for the sole purpose of supplying capital with weapons against revolts of the working-class. At the head of these in importance, stands the self-acting mule, because it opened up a new epoch in the automatic system". Lazonick (1979: 257) has examined the historical evidence in the case of Lancashire in England and concludes that the case of the self-acting mule "does not demonstrate the unfettered triumph of capital over labor through the use of the division of labor and machinery". Marx's misleading portrayal of the effects of this particular invention, it would appear, relied too much on the views of the ideological proponents of the technical change in question.

[37] Merton assumes, as do the proponents of the later de-skilling thesis, that the obsolescence of skills is irreversible. In the light of the kind of production technology used, a compensation process is presumably not considered likely. The increasing employment of labor-saving technology produces the enforced obsolescence of skills among the workers. The social and psychological consequences of discarding acquired skills are mainly connected to the demotion of status (including the possible loss of the public identity of the job) and the destruction of the positive self-image of the worker, stemming from the once confident use of those skills. In short, as Merton (1947: 80) anticipates as well, "alienation of workers from their job and the importance of wages as the chief symbol of social status are both furthered by the absence of a social meaning attributable to the task. Increased specialization of production leads inescapably to a greater need for predictability of work behavior and, therefore, for *increased discipline in the workplace*".

[38] For Adorno (1969) and Gorz ([1971] 1976: 170), for example, the answer can only be, at least as long as one examines the issue from the (interested) point of view of the owner of the means of production, that it is *not* technical progress "in the true sense" which requires hierarchy and a fragmented division of labor in industry but the effort and determination of the class of owners for maximum exploitation. In addition, such aims are not necessarily compatible with the most

Today, fascination with the constraining features of the conditions which give rise to more hierarchy and control have been replaced by equally strong convictions about a *new* technology and a logic of organizing production which is essentially permissive (cf. Hirst/Zeitlin 1991; Sabel 1991: 24). These views are linked in turn to the distinction between the declining regime of mass manufacturing ('Fordism') and the growing system of "flexible specialization" in production (cf. Piore/Sabel 1984).[39] As a result, technology is no longer seen as a dehumanizing force but as one which enables or at least holds the promise of participation in the affairs of work. In the sphere of work, the profound anxiety about the destructive ways of technology are now replaced by animated discussions about the freedoms from control. The vocabulary of intentionality and agency, thought to be obsolete, reappears in discussions of work, production and the social organization of work (e.g. Cavestro 1989). Paradoxically, the technology once feared to have become self-regulating now regulates itself in the sense of negating regulation. None the less, perspectives which emphasize the enabling features of new technology should not commit the same fallacy associated with assertions about the inherently repressive nature of production technologies, namely, to declare, as Karl Marx for example did, that technology invariably reproduces domination. Even if new technologies allow for greater flexibility and require it for greater efficiency, they do not thereby automatically also foreclose the possibility, depending on local circumstances and the nature of (economic and political) 'partnerships' that technology happens to find itself lodged in, that versatility and innovative capacity is restricted in the interest of sustaining hierarchical control of owners and managers.[40] There is not a natural role for technology as such in all of this.

But there is a counter-conception which accepts the fact that the conditions of work have changed but is not convinced that modern technology primarily is enabling and permissive. This perspective asks whether intellectual labor will not be subordinate to the same processes of rationalization and control that affected manual labor in industrial society and create a kind of "intellectual assembly line", that is, a division of labor in which the "rationality of the bureaucratic organization

efficient use of production techniques and work organization. The relations of production still dominate the forces of production.

[39] In much the same sense, Kern/Schumann (1983: 357) sum up some of the relevant findings of their research into rationalization and the conduct of industrial workers by noting that workers increasingly are seen, at least among the more flexible and enlightened segments of management, "as persons with complex abilities and varied developmental potential that one is able to employ with particular effectiveness if one endeavors to utilize their capacities on the job extensively rather than merely minimal segments by discarding the remainder of their ability, in other words, achievement expectations are not lowered but the utilization of the intellectual and motivational capacities is increased".

[40] Summing up a number of case studies in this area of research, Jones (1990: 306) arrives at such a cautionary note when he observes that "in general the prospect of using these systems [flexible specialization] to tighten hierarchical control over final operations may prove more appealing to many managers than the surrender of detailed powers to the shop floor *that is necessary* for versatile and innovative productive capability".

acquires the mechanized efficiency of the factory, and in which mental labor is subjected to both the rationalization of its knowledge and the gradual automation of its productive activity" (Perrole 1986: 111).

The deep divisions in the theoretical assessment of the nature of the impact of new technologies, in this case on work and the organization of work, attest that it is impossible to factor out effects which would indicate that these technologies by themselves are determining factors.[41] The essentially contested concept about the impact of technology also displays, in other words, the variety of *empirical* situations and interpretations about the effects of new technologies to which reference is made.

16.5 From the Employment Society to the Consumption Society

Compared to virtually all other prior historical societies, the modern capitalist societies that emerged in North America and Europe during the eighteenth and nineteenth centuries have been most concerned with the conditions of employment. The extraordinary and consistent preoccupation with paid work and the emergence of paid labor as a social activity separate from the household justify the label of "employment societies" (Keane 1988). Not surprisingly, economic discourse, too, is preoccupied with concerns directly linked to the worlds of production (of goods and services), work and incomes (in the form of rents, interest and wages). In other words, economic discourse today continues to be linked to the eighteenth-century definition of the *major* factors of production, namely, capital and labor, its mix and consequences measured in monetary units. The world of work finds its mirror image in the sphere of consumption. Society produces in order to consume and it consumes in order to produce. For many purposes, such a focus indeed may still be quite appropriate. For example, if one is concerned with the productivity of capital or labor, such an arithmetic is sufficient. Even though in modern societies "less than one sixth of the total time of the average fit adult" (Gershuny 1988: 6) is devoted to paid work, paid labor as a separately institutionalized activity continues to constitute the major social activity of large segments of the population and much of the energies of modern society are still geared toward efforts to constantly expand the production of goods and services.

However, if the focus and the implied equivalence as well as equilibrium between production and consumption shift, both in the case of those who are still part of the realm of labor and those who are not, namely, from work in the narrow sense of the term, to *forms of life* of employees and households in modern society,

[41] That is, Touraine ([1984] 1988: 108) is probably correct to infer from the essentially contested view about the status of new technologies that it is "impossible to isolate a primary cause of technological origin as the determining factor of all programmed society".

then an analysis, as Luhmann (1988: 164–166) for example has emphasized, of the *consumption* side—especially in relation to the total wealth (*Besitzstand)*—is more pertinent than is the mere income of individuals or households.[42] Consumption acquires greater independence from production. Undoubtedly, work provides meaning to consumption. As the aggregate of work shrinks and as the relative amount of time individuals devote in their lifetime to work for which they are compensated is reduced and as consumption is less immediately tied to labor, the meaning of consumption changes as well.

The growth in the total wealth and entitlements of individuals and households has accelerated enormously. Many individuals and households become, through pension funds for example, indirect owners of the means of production as well, though such ownership of course has lost much of its traditional attributes. What is meant here, therefore, is not related to the shift, although quite real, in the quantity of time spent outside the work environment and therefore a shift in central-life interests (Dubin 1956) to leisure activities. The focus still remains with the material or economic well-being of individuals and households; however, forms of life, in as much as they are dependent on material well-being, are not driven any more by considerations directly linked to the value of income but consumption patterns and their determinants, namely the *Besitzstand* of the actor(s). The determinants of the consumption patterns are related to the specific circumstances of the individual and the household unit. Structures of social inequality resonate with such circumstances as well.[43]

Among the outcomes of such a change in the circumstances which affect forms of life and material well-being is a closer link between the economic and social spheres, or a shift in the economic dependency relations of individuals and households. But even more importantly, the specificity of social conflicts in knowledge societies changes dramatically. The displacement of concerns and struggles which primarily revolve around the satisfaction of economic needs, the allocation of monetary income, interests and rents, shifts the locus of major societal conflicts to more generalized and global needs. The primary role in terms of which social struggles take place no longer involves *workers as workers* and the owners of the means of production as capitalists. The locus of the conflicts shifts to the individual as a configuration of roles, or as Touraine ([1984] 1988: 11) puts it, to the *social actor* in any one of his or her roles: "One could almost say that it is the

---

[42] Among sociologists who emphasized rather early the growing importance of the consumer position of individuals in modern society, at the expense of the importance of the occupational position, for their consciousness and status, are Schelsky (1956: 65) and Dahrendorf ([1957] 1959: 273).

[43] Luhmann (1988: 165) illustrates this proposition as follows, "whether one is married or not and whether one has children or not, whether the spouse works or not and whether, as the case may be, one may have to support divorced spouses, whether one lives in an inherited home or has to rent—all these factors contribute more significantly to the economic life chances than collectively agreed upon wage rates or, as the case may be, insurance or pension payments".

human being as living being." One of the principal axes of conflicts pits the consumer against production regimes of all sorts.[44] The generalization of issues contested in societal conflicts deprives knowledge societies of a central locus and arena in which these struggles take place.

16.6 The Emergence of the Symbolic Economy

A further major change in the structure of the economy of post-industrial societies is the emergence of an (internationalized)[45] symbolic economy[46] which 'deals' in monetary and non-monetary symbolic commodities. Drucker (1989: 127) assumes that the symbolic economy, in the form of money flows, already shapes and rivals the transnational material economy.[47]

Initially, the term "symbolic commodities" should be put into quotation marks, lest one simply assumes that the full range of symbolic commodities has *economic, legal* and *practical* qualities not unlike any other commodity, for example durable goods which have a certain utility independent of the specific context in which the product is produced, exchanged or consumed, and a legal status, especially property rights attached to it. None of these attributes applies, at least in the strict sense, to a number of the symbolic commodities. Most importantly perhaps, the identity and utility of the symbolic items are often highly context sensitive and cannot be 'understood' or estimated separate from the context in which they originated and were 'consumed'. The proximity of the context of production and utilization of symbolic commodities is often quite close; the life expectancy of symbolic commodities is fairly limited. Property rights to symbolic commodities are virtually absent. The regulative

[44] More concretely, the new social conflicts have involved consumers in quite a spectacular and many-faceted way. They have spoken out against 'schools or against the university in the name of education, against the scientific-political complex in the name of public good, against hospitals in the name of health, against urban planning in the name of interpersonal relations, against the nuclear industry in the name of ecology (Touraine [1984] 1988: 110).

[45] "This means that few countries, if any, are able to control their own currency. There is a loss of one of the main levers of power and influence" (Bell 1987: 9).

[46] In an analogous sense, Touraine ([1984] 1988: 104) records that the passage to post-industrial society takes place "when investment results in the production of symbolic goods that modify values, needs, representations, far more than in the production of material goods or even of 'services'. Industrial society had transformed the means of production; postindustrial society changes the ends of production, that is, culture".

[47] Following the "currency crisis" in the third week of September 1992, the *New York Times* (September 23, 1992, Section Cl) was prompted to observe that "on a dull day, hundreds of billions of dollars worth of marks, yen, dollars and other currencies change hands, as speculators bet on the direction of currency markets and money managers seek opportunities overseas. On a busy day, volume can top a trillion dollars. That is a lot of money. And as last week proved, the combined power of all these traders can overwhelm the power of governments, even when all of Europe is trying to act in concert. The events provided a bitter reminder to central bankers and finance ministers around the world that the power of governments to control economics and currencies has eroded".

principles which govern market exchanges and intervention into the market do not apply in full force to exchange processes involving symbolic commodities.

Symbolic 'commodities' of a monetary nature, in particular capital movements, cross-rates, exchange rates, interest-rate differentials and credit flows, are to a considerable extent "unconnected to trade—and indeed largely independent of it" and "greatly exceed trade finance" (Drucker 1986: 782); they are more important now for the world economy than the traditional flow of goods and services.[48] The gold exchange standard which operated for much of the life span of industrial society has been replaced by the electronic information system of today.

Symbolic commodities of a non-monetary nature are, for example, data ("sets of numbers"), technological trajectories, statistics, fashion regimes, programs, product marketing and organizational 'knowledge' as well as the growing flow of information within and across national boundaries. The acceleration of the flow of information increases uncertainty; more precisely, it reduces the length of those moments in which certainty appears to prevail. The rapid dissemination of symbolic goods accelerates their obsolescence. In the manufacturing industry, for example, the growing importance of symbolic commodities for the provision of products and their production raises costs and demands larger markets to absorb these expenditures. Commodities and services, to a growing extent, embody knowledge.

Developments of the monetary symbolic economy, changes in its trends and abrupt shifts occur often in response to anticipated political events or are driven by unanticipated crises in different parts of the world. Indeed, not only trade in goods and services is very much affected by the symbolic economy, the dynamics of the symbolic economy often have political repercussions. In addition, in the traditional realm of international trade and services, the movement of *symbolic commodities*, that is, knowledge, has become a more salient factor in the world economy (cf. Dickson 1984: 163–216).

16.7 The Eclipse of Time, Distance and Place

One further significant effect of the production of goods and services more dependent on knowledge is the growing irrelevance of time and place (and therefore distance) as a constraint for production[49]; that is, competitive advantages increasingly are expressed in symbolic terms and such capital is much more mobile within and across national boundaries. The potential for spatial reorganization and the redisposition of

[48] Drucker (1986: 782) provides the following figures to illustrate the claim: "World trade in goods is larger, much larger, than it has ever been before. And so is the 'invisible trade', the trade in services. Together, the two amount to around \$2.5 trillion to \$3.0 trillion a year. But the London Eurodollar market, in which the world's financial institutions borrow and lend to each other, turns over \$300 billion each working day, or \$75 trillion a year, a volume at least 25 times that of world trade."

[49] Geographers, economists, planners and other social scientists have enumerated many of the specific conditions which give rise to the greater "locational capability" of firms and enterprises (cf. Storper/Walker 1983).

time in production, distribution and consumption activities arises from the ability of information technology to 'overcome' time and distance constraints. However, the uneven mosaic of existing spatial divisions of firms and enterprises, that is, the existing high geographical concentration of industries within and among nations, will not suddenly give way to a less structured and unequal location of economic activities. Nor does the increasing emancipation of the productive process from time and place mean that production is no longer taking place within country or region-specific social and political contexts and constraints or that only decentralization effects will be observable. While constraints on place do not disappear altogether, changing spatial and time constraints allow for *many more locational configurations* than was the case under previous regimes with much more restrictive constraints (e.g. Hepworth 1986). As a matter of fact, new constraints including the compression of time are added as production becomes knowledge-based manufacturing,[50] and others, such as the existing services and the infrastructure generally in a particular location, remain significant in decisions to abandon or position enterprises.

None the less, contemporary "locational capacities" of firms and enterprises, although not equally distributed across the range of manufacturing and service industries, have multiplied considerably. Companies have more choice as to where they decide to combine mobile and relatively immobile, i.e. country or region-specific, endowments. Specific decisions of course will depend on a host of factors, for example, the reasons for investing in the first place, the product characteristics, the behavior of competitors, the regulations and policies of host countries and social and cultural factors (cf. Dunning 1989: 33–36). Most importantly, however, the relative eclipse in the importance of scarce locational features, distance and time for productive processes, and in many instances, services, represents a radical inversion of the governing calculus compared to the importance of locational configurations which count for economic production processes in industrial societies. In industrial societies production is in principle still tied closely to location (region) and/or time by virtue of the weight and cost of moving crucial productive ingredients, factors which allow for manufacture of commodities in the first place. In contrast, knowledge, in principle, is highly mobile and travels well. Under the proper conditions, especially in the presence of economic

[50] In terms of the relative contribution to any increase in the locational capacities of firms, the flexibility of production processes, the mobility of capital or other factors *not* directly related to 'labor' costs, such as the decline in transportation and communication costs, may be of particular significance. At the same time, these changes imply increases in the skill requirements of jobs and therefore a relative increase in the importance of labor to location. The locational choices of high-tech firms, for example, are constrained by the need for highly trained labor and different types of labor depending on the kind of production phase (cf. Glasmeier 1990). In a world in which labor is scarce, the increase in the importance of labor to locational decisions would mean a more powerful voice and higher rewards for labor. However, such would not appear to be the case. A leveling in unequal patterns of location would require that labor, too, becomes highly flexible and mobile. Unless one excludes labor almost entirely from production, such a development is unrealistic (cf. Storper/Walker 1983: 34). Existing empirical information, based on conventional classifications of industries, trade patterns and employment structures, are somewhat of a handicap in producing impelling and comparative empirical evidence about changes in locational constraints (cf. Krugman 1991).

incentives, knowledge not only travels well but fast. This also means that an efficient communication infrastructure will be quite important for the economy in the knowledge society (cf. Nicol 1985: 192; Henderson/Castells 1987). The choices of potential contexts for production have multiplied immeasurably and have become, as some economists begin to name it, 'global'. It also means that the kinds of considerations which enter into the determination of the location of production extend beyond those crucial in the past, namely a calculus primarily, though rarely only, driven by considerations of economic efficiency.

Historically, location theories which tried to explain the distribution of employment in manufacturing, for example, have emphasized the constraints related to the costs of transportation, access to the means of production, especially labor and the relative rigidity of production methods. At the same time, the relevant boundaries within contemporary economic discourse still presume that the decisive boundaries are those of the (sovereign) nation-state. The growing irrelevance of location and time for production, distribution and consumption also means that the link between what were once thought to be norms or 'rationalities' of *different* social systems, for example leisure and economics, converge or are confronted in decisions about production facilities. Location is more than merely an allocation problem (cf. Storper/Walker 1983: 34).

In knowledge societies, production and enterprises are, or will be, largely emancipated from the geographical features of a location. The redefinition and rearrangement in the location of enterprises or production are related, on the one hand, to the "enabling developments in the service components of goods production, and information handling and communication technologies" (Britton 1990: 536) and, on the other hand, dramatic changes in the production regimes themselves. Important consequences follow from this. In many instances, the specific location for production, while independent of a certain *natural* geographical location, as was often the case in the past, in fact has to exist or be created in the first place. In that respect at least, the choice of location remains rather closely tied to the idea of a specific context and particular locations, constraints which then account for the spatial division of labor. In the United States, "high-tech industries are likely to be found in states with traditions of innovative manufacturing, and within major metropolitan areas where business services and other urban amenities are ample" (Glasmeier 1990: 73). It is a (socially) constructed context in which these firms decide to locate, a context which can be provided for production, in principle, almost anywhere, especially if one assumes that the calculations which lead to a particular location of economic activities are not based exclusively on economic dimensions.

The efficiency of economic activities that are functionally interdependent will, in the future, not decrease despite increasing physical distance between parts of production activities (cf. Nicol 1985: 198).[51] Similarly, a decentralization of organizational activities should not seriously interfere with the ability to communicate and

[51] As Glasmeier (1990: 73) indicates, in "high-technology industries, the division of labor facilitates such decentralization. High-tech products can be segmented; firms locate technical activities in core regions, but move production to other regions where appropriate pools of labor can be found".

co-ordinate tasks. The ability to more freely divide activities spatially actually turns into an asset for economic activities. As long as the determination of location can emancipate itself from the now dominant close relationship between costs and distance, other factors such as the availability of skills and composition of the labor force will influence locational decisions. As a result, it is probably safe to assume that urban concentration, for example, will not decline; on the contrary, it may continue to increase despite the diminishing importance of the cost of transportation for goods and services.[52]

The growing irrelevance of time to production does not mean that time becomes unimportant altogether either, but rather takes on a very different kind of importance. For example, the possibility of a closer coordination of production schedules, even over a great distance, means that questions of the storage of parts and the like declines, and production is closer to actual needs. It is therefore more important now to have a specific item available at a precise instant, yet to live up to that requirement becomes easier as production schedules 'communicate' with each other.

New production technologies often imply that a major economic factor is not so much the time spent producing, but the time during which equipment is idle, including 'downtime' because of malfunctions. Thus, the irrelevance of the time of year or day in production renews the issue of total time of production and of working hours. Knowledge-based production is more flexible and allows for, some would argue requires, a much greater flexibility in working hours.

The reasons for the irrelevance of time and place for production and the provision of services have to do, on the one hand, with technologies, or better technological regimes, which 'diminish' space and 'shrink' time and, on the other hand, with the qualities of the object that need to be moved in order to produce and in order to consume. Limits to the speed and ease with which the prerequisites of production and the 'products' can be moved are increasingly disappearing. The *enabling* technologies (Dicken 1992: 103) which overcome the limits to movement in industrial society (and generate different frictions of space and time in knowledge societies) are the new media of communication and transportation. For much of human history and a considerable portion of the life span of industrialized society, the speed with which materials, products and individuals were transported was identical to the speed and obstacles faced by entities which had to be communicated across distances. In addition, the costs of both moving tangible and intangible goods was quite sensitive to the distance which had to be traveled and the volume which had to be moved. Today, the mobility of 'information' and the speed with which tangible goods are moved are increasingly at odds. Much of the cost of communication is virtually independent of distance (cf. de Sola Pool 1990: 34–39) and volume, while the cost of the transport of goods is still contingent on distance and volume. Moreover, the cost of communication has fallen sharply. The gap in

---

[52] Florence (1948: 128, 136–140), in his study of *Investment, Location, and Size of Plant,* already anticipates that a decline in transport and communication costs will probably exercise a considerable influence on the mobility of location but these and other technical changes continue to favor a further growth of large cities.

the time, ease and cost it takes to move information and tangible products represents one of the constraints on production or incentives to reduce the amount of tangible entities used in production. In addition, some of the same enabling technologies have altered the rigidities of production regimes and corporate organization, making both potentially more divisible and adding further to the process of emancipating production from constraints of time and place.

16.8 New Limits to Growth

Dahrendorf (1988: 123) makes the point that the 1970s were a time of "enormous exaggeration. The exaggeration of gloom and doom." Not since José Ortega y Gasset's *The Revolt of the Masses* and Oswald Spengler's *Decline of the West* in the late 1920s and early 1930s have so many books been written about the pending descent and dissolution of a way of life. But no title better reflected and symbolized the spirit of the discussion and concerns than the study of the Club of Rome on the *Limits to Growth* (Meadows et al. 1972). The despairing prognosis of the 1972 Report was that present growth trends in world population, industrialization, food production, environmental decay (in particular, pollution) and the exhaustion of natural resources have to come to a halt within the next century.

The thesis that the world will reach the limits of resource availability on a global scale is self-evident or a tautology. For practical purposes, what is relevant is the time scale. And in this respect at least, discussion about the limits of growth, both then and probably now, continues to suffer from simply extrapolating established trends into the future. The limits-to-growth discussion of the early 1970s was, of course, based on certain premises about the nature of the modern productive and distributive process, with trends extended into the (near) future given specific assumptions, especially about not only scarce but finite resources and a growing world population (cf. Meadows et al. 1972). The outcome of such reflections was the conviction that continued economic growth in industrial societies, and efforts of Third World pre-industrial economies to catch up, is not sustainable and will, in fact, soon lead to catastrophe. But these predictions were soon contradicted by competing analyses (e.g. Leontief 1977) and events. But at issue here is not whether economic growth is desirable[53] or whether the ratio of resources to population

[53] One of the consequences of the 1972 report was to stimulate a discussion about the socio-economic principle and desirability of economic growth, in the first instance, as well as debate about the ways in which growth ought to be conceptualized. Vigorous voices reject current economic theory and policy if its orientation boils down to mere efforts to ensure increases in the gross domestic product (GDP). The denial of a growth-orientation is usually based on three considerations: (a) conventional assumptions about economic growth confuse means and ends; (b) they fail to take the reality of the finite state of the planet into account; and (c) the pursuit of such a perspective, paradoxically, assures that some of the very problems, such as unemployment and inflation, that it hopes to cure will actually become worse (e.g. Elkins 1986).

trends and the impact of economic growth on the environment will lead to a sudden reversal in secular advances in economic well-being in the near future, but, rather, the changes in the nature of the productive process itself (not only driven by economic considerations) and the political agenda on any discussion about the limits to economic growth.

One of the crucial deficiencies of the Meadows Report is not so much the notion of constraints on economic activity and patterns of growth, or even 'limits', but concerns the determination of such limits, namely the simple extrapolation of existing trends into the future. Mere extrapolation ignores a whole range of dynamic economic, social and political processes which determine future outcomes, including self-fulfilling and defeating conduct.

The growing centrality of knowledge to the productive process alters the import of certain resources and accelerates the significance of others with different limits. The outcome is that new or different but not necessarily no limits to growth become relevant. One of the commendable outcomes of the *Limits to Growth* has been to affect the agenda of political discourse and policy. The issue of environmental consequences of human activity is now part of the political agenda in many countries.

The changing limits to the growth of national economies or to the global economy also raise the question of the contribution of 'knowledge' to production and increases in output. Available aggregate estimates from economists tend to be fairly imprecise as well as ambivalent; perhaps such figures will never be very precise. One estimate available for the United States credits 'knowledge', which in this instance includes advances in technological, managerial and organizational knowledge, as a source for 54 percent of the total gain in economic growth during the period of 1948–1973 (Denison 1979: 2), while knowledge accounted for only 26 percent of the growth in the years between 1929 and 1948 (Denison 1979). But as the author of these figures points out himself, these percentages are obtained as residual figures "because there is no way to estimate it directly" (Denison 1979: 131). In fact, the economic growth due to knowledge is therefore, following the advice of Solow (1957), merely that "percentage of the measured growth rate in output that cannot be explained by the growth rate of total factor inputs and by other adjustments made for other types of productivity increases" (Feller 1987: 240). Since the different 'variables' typically taken into consideration in these estimates tend to be interrelated, but no theory about their interdependence is available in economic discourse, decompositions of the relative contributions to economic growth only constitute mere illustrations of the growth process (cf. Nelson 1981). For the most part, estimates of the contribution of knowledge (or technology) to economic growth in the long term are just beginning to be researched more comprehensively (e.g. Fagerberg 1988, 1991). For the time being, many dimensions of the use and change produced by knowledge in the economy are not taken into consideration in these estimates. It is, therefore, quite possible that the contribution of knowledge is systematically under-represented to date. And since the estimates are aggregate figures, it is far from clear which sectors and what commodities are knowledge-intensive and which are not; at least these numbers do not allow for any inferences about such questions. In addition, the increased importance of the knowledge factor does not imply that the 'welfare' of society benefits, assuming one has a definite notion of what constitutes a contribution to the welfare of society. But it is entirely possible that much of the growth attributed to knowledge occurs as a result of the production of weapons, other destructive means, commodities which have detrimental environmental impact, nuclear energy or reflect work done in the area of space exploration, all with dubious social utility. In short, the figures need to be much more carefully dissected, although the question of the social utility of economic growth raises difficult, contentious questions (cf. Heilbroner 1973).

16.9 The Fragility of the Future

Although much effort has been invested in the reduction of the contingencies of economic affairs and in the improvement of the possibilities of planning and forecasting, the economy of the knowledge society is, as much as the rest of global society, increasingly subjected to a rise in indeterminacy. While success may at times justify the high hopes of many that techniques and technologies will be developed to reduce if not eliminate much of the uncertainty from economic conduct, sudden and unexpected events almost invariably disconfirm, almost cruelly, such optimistic forecasts about the possibility of anticipating and therefore controlling future events. As a matter of fact, and paradoxically, one of the sources of the growing indeterminacy can be linked directly to the nature of the technological developments designed to achieve greater certainty. The new technology contributes to and accelerates the malleability of specific contexts because of its lower dedication (limitation) to particular functions. Technological developments add to the fragility of economic markets and the need of organizations operating in such a context to become more flexible in order to respond to greater mutability in demand and supply. In the sphere of production, as a result, a new utopian vision arises, a vision which Sabel (1991: 24)[54] sketches in the following and deliberately enabling terms:

> Universal materializing machines replace product-specific capital goods; small and effortlessly re-combinable units of production replace the hierarchies of the mass-production corporation; and the exercise of autonomy required by both the machines and the new organizations produces a new model producer which view of life confounds the distinction between the entrepreneurial manager and the socialist worker-owner.

Much of the standard discussion of these matters, at least until recently, has been animated by opposite expectations. Bell (1973a: 26), for example, confidently

[54] Sabel acknowledges that he has been seen as the major author and therefore responsible for this utopian vision; however, he prefers to subscribe to a more "prudent version of these caricatures". This is a perspective which accounts for the "diversity and similarity of efforts to adjust to the new competitive environment" (Sabel 1991: 24–25).

asserts that the "development of new forecasting and 'mapping' techniques makes possible a novel phase in economic history—the conscious, planned advance of technological change, and therefore the *reduction of indeterminacy* about the economic future" (emphasis added).

But the factor of greater fragility, malleability and volatility is not confined to the economy, the labor market and the social organization of work and management, nor does it merely have 'positive' effects on social relations and individual psyches. Greater vulnerability corresponds to greater fragility and greater flexibility is linked to new regimes of exclusion.

References

Adorno, Theodor W. 1969. "Spätkapitalismus oder Industriegesellschaft", in: Theodor W. Adorno (ed.), *Spätkapitalismus oder Industriegesellschaft*. Verhandlungen des 16. Deutschen Soziologentages. Stuttgart: Ferdinand Enke: 12–26.

Arendt, Hannah. [1960] 1981. *Vita activa, oder vom tätigen Leben*. München: Piper.

Baumol, William J. 1967. "The Macroeconomics of Unbalanced Growth: The Anatomy of Urban Crisis", *American Economic Review,* 57: 415–426.

Baumol, William J., Sue Anne Batey Blackman and Edward N. Wolff. 1985. "Unbalanced Growth Revisited: Asymptotic Stagnancy and New Evidence", *American Economic Review,* 75: 806–817.

Bean, Charles. 1990. *European Unemployment: A Survey*. Centre for Economic Performance Working Paper No. 35. London: London School of Economics.

Bell, Daniel. 1973a. *The Coming of Post-Industrial Society. A Venture in Social Forecasting*. New York: Basic Books.

Bell, Daniel. 1976. *The Cultural Contradictions of Capitalism*. New York: Basic Books.

Bell, Daniel. 1979a. "The Social Framework of the Information Society", in: Michael L. Dertouzos and Joel Moses (eds.), *The Computer Age: A Twenty-Year View*. Cambridge, MA: MIT Press: 163–211.

Bell, Daniel. 1987. "The World and the United States in 2013", *Daedalus,* 116: 1–31.

Block, Fred. 1985. "Postindustrial Development and the Obsolescence of Economic Categories", *Politics and Society,* 14: 416–441.

Block, Fred. 1987. *Revising State Theory: Essays in Politics and the Contradictions of Contemporary Capitalism*. Philadelphia: Temple University Press.

Block, Fred and Larry Hirschhorn. 1979. "New Productive Forces and the Contradictions of Contemporary Capitalism", *Theory and Society,* 17: 363–395.

Bluestone, Barry and Harrison, Bennett. 1982. *The Deindustrialization of America*. New York: Basic Books.

Böhme, Gernot et al. 1978. "The 'Scientification' of Technology", in: Gernot Böhme et al. (eds.), *The Dynamics of Science and Technology*. Dordrecht: D. Reidel: 219–250.

Braverman, Harry. 1974. *Labor and Monopoly Capital: The Degradation of Work in the Twentieth Century*. New York: Monthly Review Press.

Britton, Stephen. 1990. "The Role of Services in Production", *Progress in Human Geography,* 14: 529–546.

Cavestro, William. 1989. "Automation, New Technology and Work Content", in: Stephen Wood (ed.), *The Transformation of Work*? London: Unwin Hyman: 219–234.

Channell, David F. 1982. "The Harmony of Theory and Practice: The Engineering Science of W.J. M. Rankine", *Technology and Culture,* 23: 39–52.

Cohen, Stephen S. and John Zysman. 1987. *Manufacturing Matters. The Myth of the Post-Industrial Economy*. New York: Basic Books.
Cutler, Tony, Karel Williams and John Williams. 1986. *Keynes, Beveridge and Beyond*. London: Routledge and Kegan Paul.
Dahrendorf, Ralf. 1987. "Changing Perceptions of the Role of Government", in: *Organisation for Economic Co-Operation and Development, Interdpendence and Co-Operation in Tomorrow's World*. Paris: OECD: 110–122.
Dahrendorf, Ralf. [1957] 1959. *Class and Class Conflict in Industrialized Society*. Stanford: Stanford University Press.
Dahrendorf, Ralf. 1988. *The Modern Social Conflict. An Essay on the Politics of Liberty*. Berkeley: University of California Press.
de Sola Pool, Ithiel. 1990. *Technologies Without Boundaries. On Telecommunications in a Global Age*. Cambridge, MA: Harvard University Press.
Denison, Edward. 1979. *Accounting for Slower Economic Growth*. Washington, DC: Brookings Institution.
Dicken, Peter. 1992. *Global Shift. The Internationalization of Economic Activity*. Second Edition. New York and London: Guilford Press.
Dickson, David. 1984. *The New Politics of Science*. New York: Pantheon Books.
Dosi, Giovanni, Christopher Freeman, Richard Nelson, Gerald Silverberg, and Luc Soethe (eds.). 1988. *Technical Change and Economic Theory*. London: Pinter.
Drucker, Peter F. 1986. "The Changed World Economy", *Foreign Affairs*, 64: 768–791.
Drucker, Peter F. 1989. *The New Realities: In Government and Politics/In Economics and Business/In Society and World View*. New York: Harper & Row.
Dubin, Robert. 1956. "Industrial Workers' Worlds: A Study of the "Central Life Interests' of Industrial Workers", *Social Problems*, 3: 131–142.
Dunning, John H. 1989. "Transnational Corporations and the Growth of Services: Some Conceptual and Theoretical Issues", *United Nations Centre on Transnational Corporations*. Current Studies Series A, No. 9. New York: United Nations.
Durkheim, Emile. [1893] 1964. *The Division of Labor in Society*. New York: Free Press.
Elkins, Paul. (ed.). 1986. *The Living Economy. A New Economics in the Making*. London: Routledge and Kegan Paul.
Ellul, Jaques. 1954. *The Technological Society*. New York: Vintage.
Esping-Anderson, Gosta. 1992. "Three Postindustrial Employment Regimes", in: Jon E. Kolberg (ed.), *The Welfare State as Employer*. Armonk, NY: M.E. Sharpe: 149–188.
Fagerberg, Jan. 1988. "Why Growth Rates Differ?", in: Giovanni Dosi, Christopher Freeman, Richard Nelson, Gerald Silverberg, and Luc Soethe (eds.), *Technical Change and Economic Theory*. London: Pinter: 432–457.
Fagerberg, Jan. 1991. "Innovation, Catching-Up and Growth", in: *Organisation for Economic Co-Operation and Development, Technology and Productivity*. The Challenge for Economic Policy. Paris: OECD: 37–46.
Feller, Irwin. 1987. "The Economics of Technological Change Filtered Through a Social Knowledge System Framework", *Knowledge*, 9: 233–253.
Florence, P. Sargant (assisted by W. Baldamus). 1948. *Investment, Location, and Size of Plant*. Cambridge: Cambridge University Press.
Floud, Jean. 1971. "A Critique of BELL", *Survey*, 16: 25–37.
Forstner, Helmut and Robert Ballance. 1990. *Competing in a Global Economy. An Empirical Study of Specialization and Trade in Manufactures*. Prepared for the United Nations Industrial Development Organization. London: Unwin Hyman.
Fourastié, Jean. 1950. *Le Grand Espoir du XX siècle*. Paris: PUF.

Freeman, Christopher. 1979. "The Kondratiev Long Waves, Technical Change and Unemployment", in: *Structural Determinants of Employment and Unemployment*. Paris: OECD.

Freeman, Christopher, John Clark, and Luc Soete. 1982. *Unemployment and Technical Change. A Study of Long Waves and Economic Development*. Westport, CT: Greenwood Press.

Friedmann, George. [1956] 1992. *The Anatomy of Work. Labor, Leisure and the Implications of Automation*. With a New Introduction by Donald C. King. New Brunswick, NJ: Transaction Books.

Gershuny, Jonathan I. 1988. *The Social Economics of Post-Industrial Societies*. A Report to the Joseph Rowntree Memorial Trust. Bath: University of Bath.

Glasmeier, Amy. 1990. "High-Tech Policy, High-Tech Realities: The Spatial Distribution of High-Tech Industry in America", in: Jürgen Schmandt and Robert Wilson (eds.), *Growth Policy in the Age of High Technology. The Role of Regions and States*. Boston: Unwin Hyman: 67–96.

Gorz, André. [1971] 1976. "Technology, Technicians and Class Struggle", in: André Gorz (ed.), *The Division of Labour. The Labour Process and Class Struggle in Modern Capitalism*. Hassocks: Harvester: 160–189.

Granovetter, Mark. 1990. "The Old and the New Economic Sociology: A History and an Agenda", in: Roger Friedland and A.F. Robertson (eds.), *Beyond the Marketplace. Rethinking Economy and Society*. New York: Aldine de Gruyter: 89–112.

Gray, Jon. 1988. "Hayek, the Scottish School, and Contemporary Economics", in Gordon C. Winston and Richard F. Teichgraeber III (eds.), *The Boundaries of Economics*. Cambridge: Cambridge University Press: 53–70.

Grilli, Enzo R. and Maw Cheng Yang. 1988. "Primary Commodity Prices, Manufactured Good Prices, and the Terms of Trade of Developing Countries: What the Long Run Shows", *The World Bank Economic Review*, 2: 1–47.

Hayek, Friedrich A. [1945] 1948. "The Use of Knowledge in Society", in Friedrich A. Hayek (ed.), *Individualism and Economic Order*. Chicago: University of Chicago Press: 77–91.

Heilbroner, Robert L. 1973. "Economic Problems of a 'Post-industrial' Society", *Dissent*, 20: 163–176.

Henderson, Jeffrey and Manual Castells (eds.). 1987. *Global Restructuring and Territorial Development*. London: Sage.

Hepworth, Mark. 1986. "The Geography of Technical Change in the Information Economy", *Regional Studies*, 20: 407–424.

Hinrichs, Karl. 1991. "Working-Time Development in West Germany: Departure to a New State", in: Karl Hinrichs, William Roche and Carmen Sirianni (eds.), *Working Time in Transition. The Political Economy of Working Hours in Industrial Nations*. Philadelphia: Temple University Press: 27–59.

Hinrichs, Karl, Claus Offe and Helmut Wiesenthal. 1988. "Time, Money, and Welfare-State Capitalism", in John Keane (ed.), *Civil Society and the State. New European Perspectives*. London and New York: Verso: 221–243.

Hirst, Paul and Jonathan Zeitlin. 1991. "Flexible Specialization Versus Post-Fordism: Theory, Evidence and Policy Implications", *Economy and Society*, 20: 1–56.

Hobson, John A. 1910. *The Industrial System. An Inquiry into Earned and Unearned Income*. London: Green.

International Monetary Fund. 1992. *World Economic Outlook*. May 1992. Washington, DC: The Fund.

Jackman, Richard and S. Roper. 1987. "Structural Unemployment", *Oxford Bulletin of Economic and Statistics*, 49: 9–36.

Jones, Bryn. 1990. "New Production Technology and Work Roles: A Paradox of Flexibility Versus Strategic Control", in: Ray Loveridge and Martyn Pitt (eds.), *The Strategic Management of Technological Innovation*. New York: Wiley: 293–309.

Kalmbach, Peter. 1988. "Der Dienstleistungssektor: Noch immer die große Hoffnung des 20. Jahrhunderts?", in: Werner Süß und Klaus Schroeder (eds.), *Technik und Zukunft. Neue Technologien und ihre Bedeutung für die Gesellschaft*. Opladen: Westdeutscher Verlag: 166–181.

Keane, John. 1988. *Democracy and Civil Society. On the Predicaments of European Socialism, the Prospects for Democracy, and the Problem of Controlling Social and Political Power*. London: Verso.

Kern, Horst and Michael Schumann. 1983. "Arbeit und Sozialcharakter: Alte und neue Konturen", in: Joachim Mattges (ed.), *Krise der Arbeitsgesellschaft*. Verhandlungen des 21. Deutschen Soziologentages in Bamberg 1982. Frankfurt am Main: Campus: 353–365.

Keynes, John M. 1936. *The General Theory of Employment, Interest and Money*. London: Macmillan.

Kolberg, Jon E. and Gosta Esping-Anderson. 1992. "Welfare States and Employment Regimes", in: Jon E. Kolberg (ed.), *Between Work and Citizenship*. Armonk, NY: M. E. Sharpe: 3–35.

Kolberg, Jon E. and Arne Kolstad. 1992. "Unemployment Regimes", in: Jon E. Kolberg (ed.), *Between Work and Citizenship*. Armonk, NY: M. E. Sharpe: 171–192.

Krugman, Paul. 1991. *Geography and Trade*. Cambridge, MA: MIT Press.

Layard, Richard and Stephen Nickell. 1985. *The Causes of British Unemployment*. London: National Institute of Economic and Social Research.

Layard, Richard, Stephen Nickell and Richard Jackman. 1991. *Unemployment. Macroeconomic Performance and the Labour Market*. Oxford: Oxford University Press.

Layton, Edwin T. 1976. "American Ideologies of Science and Technology", *Technology and Culture*, 17: 688–701.

Lazonick, William. 1979. "Industrial Relations and Technical Change: The Case of the Self-acting Mule", *Cambridge Journal of Economics*, 3: 231–262.

Lederer, Emil. [1931] 1938. *Technical Progress and Unemployment: An Inquiry into the Obstacles of Economic Expansion*. London: P.S. King & Son.

Leontief, Wassily. 1985. "The Choice of Technology", *Scientific American*, 252: 37–45.

Leontief, Wassily et al. 1977. *The Future of the World Economy*. New York. Oxford University Press.

Lowe, Adolph. 1986. "The Specter of Technological Unemployment", Workings Papers: Forschungsgruppe "Technologischer Wandel und Beschäftigung". Bremen: Universität Bremen.

Luhmann, Niklas. 1988. *Die Wirtschaft der Gesellschaft*. Frankfurt am Main: Suhrkamp.

Mannheim, Karl. [1935] 1940. *Man and Society in an Age of Reconstruction. Studies in Modern Social Structure*. London: Routledge and Kegan Paul.

Marx, Karl. [1867] 1967 *Capital*. Volume 1. New York: Vintage Books.

Marx, Karl. [1939–1941] 1973. *Grundrisse. Introduction to the Critique of Political Economy*. New York: Vintage Books.

Marx, Karl and Friedrich Engels. [1932] 1960. *Die deutsche Ideologie*. Berlin: Dietz.

Meadows, Donella H., Dennis L. Meadows, Jürgen Randers and William W. Behrens III. 1972. *The Limits to Growth*. New York: Universe Books.

Merton, Robert K. 1947. "The Machine, the Worker and the Engineer", *Science*, 105: 79–84.

Mingione, Enzo. 1991. *Fragmented Societies. A Sociology of Economic Life Beyond the Market Paradigm*. Oxford: Basil Blackwell.

Nelson, Richard R. 1981. "Research on Productivity Growth and Productivity Differentials: Dead Ends and New Departures", *Journal of Economic Literature*, 19: 1029–1064.

Nicol, Lionel. 1985. "Communications Technology: Economic and Spatial Impacts", in: Manuel Castells (ed.), *High Technology, Space and Society*. Beverly Hills, CA: Sage: 191–209.

Offe, Claus. 1984. *Arbeitsgesellschaft: Strukturprobleme und Zukunftsperpektiven*. Frankfurt am Main: Campus Verlag.

Organisation for Economic Co-Operation and Development. 1991. *Employment Outlook*. Paris: OECD.

Organisation for Economic Co-Operation and Development. 1992a. *Economic Outlook. Historical Statistics, 1960–1990*. Paris: OECD.

Organisation for Economic Co-Operation and Development. 1992b. *Employment Outlook*. Paris: OECD.

Organisation for Economic Co-Operation and Development. 1992c. *Structural Change and Industrial Performance. A Seven Country Growth Decomposition Study*. Paris: OECD.

Organisation for Economic Co-Operation and Development. 1992d. *Quarterly Labour Force Statistics*. Number 3. Paris: OECD.

Organisation for Economic Co-Operation and Development. 1992e. *National Accounts: Detailed Tables, 1978–1990*. Volume 2. Paris: OECD.

Perrole, Judith A. 1986. "Intellectual Assembly Lines: The Rationalization of Managerial, Professional, and Technical Work", *Computers and the Social Sciences*, 2: 111–121.

Piore, Michael J. and Charles F. Sabel. 1984. *The Second Industrial Divide*. New York: Basic Books.

Rahman, Syed S. and Surendra Gera. 1990. *Long-Term Unemployment: The Canadian Experience*. Working Paper No. 12. Ottawa: Economic Council of Canada.

Roach, Stephen S. 1991. "Pitfalls on the 'New' Assembly Line: Can Services Learn from Manufacturing?", in: *Organisation for Economic Co-Operation and Development, Technology and Productivity. The Challenge for Economic Policy*. Paris: OECD: 119–129.

Sabel, Charles F. 1991. "Moebius-Strip Organizations and Open Labor Markets: Some Consequences of the Reintegration of Conception and Execution in a Volatile Economy", in: Pierre Bourdieu and James S. Coleman (eds.), *Social Theory for a Changing Society*. Boulder: Westview Press: 23–54.

Salz, Arthur. 1932. "Die Kontrolle des technischen Fortschritts: Technischer Fortschritt und Arbeitslosigkeit", *Der Deutsche Volkswirt*, 6: 1607–1609.

Scharpf, Fritz W. 1988. "Strukturen der post-industriellen Gesellschaft oder: Verschwindet die Massenarbeitslosigkeit in der Dienstleistungs- und Informations-Ökonomie?", *Soziale Welt*, 38: 3–24.

Schelsky, Helmut. 1954. "Zukunftsaspekte der industriellen Gesellschaft", *Merkur*, 8: 13–28.

Schelsky, Helmut. 1956. "Gesellschaftlicher Wandel", *Offene Welt*, 41: 65.

Simonis, Udo E. 1989. "Ecological Modernization of Industrial Society: Three Strategic Elements", *International Social Science Journal*, 41: 347–361.

Sohn-Rethel, Alfred. 1978. *Intellectual and Manual Labour. A Critique of Epistemology*. Atlantic Highlands, NJ: Humanities Press.

Solow, Robert. 1957. "Technical Change and the Aggregate Production Function", *Review of Economics and Statistics*, 39: 312–320.

Stehr, Nico. 1992. *Practical Knowledge. Applying the Social Sciences*. London: Sage.

Stigler, George J. 1961. "The Economics of Information", *The Journal of Political Economy*, 69: 213–225.

Storper, Michael and Richard Walker. 1983. "The Theory of Labour and the Theory of Location", *International Journal of Urban and Regional Research*, 7: 1–41.

Swedberg, Richard. 1987. "Economic Sociology: Past and Present", *Current Sociology*, 35: 1–221.

Therborn, Göran. 1986. *Why Some People Are More Unemployed Than Others. The Strange Paradox of Growth and Unemployment*. London: Verso.

Touraine, Alain. [1984] 1988. *Return of the Actor. Social Theory in Postindustrial Society.* Minneapolis: University of Minnesota Press.

United Nations. 1991. *Economic Survey of Europe in 1990–1991.* New York: United Nations Publications.

United States Department of Labor. 1984. *Technological Change and its Labor Impact in Four Industries.* Washington, DC: U.S. Department of Labor.

United States National Commission on Technology, Automation and Economic Progress. 1966. *Technology and the American Economy.* Washington, DC: United States Government Printing Office.

Walsh, Kenneth. 1987. *Long-Term Unemployment: An International Perspective.* London: Macmillan Press.

Whitley, Richard. 1988. "The Transformation of Expertise by New Knowledge: Contingencies and Limits to Skill Scientification", *Social Science Information,* 27: 391–420.

Wolff, Edward N. and William J. Baumol. 1989. "Sources of Postwar Growth of Information Activity in the United States", in: Lars Osberg, Wolff, Edward N. and William J. Baumol (eds.), *The Information Economy: The Implications of Unbalenced Growth.* Halifax: The Institute for Research on Public Policy: 17–46.

World Bank. 1992. *World Development Report 1992. Development and the Environment.* New York: Oxford University Press.

Yuchtman-Yaar, Ephraim. 1987. "Economic Culture in Post-industrial Society: Orientation Toward Growth, Work and Technology", *International Sociology,* 2: 77–101.

zur Nedden, F. 1930. "Technischer Fortschritt und Weltwirtschaftskrise", *Die Hilfe,* 36: 899–903.

Chapter 17
The Productivity Paradox: ICTs, Knowledge and the Labor Market

Nico Stehr

17.1 Introduction

When it comes to the economic affluence of a nation and the ability of a country's economy to improve the standard of living of its citizens and compete internationally, social scientists are in an unusual agreement that productivity "in the long run is almost everything" (Krugman 1994: 13). Manuel Castells (1996: 80) throughout his extensive study of modern society as a network society seconds this observation and concludes that "productivity is the source of the wealth of nations".[1] Not only only shifts in standards of living follow from changes in productivity performance. Less immediately related non-economic transformations in response to unequal national productivity gains occur, including major changes in the balance of global power relations. In this light, Krugman (1994: 17) comments somewhat despairingly, the slowdown of "American productivity growth since the early 1970s becomes the most important single fact about our economy." Global competition represents the linkage path that combines information and communication technology (ICT),

---

This text was first published as: Stehr, Nico. 2000. "The Productivity Paradox: ICTs, Knowledge and the Labour Market", in: John de la Mothe and Gilles Paquet (eds.), *Information, Innovation and Impacts*. Norwell, MA: Kluwer Academic Publishers: 255–271. The permission to republish this text here was granted by Springer.

[1] The reasons Castells (1996: 468) offers in support of his label of network society for the emerging social structures throughout modern society refer to a historical trend in which the "dominant functions and processes in the information age are increasingly organized around networks. Networks constitute the social morphology of our societies, and the diffusion of networking logic substantially modifies the operation and outcomes in processes of production, experience, power, and culture." For a critique of aspects of Castells theory of society see Stehr (1999).

M. T. Adolf (ed.), *Nico Stehr: Pioneer in the Theory of Society and Knowledge*, Pioneers in Arts, Humanities, Science, Engineering, Practice 16,
https://doi.org/10.1007/978-3-319-76995-0_17

organizational change and productivity growth. Productivity gains are stratified. Productivity improvements or the lack thereof do not occur in a linear fashion throughout the economy nor among all divisions within a firm or segments of the work force. In the advanced economies the challenge is no longer "to make manual work productive [...] the central challenge is to make knowledge workers productive" (Drucker 1999: 141). And the tools modern knowledge workers increasingly deploy are information and communication technologies.

A similar equally widespread agreement among economists, policy makers, sociologists and educators extends to the notion that "structural changes" in the modern economy transform the labor market and generate demands for specific skills. Changes associated with the emergence of the knowledge-intensive economy are seen to engender an increased need for skilled workers (e.g. Baldwin/Gellatly 1998: v) or a growing presence of knowledge workers in the economy.[2]

Three decades ago, Daniel Bell and others began to explicate their idea that industrial society is giving way to a post-industrial society. One of their central tenets was to refer to the centrality of theoretical knowledge as constitutive for post-industrial society. According to Bell (1968: 158), the crucial political questions post-industrial society will face "deal with education, talent, and science policy. The rapid expansion of a professional and technical class, and the increased dependence of the society on scientific manpower, suggest a new and unique dimension in social affairs: i.e., that the economic growth rate of a postindustrial society will be less dependent on money than on 'human capital'."

In the discussion and controversies surrounding the productivity paradox, these and related observations about the decline in the economic importance of the forces of production that shaped industrial society are rarely moved toward the center of the analysis or linked to observations about the nature of the productivity puzzle even though they now form the premise of many general analyses of the modern economy as a knowledge-based economy.

I will try to link the analysis of the productivity paradox in this paper to the emergence of knowledge as a source of economic growth and changes. I will argue that the productivity paradox lean be better understood if one recognizes three empirical facts. First, highly skilled labor appears on the scene before information

[2] I will only cite one more example of the by now rather orthodox claim in science, in education, the economy and in much of the public realm that there is a technology driven demand for highly workers. That claim also resonates strongly with technological determinism: Davenport (1997: 2) deferring to Peter Drucker's ideas on the knowledge society explicates the term knowledge-based economy and indicates that the characteristic technological basis of the knowledge-intensive economy, namely information and communication technology, biotechnology, and new materials have created a "remarkable demand for highly educated workers, not only to advance and manage the technologies themselves, but to serve as experts in the finance, production, and marketing of the new products and services which the technologies produce." What is remarkable about Davenport's assertion is that Drucker is one of the few economists who has indicated that the really intriguing dynamics of growth of knowledge workers is the extent to which such gains may not be demand but supply side driven (see Drucker [1968] 1992: 279). I will return to this point in detail later in the paper.

technology. Second, the increasing importance of highly skilled labor is not a reaction to demand for such labor, but rather there is an autonomous (i.e. societally driven) supply shift. And third, information technology actually helps entrepreneurs and managers to catch up with and reverse the rising labor costs implied by this supply shift. Therefore, the productivity paradox can help us to understand that we are not faced with a technology driven transition from industrial to informational society but rather with a societally driven transition from an industrial to a knowledge society. In this sense, then, we have entered a new modernity.

17.2 The Productivity Paradox

In the last two decades economists in particular have been puzzled and even irritated about the apparent lack of measurable productivity gains in goods producing and services industries in OECD countries in response to or in conjunction with the immense investments in recent years in information and communication technologies. The choice of labelling this phenomenon the "productivity paradox" results from the disjuncture between the immense economic expectations and promises that have been engendered by the "computer age", on the one hand, and the apparent lack of sustainable economic payoffs resulting from the enormous investments by corporations and the state in information and communication technologies, on the other hand.

In 1990 alone, U.S. businesses invested $61 billion in hardware, $18 billion in software and $75 billion in data processing and computer services (U.S. Department of Commerce 1991). Attewell (1994: 24) sums up previous research on the productivity paradox affirming its existence and comments that "no study documents substantial IT effects on productivity." Although conceptual, methodological and data difficulties that extend to the very definition of productivity do exist in the information that are typically utilized in generating these findings, they do not appear to invalidate the results completely.[3] Given the excessive and often repeated claims about the transformative capacity of information technologies, one could even be tempted to ask, why have productivity gams that can be attributed to ITCs not been even more spectacular?

In an intellectual version of the continental drift, the social sciences disciplines are moving farther and farther apart (cf. Luhmann [1991] 1993: 2). Given the socio-cognitive state of contemporary social science, it is not surprising that the

[3] Castell's (1996: 78) expresses the suspicion that the poor validity of the economic statistics might be responsible for the productivity puzzle and therefore may not even be real: "It may well be that a significant proportion of the mysterious productivity slowdown results from a growing inadequacy of economic statistics to capture movements of the new informational economy, precisely because of the broad scope of its transformations under the impact of information technology and related organizational changes." However, he does not indicate how one might be able to specifically 'heal' the deficiencies of the current statistical regime.

growing literature on conditions for the possibility of "productivity growth", let alone the nature and consequences of the "productivity paradox" is not accompanied by a common conception of its meaning and empirical referents. There is no agreement on why productivity varies, let alone on what and how one might account for the essentially contested observation that the growing diffusion of information and communications technologies at work have not improved the productivity of firms as measured by official statistics. For example, is the productivity paradox in the field of ICT a signal that mirrors a more global signal that displays a secular decline in productivity gains? Distinctive research cultures and networks have: grown up around the notion of the productivity paradox; networks that do not communicate with each other and pursue their own strategies in examining the issue of the productivity payoff of modern information and communication technologies. These points have been elaborated in the proceedings of a recent OECD conference, *The Social Sciences at a Turning Point?*, and particularly in the contribution of Langenhove (1999).

The variety of research and accounts of the productivity paradox exemplify the growing and deepening division of labor in social science and its essentially contested nature. For some observers, the productivity paradox does not exist in reality. The productivity puzzle is a measurement construct or indicative of a mismeasurement of outputs that conceals real gains that are made (cp. Quinn 1996; Diewert/Fox 1997). But even if the puzzle should exist, the magnitude of the problem is small upon first examination since investments into computers form a relatively minor part of all capital input. For others although the paradox is real its represents but a transitory phase not unlike the productivity lag produced by the transition to technological systems in the past such as the diffusion of electric power. And as is the case for other learning processes, it takes a protracted period of time before the economic benefits show up (cf. David 1990; Petit/Soete 1997; Davenport 1997). Still other observers see the productivity paradox not as a gap that reflects economic realities. Its persistence is rather an indicator of intellectual or theoretical deficits in economic discourse (cp. Jorgenson 1997).[4] Last but not least, some economists have signaled that the danger has passed and that the productivity paradox disappeared by 1991 (e.g. Brynjolfsson/Hitt 1996; Sichel 1999).

Sociologists and scholars from other disciplines who have paid attention to the issue of the productivity paradox have attempted to explain (deaggregate) the paradox in a great variety of ways. In particular, they have tried to identify various social and organizational mechanisms that undercut or stretch out the potential

[4] Jorgenson (1997: 4) sees the productivity paradox as arising from the prevailing identification of "productivity growth with technological change". Technological change and productivity gains are distinct. Productivity growth is but a minor component to growth. Technological change occurs, he argues, as a result of investments; economic growth also is due to capital investment. Capital investments can be categorized into investments into tangible assets, human and intellectual capital. The purchase of computers constitutes an investment into tangible assets. But the key concept in this context, intellectual capital remains but a vague and perhaps even more irritating to economists an unmeasured and unmeasurable concept.

productivity payoffs from ICTs in firms. Attewell (1994: 48) suggests that productivity payoffs from investments in information technology may result from a series of trade-offs within firms at both the individual as well as collective level. For example, the "potential benefits of the technology may be channeled into alternative directions—either doing the original work more efficiently (productivity enhancing) or doing a different kind of activity or the same activity more often." Pinsonneault (1998) has pointed to the differential patterns of association between the usage of information technologies and the nature of managerial work in different firms.

17.3 Employing Information and Communication Technologies

I would like to frame the issue of the productivity paradox in a radically different manner leaving neoclassical conceptions of rational economic conduct and of work behind. Perhaps the most fundamental yet unexamined, assumption of the essentially contested accounts of the productivity paradox is the idea that the deployment of information and communication technologies entirely follows the logic of all economic behavior. Investments into tangible and intangible capital is motivated up front and primarily driven by the desire to enhance the profitability, competitiveness, and productivity of firms that decide to embark on such investments.

There is no reason to doubt that such motives play a key role in investment behavior and that expectations of senior managers who invest in information technologies are that robust gains in productivity will be commensurate with the cost of modern ITCs. In the absence of such expectations it makes no sense to speak of a productivity puzzle.

The question therefore becomes whether conventional economic motives and forces are the exclusive or even main reason for the heavy commitment of funds into ICTs. It is not only the possible lack of any visible and robust payoffs in terms of the orthodox motives of economic conduct, or the kinds of everyday frustrations that accompany the use of information technologies (cp. Landauer 1995) and often speedy obsolescence of skills that generate doubt about the persisting force or viability of such normative dependencies and asymmetries in the relation between the employment of new technologies, the nature of the world of work, production processes and profits.

In a knowledge-based economy other, and as I want to stress, *additional* processes[5] are at work that account for the transformation of the work place and the

[5] My cautious observations about a gradual rather than an abrupt, decisive transformation of the economy of industrial society into a knowledge-based economy that fully displaces and renders obsolete the earlier economic formations resonate with Sombart's ([1927] 2000) reflections about erroneous expectations that economic systems can change in dramatic even violent ways: "All those opinions are mistaken which expect a violent upset of the existing economic constitution and a sudden change of the, bases of economic life. This opinion too misjudges the nature of economic

world of work and therefore for the growth and employment of information and communication technologies even though global and/or local empirical signals as far as the payoff, as measured by conventional economic indicators, are difficult if not impossible to obtain.

If it can be shown that such additional reasons also and increasingly account for the widespread marshalling of ITCs at work, it follows that the productivity paradox assuming that it is real can be evaluated in a very different light and has to say the least, multiple causes.

The main weakness of neoclassical perspectives of work is the failure to attend to the social context of work. Key qualitative features of the world of work such as job satisfaction, social networks, management style and the quality of the products and services, are left out. In fact, "social context and social connections envelop the world of work" (Tilly/Tilly 1998: 13). In knowledge-intensive economies, I would want to maintain the relevant social context of work acquires even greater significance for the competitiveness of a firm, its innovative capacity and profitability. Differences in social context that should not be simply stripped away are of particular importance as one explores differences in the world of work across time. I want to explore the possibility that the productivity paradox constitutes a global signal for a dramatic transformation of the economy generally and the world of work characteristic of that economy in particular.

17.4 The Loss of Agency or the World of Work in Industrial Society

The meaning associated with the term 'labor' or 'work' today is a product of industrial society.[6] The dominant perspective of industrial society sees its technological regime closely linked to mass production systems, intensive productivity gains and the capacity to produce an abundance of goods as well as hierarchical forms of work organization and control. In the end, the (capitalist) logic always contributes to a massive alienation of workers, or as argued more recently, to an extensive de-skilling of the work force (e.g. Braverman 1974; Gill 1985). Although the lament about the deskilling of the individual worker often is associated with the work of Hairy Braverman, there are numerous predecessors. The predecessors include Helmut Schelsky. Schelsky (1954: 20) asserts: "The closer we approach automation, though without ever fully reaching it, the greater the degree to which

development, which always proceeds in the form of a gradual, 'organic' reshaping of existing conditions. A new economy 'grows', like a plant, or an animal. Forcible interventions may well destroy, but they build nothing. All previous history confirms the accuracy of this observation."

[6] It is worth noting that the meaning of the terms work and labor now are used almost interchangeably. Labor in contrast to work may or was seen as based in circumstances in another person's direction and control while work refers to what directly maintain one's existence and is earned with relative autonomy.

work becomes spiritless and stressful and the lesser the extent to which it requires interest in technical matters and skills or, even initiative of any sort." Even earlier, in an essay entitled "The machine, the worker and the engineer", Merton (1947: 180) refers to numerous social implications of labor-saving technology including the "enforced obsolescence of skills" and the loss of status as well as self-image that accompanies the de-skilling process of workers.

Merton assumes, as does the later de-skilling thesis, that the obsolescence of skills is irreversible. In light of the kind of production technology used, a compensating process presumably is not considered likely. The increasing employment of labor saving technology produces the enforced obsolescence of skills among the workers. The social and psychological consequences of discarding acquired skills are mainly connected to the demotion of status (including the possible loss of the public identity of the job) and the destruction of the positive self-image of the worker, stemming from the once confident use of those skills. Merton (1947: 80) anticipates an "alienation of workers from their job and the importance of wages as the chief symbol of social status are both furthered by the absence of asocial meaning attributable to the task. Increased specialization of production leads inescapably to a greater need for predictability of work behavior and, therefore, for *increased discipline in the workplace*."

Massive power asymmetries in work organizations exist and the struggle for power is a zero-sum game. The ability of management to preserve and exercise domination is assured by virtue of holding on to or monopolizing *knowledge* about the conceptions on which production is based. Knowledge is located in specialized departments only. That is, the successful separation of execution and conception is the key to the control and the persistent degradation of the worker. Employees are mere executioners of tightly prescribed and increasingly fragmented tasks. The new version of the oppression thesis also generalizes about the work place without any credit to the imagination of the worker and specific conditions of work. Any resistance by employees that may be evident is merely in response to the oppressive control exercised by management. The thesis minimizes, consistent with Marx's portrait of the labor process in capitalist society, the active ability of the worker to affect his or her working conditions. Technological developments simply reproduce the domination of capital over labor, often on a more repressive scale contributing to what Merton (1947: 80) as indicated called an "enforced obsolescence of skills." The depressing conclusion therefore can only be that work in the modern factory continues to be trivialized and that the "Taylorist philosophy is in many cases still being earned over to the era of microelectronics systems in manufacturing" (Gill 1985: 87).[7]

[7] Several authors have adopted the term 'neo-Fordism' to describe the convergence of information technology and the managerial orientation of 'Fordism' (cp. Massey 1984; also Jaeger/Ernste 1989).

17.5 Agency or the World of Work in Knowledge Societies

It is by no means a novel observation that the social organization of work is changing and that the nature of the change has to do with what originally constituted, at least according to Marx and Engels, the condition for the possibility of the division of labor in society: the separation of labor into manual and intellectual labor (cf. Marx/Engels [1932] 1960: 28). Although physical labor and the expenditure of human energy in such tasks will not disappear entirely in knowledge societies, the dominant trend is away from manual labor to intellectual labor, and therefore to a corresponding increase in the role knowledge and learning play in shaping work, the ability to work and continued employment.

More recently, fascination with the constraining features of the social and material conditions which allegedly give rise to persistent hierarchy and isolation of the individual in the workplace, have been somewhat replaced by equally strong convictions about a *new* division of labor, a new technology, and new logics of organizing production that are described as permissive forms of domination (cf. Sabel 1991: 24; Hirst/Zeitlin 1991; Sabel 1995). The social distribution of power in the workplace may, as a result, no longer constitute a zero-sum game. These views are linked in turn to the distinction between the declining regime of mass manufacturing ('Fordism')[8] and the growing system of "flexible specialization" in production (cf. Piore/Sabel 1984). As a result, technology is not seen as an essentially dehumanizing force but as one that enables or at least holds the promise of participation in the affairs of work. This in turn moves the notion of the growing condition for the possibility and importance of "subjectivized conduct of work" (Böhle 1998: 241) into the center of the analysis of the world of work.

The extensive use of technology does displace least-skilled work. For example, in a case study of the impact of computer-aided design on skills m U.S. aircraft and automobile manufacturing industry, Salzman (1989: 260) finds that "the technology is relatively effective at [...] automating the least skilled work (the simplest connections in this case) leaving only the most skilled aspects for the designer." Whether one should really be concerned about such a change is a different matter. It certainly does not represent an instance of de-skilling. Computer-aided design is not the "Trojan Horse of Taylorism" but its opposite, a counter-tendency to a Taylorist organization of work.

In the sphere of work, the profound anxieties about the destructive ways of technology are now replaced by animated discussions about freedoms from control. The vocabulary of intentionality and agency or collaboration and working together, thought to be obsolete, reappears in discussions of work, production and the social organization of work (e.g. Cavestro 1989). Paradoxically, the technology once

[8] A brief description of the emergence, nature and recent challenges to mass production in the automobile industry in North America and Europe described as Fordism may be found in Dankbaar (1995).

feared to have become self-regulating now (de-)regulates itself in the sense of negating regulation.

The distribution and utilization (deployment) of knowledge is not necessarily as one-sided as accounts have stressed the ease with which superiors in modern factories and offices manufacture consent and impose control. Knowledge as a capacity for action is not easily monopolized. Nor is knowledge a one-dimensional and static phenomenon. Knowledge is contingent, contestable, multiple and shifting. Subordinates do not always need to control as much knowledge as their superiors perhaps do in distinct instances or precisely the same forms[9] in order to effectively resist and override controls imposed by management and owners.

If it is indeed the case that modern technology enlarges the capacity to act not only among management, but also among its employees, it would follow that employment regimes that rely on command, control and coercion to ensure performance constitute an increasingly ineffective basis for the coordination of production. Social relations in the workplace could therefore be best shaped by what Weber ([1913] 1981) has called *Einverständnishandeln,* or social action that derives from mutual trust. Social relations in the workplace that are based on trust may displace those linked to its opposite, namely distrust. Distrust is, of course, at the heart of the social organization of work that relies on the principles developed at the beginning of the century by Frederick W. Taylor or Henry Ford. Mutual expectations that invoke trust rather than distrust may not only be more typical in organizations that require considerable flexibility, initiative, and autonomy (cf. Heisig/ Littek 1995), but could also be a 'rational' response by management to enlarged capacities of action, including the employees' abilities to resist.

However, in spite of the discovery of the growing role of agency in the world of work in knowledge societies, another leading assumption about the labor market that remains unaffected to this day are that skills the individual is expected to bring to the world of work are those which the pre-existing system of work demands and imposes. In an even more general sense, Bourdieu (1973: 72) describes the assumptions (or laws as he calls it) that govern such asymmetric social relations as the tendency of structures "to reproduce themselves by producing agents endowed with the system of predispositions which is capable of engendering practices adapted to the structures and thereby contributing to the reproduction of the structures" of labor. In other words, it is a widely shared assumption that growth of the number of knowledge workers in the modern economy occurs in response to the *requirements the world of work,* especially those its modern technological regime imposes. And as long as one holds on to this assumption the productivity puzzle becomes an even more severe enigma. The share of highly paid and highly skilled workers grows yet a consummate payoff is not in sight.

[9] The so-called scientific management principles that Taylor developed were, for example, designed to limit and restrict the exceptional control workers often had with respect to forms of knowledge that related directly to technical and performance-related knowledge on the shop floor.

17.6 The Growth of Knowledge Workers

Peter Drucker presents a much more surprising and perhaps plausible hypothesis about the reasons for the growth of the number of knowledge workers in modern society. He suggests that the stimulus for the increasing demand for knowledge-based work has to do less with more difficult and complex job skills, changing technological regimes, the growing complexity and specialization of the economy or enhanced functional steering and coordination needs. The growth of knowledge workers has more to do with the substantial extension in the working life span of individuals and the enhanced knowledge with which individuals come to the labor market in the first place. The observed shift toward knowledge-based jobs should not be seen to result from changing occupational preference and choices of new entrants to the labor market, although changes in preferences of job content and ways of achieving job satisfaction certainly are part of the transformation. In the United States for example full scale labor force entry of adult college trained baby boomers coupled with steady retirements of less-educated persons who reached working age during the 1930s and 1940s resulted in quantum increases in the average educational endowment of the labor force. Prime working age participants with four years or more of college training rose by 64 percent to 25.5 million during the 1980s and those with one to three years rose by 58 percent to 20.8 million (Wetzel 1995: 60).

If we follow Drucker, it is not so much the *demand* for labor and particular skills as the result of more complex and exacting jobs, but the *supply* of highly skilled labor that underlies the transformation of the world of work. More specifically the "direct cause of the upgrading of the jobs is [...] the upgrading of the educational level of the entrant into the labor force" (Drucker 1968: 279).[10] Whether the transformation of work in a world of knowledge work by knowledge workers constitutes a historically unique development or will continue to be major attribute of work in the future is an open question. It is likely once the world of work has been thoroughly transformed that 'demand' attributes will become more prominent attributes influencing the texture of the labor market.[11]

[10] Among the surprising even amazing properties of the transformation of the labor market is that the American economy was able to "satisfy the expectations of all these people with long years of schooling [...]. As a result of the change in supply, we now have to create genuine knowledge jobs, whether the work itself demands it or not. For a true knowledge job is the only way to make highly schooled people productive [...]. That the knowledge worker came first and knowledge work second—that indeed knowledge work is still largely to come—is a historical accident. From now on, we can expect increasing emphasis on work based on knowledge, and especially skills based on knowledge" (Drucker 1968: 285).

[11] John Kenneth Galbraith dismisses Drucker's argument out of hand. He affirms the orthodox view about the relation between education and the labor market and considers Drucker's perspective as evidence for the typical self-complacency and pretension of the educator misreading the real power balance in society in the process. Galbraith (1967: 238) suggests it is the "vanity of educators that they shape the educational system to their preferred image. They may not be without influence but the decisive force is the economic system. What the educator believes is latitude is

The much more common perspective is that rapid diffusion of information and communication technologies has altered the production process and the delivery of services and that the employment structure has changed as a result. The new technologies increase the demand for highly skilled workers. Put more technically, most observations "model changes in workforce skill as a function of changes in industry capital intensity and industry-level investment in computer equipment" (Doms et al. 1997: 254).

At the present time, there are no empirical studies, to my knowledge, that explicitly examine the relationships stipulated by Drucker. However, two recent intensively discussed and researched economic issues that arrive at largely unanticipated results may offer at least an indirect measure of confirmation of the Drucker's thesis. First, a large volume of empirical studies of the US labor market triggered by the observation that income inequalities between well-paid and more poorly paid jobs have risen in recent years considerably, prompted economist to ask what developments may be responsible for these income trends. In particular, the relations between technological change, skill level and income have been studied. The primary assumption examined in these studies is that the increasing polarization of the labor market may be linked technical changes that in turn cause firms to hire more highly skilled labor (Gottschalk 1997). Second, there are research designs directed toward analyses of the causes of the "productivity puzzle" that could be relevant.

For the researchers concerned with the growing polarization of income levels, two explanations are of particular interest. First of all, the growing differentiation of pay for skilled and lesser skilled labor may be caused by technological change. More precisely, the increase in *demand* for skilled labor and a growing proportion of skilled workers in the labor force is seen as induced by technical developments (Johnson 1997). The second explanation has a family resemblance to the first account: the *demand* for technologically more sophisticated products and services has triggered a growing need for a highly skilled work force (cf Bernard/Jensen 1997: 5). In short, changes that can be attributed in one way or the other to demand induced forces provoke a change in the balance of skilled to lesser skilled workers. And, as a result, these transformations in the nature of demand activate and accelerate a growing inequality in incomes by skill levels.

In a broadly-based, cross-sectional, empirical study at the level of individual manufacturing firms and using individual rather than aggregate data for the American economy, Doms et al. (1997) for example have examined the relationship between technology use,[12] the education, occupation and wages of the employees

usually latitude to respond to economic need." In other words, Galbraith insists that the demand-side explanation generally favored by economists (as well as employers, educators and educational policy makers one should add) primarily accounts for the increase in skilled work.

[12] The degree of technology intensity in individual firms was measured by the authors of the study by counting the number of technical processes or devices such as computer driven machines, robots and so on found in the plants (see Doms et al. 1997: 287–288 for a detailed description of the different processes and devices).

in the manufacturing sector. As the data reproduced in Table 17.1 show, one is able to conclude that there is a growing covariance between the degree of technology use in firms, that is to say, the progress made in automating both the development and the production process, and the educational level of the employees. The conclusion therefore not only is that "skilled workers and advanced manufacturing technologies are complements" but also that the proportion of employees "in skilled occupations rises significantly with the number of technologies employed" (Doms et al. 1997: 261, 263; see also Berman et al. 1994) *as well as* the proportion of employees not directly active m the production process.[13] A variety of controls confirm these findings. In addition, the authors report that employees in firms with extensive technology deployment earn higher wages and salaries.

The cross-sectional data however cannot offer an answer to questions about the timing of the observed marked substitution in favor of skilled labor in the manufacturing firms that are technology intensive. As a result, Doms, Dunne and Troske attempt to extend their analysis by relating the utilization or adoption of different technologies over time in these firms to changes in the different variables such as the wage levels, the proportion of employees not directly involved in the production process etc. Aside from methodological problems such a procedure may have as the result of the absence of valid longitudinal data, the overall result of their efforts to operationalize "technology adoption" is that "technology adoption is relatively uncorrelated with the changes in the nonproduction labor share, workers wages, or labor productivity" (Doms et al. 1997: 277).

One possible 'explanation' for the 'negative' finding would be that the firms the author included in their study already employed or hired a large number of highly skilled employees *prior* to the adoption of new technological means. Thus, "if plants that adopt technologies have more skilled workforces prior to adoption, then we would expect that the pre-adoption wages and labor productivity should be correlated with future technology use" (Doms et al. 1997: 277). The results of their study once more are far from transparent. The authors sum up the relation they are able to document as follows: "Plants that adopt a large number of new technologies have more skilled workers both pre- and postadoption" (Doms et al. 1997: 279). As Drucker assumes, one cannot preclude the possibility in other words that highly skilled employees force the modernization of their work places, in the first instance. It is likely therefore that the supply of skilled workers rather than the demand for

[13] In concrete terms, "the positive relationship between technology use and the percent of skilled workers is primarily due to a dramatic increase in the percent of scientists and engineers in the most technologically advanced plants" (Doms et al. 1997: 263).

Table 17.1 Technology use, education, wages and occupation at US Manufacturing Plants, 1988–1990

	Percent					
	of workers with at least a college degree	of nonproduction workers with at least a college degree	of production workers with at least some college	of workers in managerial, scientific, engineering, or precision-craft occupations	of non-production workers	of total wages paid to nonproduction workers
	(1)	(2)	(3)	(4)	(5)	(6)
Plants using less than 4 technologies	9.4	24.1	21.2	33.7	32.7	41.0
Plants using 4–6 technologies	12.2	31.2	24.2	35.6	33.3	42.2
Plants using 7–8 technologies	14.0	34.5	27.1	36.6	34.9	39.5
Plants using 9–10 technologies	16.2	34.9	27.7	37.4	40.7	46.2
Plants using 11–13 technologies	15.2	37.5	29.7	33.1	34.3	37.3
Plants using more than 13 technologies	33.1	53.9	34.9	48.6	56.9	62.4
Full sample	18.3	40.1	23.9	38.5	40.5	47.2

There are 3251 workers in the sample that work in plants using less than 4 technologies, 4690 (4–6 technologies), 6403 (7–8), 5914 (9–10), 5931 (11–13) and 7844 workers in plants using more than 13 technologies

Sources Doms et al. (1997: 262)

workers with such skills that constitutes the motor of the rapid and radical transformation of the world of work.[14]

A further but confirmation of Drucker's thesis of a supply induced transformation of the world of work is aggregate data about the growing "skill level" of the population in most OECD countries. According to Johnson (1997: 42), the relative skill supply measured as a ratio in the population of high school to college equivalency in the US has risen from 0.105 in 1940 to 0.496 in 1993. In five decades, the proportion of the population with a college education has increased fivefold. The increase of college educated labor is particularly strong in 1970s, reflecting a growing proportion of college students in the latter half of the sixties. It would be far too simple to suggest that the tremendous increase in the collective skill level of the work force simply occurred in direct response to market forces. Although individuals will respond to perceived market opportunities, the fit not only in terms of time between perceived market opportunities and education could hardly be expected to be exceptionally close. Too many other factors and forces impinge upon those choices and after many years of education perhaps result in "higher skill levels".

In the modern economy, knowledge is the most important resource. As a result the production of knowledge and learning are the most significant processes in the knowledge society. Public policy in turn must be attuned and attend to these features of modern society (cf. Alexander 1997).[15] Knowledge societies are changing with rapid speed. For this reason alone it is not sensible to adopt a strict demand or storage model of the kinds of skills and competencies that schools, universities and other educational institutions ought to deliver. Future occupational requirements of the world of work and their obsolescence are difficult if not impossible to anticipate or predict. The close even intimate linkage between prior curriculum and subsequent occupational tasks that is often expected and demanded cannot be accomplished. A decisive feature of the labor market in knowledge societies is its unpredictability and the insecurities that the substance of educational needs can be determined by future features of the labor market and the world of work. The storage model—schools and universities supply those skills and competencies that can be immediately utilized at work—has to be replaced by a model that couples work and education under conditions of uncertainty and agency.

---

[14] In response to the question of the reasons for the immense growth of the service sector in recent decades, Landauer without referring to Drucker, also offers an account that stresses factors induced by the demand for jobs. Thus, new jobs were needed, so new services were invented. Many new or expanded services depended on computers: a plethora of investment instruments—complex new mutual funds and trading schemes, a deluge of new insurance policies and options, a myriad of debit and credit cards, dozens of new kinds of bank accounts and novel banking services offered from widely dispersed branches and machines, multitudes of new medical techniques and therapies, fast food restaurants, fast copy stores, fully filled planes with frequent flyer plans, mom and pop mail order firms, direct marketing, PC maintenance, and so forth (Landauer 1995: 74–75).

[15] Concerns that the quality of the available jobs may not be compatible with rising educational levels (Harman 1978: 209) correspond to exactly the opposite perspective, namely that the quality of the worlds of work is primarily driven by the nature of the demand.

17.7 Outlook

Economic theories and research that deal with the productivity paradox for the most part continue to be linked in a rather close fashion to the production process of industrial society, its organization and outcomes. Moreover, the lingua franca of economic theories remains tied to an imagery which displays agents as inflexible creatures entangled in single-purpose, dedicated structures of purposes which imposes its rhythm, often over generations, on passive individuals and groups. As long as economic organizations and regimes such as the labor market are reconstructed as inflexible and largely one-dimensional, the gradients of inequality will be steep and the effects of its logic long lasting. Discourse on the role of economic agents will be concerned with the coercion, that is, the constraints, the vulnerability and effective impotence of individuals and groups in the face of power exercised by the powerful. But new realities require a new language of the world of work. The new language should, in contrast, stress agency, malleability, flexibility, multi-purpose resources, volatility, heterogeneity etc. of social structures and the extent to which individuals and social groups as embedded in their relationships with others have the capacity to employ and transform these structures, once the generalized vulnerability to the forces of the conventional logic of economic conduct has been reduced to a significant degree. The condition for the possibility of greater and more broadly based agency is knowledgeability or a bundle of more widely accessible social competencies and their impact on social structures of inequality accelerating chances for actors to re-fashion social constructions.

References

Alexander, T.J. 1997. "Human Capital Investment: Building the 'Knowledge Economy'", *Policy Options*, 18: 5–8.

Attewell, Paul. 1994. "Information Technology and the Productivity Paradox", in: Douglas H. Harris (ed.), *Organizational Linkages. Understanding the Productivity Paradox*. Washington, DC: National Academy Press: 13–53.

Baldwin, John R. and Guy Gellatly. 1998. *Are There High-Tech Industries or Only High-Tech Firms? Evidence from New Technology Based Firms*. Ottawa: Statistics Canada, Analytical Studies Branch.

Bell, Daniel. 1968. "The Measurement of Knowledge and Technology", in: Eleanor B. Sheldon and Wilbert E. Moore (eds.), *Indicators of Social Change. Concepts and Measurements*. Hartford, CT: Russel Sage Foundation: 145–246.

Berman, Eli, John Bound and Zvi Geliches. 1994. "Changes in the Demand for Skilled Labor Within U.S. Manufacturing Industries: Evidence from the Annual Survey of Manufacturing", *Quarterly Journal of Economics*, 59: 367–398.

Bernard, Andrew B. and Bradford Jensen. 1997. "Exporters, Skill Upgrading and the Wage Gap,", *Journal of International Economics*, 42: 3–31.

Böhle, Fritz. 1998. "Technik und Arbeit – neue Antwoten auf 'alte' Fragen", *Soziale Welt*, 49: 233–252.

Bourdieu, Pierre. 1973. "Cultural Reproduction and Social Reproduction", in: Richard Brown (ed.), *Knowledge, Education, and Cultural Change*. London: Tavistock: 71–112.

Braverman, Harry. 1974. *Labor and Monopoly Capital: The Degradation of Work in the Twentieth Century*. New York: Monthly Review Press.

Brynjolfsson, Erik and Lorin Hitt. 1996. "Paradox Lost? Firm-Level Evidence on the Returns to Information Systems Spending", *Management Science*, 42: 541–558.

Castells, Manuel. 1996. *The Information Age: Economy, Society and Culture. Volume I: The Rise of the Network Society*. Oxford: Blackwell.

Cavestro, William. 1989. "Automation, New Technology and Work Content", in: Stephen Wood (ed.), *The Transformation of Work?* London: Unwin Hyman: 219–234.

Dankbaar, Ben. 1995. "The Crisis of Fordism: Restructuring in the Automobile Industry", in: Rien Hiuskamp van Ruysseveldt and Jacques van Hoff (eds.), *Comparative Industrial and Employment Relations*. London: Sage: 293–314.

Davenport, Paul. 1997. "The Productivity Paradox and the Management of Information Technology", Paper presented to the Centre for the Study of Living Standards Conference on Service Sector Productivity and the Productivity Paradox, April 11–13, 1997, Ottawa, Canada.

Diewert, Erwin and Kevin Fox. 1997. "Can Measurement Error Explain the Productivity Paradox?", Paper presented to the Centre for the Study of Living Standards Conference on Service Sector Productivity and the Productivity Paradox, April 11–13, 1997, Ottawa, Canada.

Doms, Mark, Timothy Dunne and Kenneth Troske. 1997. "Workers, Wages and Technology", *The Quarterly Journal of Economics*, (February): 253–290.

Drucker, Peter. [1968] 1992. *The Age of Discontinuity. Guidelines to Our Changing Society. With a New Introduction by the Author*. New Brunswick, NJ: Transaction Books.

Drucker, Peter. 1999. *Management Challenges for the 21st Century*. New York: HarperBusiness.

Galbraith, John Kenneth. 1967. *The New Industrial State*. Boston: Houghton Mifflin.

Gill, Colin. 1985. *Work, Unemployment and the New Technology*. Oxford: Polity Press.

Gottschalk, Peter. 1997. "Inequality, Income Growth and Mobility", *Journal of Economic Perspectives*, 11: 21–40.

Harman, Willis W. 1978. "Chronic Unemployment: An Emerging Problem of Postindustrial Society", *Futurist*, 12: 209–214.

Heisig, Ulrich and Wolfgang Littek. 1995. "Trust as a Basis of Work Organisation", in: Wolfgang Littek and Tony Charles (eds.), *The New Division of Labour: Emerging Forms of Work Organisation in International Perspective*. Berlin: Walter de Gruyter: 17–56.

Hirschhorn, Larry and Joan Mokray. 1992. "Automation and Competency Requirements in Manufacturing: A Case Study", in: Paul S. Adler (ed.), *Technology and the Future of Work*. Washington, DC: The Brookings Institution: 15–45.

Hirst, Paul and Jonathan Zeitlin. 1991. "Flexible Specialization Versus Post-Fordism: Theory, Evidence and Policy Implications", *Economy and Society*, 20: 1–56.

Jaeger, Carlo and Huib Ernste. 1989. "Ways Beyond Fordism?", in: Huib Ernste and Carlo Jaeger (eds.), *Information Society and Spatial Structure*. London: Belhaven Press: 159–185.

Johnson, George F. 1997. "Changes in Earnings Inequality: The Role of Demand Shifts", *Journal of Economic Perspectives*, 11: 41–54.

Jorgenson, Dale. 1997. "Computers and Productivity", Paper presented to the Centre for the Study of Living Standards Conferrence on Service Sector Productivity and the Productivity Pradox, April 11–13, 1997, Ottawa, Canada.

Kegan, Robert. 1994. *In Over Our Heads. The Mental Demands of Modern Life*. Cambridge, MA: Harvard University Press.

Krugman, Paul. 1994. *The Age of Diminished Expectations. U.S. Economic Policy in the 1990s*. Revised and Updated Edition. Cambridge, MA: MIT Press.

Landauer, Thomas K. 1995. *The Trouble with Computers. Usefulness, Usability, and Productivity*. Cambridge, MA: MIT Press.

Luhmann, Niklas. [1991] 1993. *Risk: A Sociological Theory*. New York: Aldine de Gruyter.

Marsick, Victoria J. 1998. "Transformative Learning from Experience in the Knowledge Era", *Daedalus,* 127: 119–136.

Marx, Karl and Friedrich Engels. [1932] 1960. *Die deutsche Ideologie.* Berlin: Dietz.

Massey, Doreen B. 1984. *Spatial Divisions of Labor: Social Structures and the Geography of Production.* London: Macmillan.

Merton, Robert K. 1947. "The Machine, the Worker and the Engineer", *Science,* 105: 79–84.

Nelson, Richard R. and Gavin Wright. 1992. "The Rise and Fall of American Technological Leadership: The Postwar Era in Historical Perspective", *Journal of Economic Literature,* 30: 1931–1964.

Petit, Pascal and Luc Soete. 1997. "Is a Biased Technological Change Fuelling Dualism", Paper presented to the Centre for the Study of Living Standards Conferrence on Service Sector Productivity and the Productivity Pradox, April 11–13, 1997, Ottawa, Canada.

Pinsonneault, Rivard S. 1998. "Information Technology and the Nature of Managerial Work: From the Productivity Paradox to the Icarus Paradox?", *MIS Quarterly,* 22: 287–311.

Piore, Michael J. and Charles F. Sabel. 1984. *The Second Industrial Divide.* New York: Basic Books.

Pryor, Frederic L. and David L. Schaffer. 1999. *Who's Not Working and Why. Employment, Cognitive Skills, Wages, and the Changing U.S. Labor Market.* Cambridge: Cambridge University Press.

Quinn, James B. 1996. "The Productivity Paradox Is False: Information Technology Improves Service Performance", *Advances in Services Marketing and Management,* 5: 71–84.

Sabel, Charles F. 1991. "Moebius-Strip Organizations and Open Labor Markets: Some Consequences of the Reintegration of Conception and Execution in a Volatile Economy", in: Pierre Bourdieu and James S. Coleman (eds.), *Social Theory for a Changing Society.* Boulder: Westview Press: 23–54.

Schelsky, Helmut. 1954. "Zukunftsaspekte der industriellen Gesellschaft", *Merkur,* 8: 13–28.

Sichel, Daniel E. 1999. "Computers and Aggregate Economic Growth: An Update", *Business Economics,* 34: 18–24.

Sombart, Werner. [1927] 2000. "The Economic Life of the Future", in: Werner Sombart (ed.), *Economic Life in the Modern Age.* Edited and Introduced by Nico Stehr and Reiner Grundmann. New Brunswick: Transaction Books.

Stehr, Nico. 1999. "Deciphering Information Technologies: Modern Societies as Networks", *European Journal of Social Theory,* 3: 84–93.

Tilly, Chris and Charles Tilly. 1998. *Work Under Capitalism.* Boulder, CO: Westview Press.

United States Department of Commerce 1991 *U.S. Industrial Outlook.* Washington, DC: U.S, Department of Commerce.

Van Langenhove, Luk. 1999. "Rethinking the Social Sciences? A Point of View", in: *The Social Sciences at a Turning Point?* Paris: OECD: 43–51.

Weber, Max. [1913] 1981. "Some Categories of Interpretive Sociology", *The Sociological Quarterly,* 22: 151–180.

Wetzel, James R. 1995. "Labor Force, Unemployment, and Earnings", in: Reynolds Farley (ed.), *State of the Union. America in the 1990s.* Volume One: *Economic Trends.* New York: Russell Sage Foundation: 59–105.

Chapter 18
Knowledge Politics: Policing Knowledge

Nico Stehr

18.1 Introduction

Science has not only led to the mass production of knowledge but also has it invaded society with multifarious effects: Consequently, today one talks about knowledge in the plural, for wherever knowledge is produced counter-knowledges occur. Therefore, science studies has put a novel issue called knowledge society on the agenda: Scholars inquire into its texture (Böhme/Stehr 1986; Stehr 1994) as well as into its type of knowledge production (Gibbons et al. 1994; Willke 1998, 1999). While it is not as yet decided what a society based on knowledge will eventually look like it seems to be certain that we face up to some fundamental dilemmata of knowledge: Implementing knowledge inevitably means to adapt it to local conditions, thereby changing it. What is more, knowledge may prove not only useful and profitable but also risky. While societies promote systematic production of knowledge so as to improve individual well-being and collective standards of prosperity, health, and freedom, neither the quality of knowledge thus produced nor its effects once it has become implemented can be adequately foreseen. At issue is nothing less but the control of the unforeseeable.

Since technology and science, far more than economy, have become the real motor of societal change, institutions became established that debate and assess their potential or real effects before or while implementing them. So-called

This text was originally published as Stehr, Nico. 2001. "Knowledge Politics—The Paradox of Regulating Knowledge Dynamics—Policing Knowledge", in: Sabine Maasen and Matthias Winterhager (eds.), *Science Studies. Probing the Dynamics of Scientific Knowledge*. Bielefeld: Transcript Verlag: 257–290. Permission to reprint was granted on behalf of Transcript Verlag by Stefanie Hanneken on 21 July 2017.

M. T. Adolf (ed.), *Nico Stehr: Pioneer in the Theory of Society and Knowledge*,
Pioneers in Arts, Humanities, Science, Engineering, Practice 16,
https://doi.org/10.1007/978-3-319-76995-0_18

technology assessments, mediations, hearings or round-tables are designed to control or police knowledge, the main strategies being to minimize dangerous effects and to maximize public acceptance. Whereas two decades before nuclear energy or military research has been the primary concern of such interventions, today the attention has shifted to the biological and environmental research: Medicine, food, and nature are conceived as key issues deeply affecting individual lives and societies at large. In the light of "genetic engineering," for instance, the issue is about making individual choices, privatize knowledge, and legislate its accessability. Thus, what is at stake today is the intricate relationship between the individual, economy, and the state: If anything, they share a common interest in regulating knowledge so as to keep the ideological, cultural, and moral effect of science and technology under control. While Merton's norms are still part of the game named quality-control of knowledge, its regulation from within science does no longer seem sufficient. External regulation is sought to heighten the efficacy of policing it: Drug regulation, intellectual property, and copyright protection are examples of the ways in which the distribution and implementation of knowledge becomes a domain of explicit legislation and a target of political and economic decisions. To be sure, regulating knowledge is not about 'reducing' it (though shortage of availability and accessability are forms of policing knowledge). On the contrary: Regulating knowledge will enforce the significance of knowledge, thereby disseminating the places where knowledge becomes implemented, disputed and adapted. Policing knowledge, thus Stehr as well as Weingart (2001) concur, inevitably increases the dynamics of a knowledge-based society.

18.2 Policing Knowledge

Early one morning in late July of 1999, Lord Melchett, the head of Greenpeace in Britain, was detained for questioning by the police after he and about 30 Greenpeace members raided a field of genetically modified maize near Norfolk. The protest came to an abrupt end after the farmer called the police and they arrested the protesters. According to *The Times* (July 27, 1999) the raid left government trials of seed crops that had been genetically modified in disarray. The farm on which the protest took place was one of seven test sites damaged or destroyed within months. The protest by Greenpeace followed a recommendation by the Association of Local Governments to its 170 members in England and Wales to phase out genetically manipulated foods (or GM food) until they are proven safe. A number of councils followed the recommendation. Major food manufacturers and supermarket chains as well as fast-food chains in Britain had already announced that they will not carry any products that contain genetically modified ingredients. A poll in the summer of 1999 found that 79 percent of the British public agrees that GM crop testing should be stopped. In Canada and the United States, genetic modification of foodstuff has hardly been questioned by the public. A major political battle on this front between North America and Europe is likely.

In January of 1999, the *Daily Telegraph* (January 22, 1999: 9) reported that the British Medical Association warns, in a report entitled *Biotechnology Weapons and Humanity,* that rapid advances in genetics will "soon transform biological weapons into potent tools of ethnic cleansing and terrorism". The British Medical Association urged that the regulations of the 1972 International Biological and Toxic Weapons Convention should be tightened and improved, anticipating the possibility of genetic warfare which is a practical possibility today.

The so-called "genetic protection initiative" (the petition for a referendum "for the protection of life and environment from genetic manipulation") in Switzerland was clearly rejected in June 1998 in a plebiscite in all the cantons, to the great "relief of the pharmaceutical industry" (*Neue Zürcher Zeitung,* June 8, 1998). With a voter turnout of 40.6 percent, 66.6 percent of voters opposed the petition, which according to its advocates would have declared Switzerland to be a great, unified "genetic protection area". The petition demanded, among other things, changes to the Swiss constitution forbidding the production, purchase and sale of genetically modified animals, the release of genetically modified organisms into the environment and the granting of patents for genetically modified animals and plants.

The fact that all of my examples of recent attempts to regulate the application of knowledge deal with genetic research—and the list could easily be extended—is, of course, a result of the fears and/or nightmares which have lately been prompted by just this area of research.

18.3 Knowledge Politics

In this contribution I plan to discuss what may well become one of the most significant and contentious issues for intellectual, legal, public, scientific and political discourse in the coming century: the growing pressure to police *novel* knowledge—or in other words, the emergence of a new field of political activity, namely knowledge politics.[1] In democratically organized societies, it is a legitimate role of political discourse and action to contribute to and take part in decisions that effect the ways in which scientific knowledge and possible technological artifacts are deployed in society or not.

During the early post-war decades of rapid economic growth, the application of scientific and technical knowledge in developed societies was not necessarily unanimously and uncritically advocated, to be sure, but there was a considerable degree of silent assent.

[1] By the same token, a report issued by the Rand Corporation (Fukuyama/Wagner 2000: 1) anticipates that in the early part "of the 21st century, the technologies emerging from the information and biotechnology revolutions will present unprecedented governance challenges to national and international political systems." The report deals with the governance of both research and knowledge policies.

Such headlines of recent times as "We know too much" and "How much genetic self-knowledge is good for us?", or keywords from ever more vehement disputes, such as "We dare not make use of everything we know", are part of the background and the environment of the current increasingly urgent demands for the regulation of knowledge in modern societies.[2] These science and technology controversies open a window on modern struggles over meaning and morality, economic benefits and damages, as well as the emerging and shifting locations of social power and control in knowledge societies.[3]

More specifically, it is the shift from regulating and policing normality or identity (Foucault) to the growing concern in knowledge societies with efforts to police novelty and differences. As I have indicated, the examples that come to mind, and that have captured the attention of the media and the public recently, are numerous and growing.[4] For example, the United Nations, provoked by advances in ocean exploration, is drafting a treaty that attempts to regulate marine archaeology and commercial efforts to hunt for and reclaim lost cultural treasures—and therefore the knowledge about ancient civilizations, such as the empire of the Phoenicians, that may come with their discovery (cf. *New York Times,* October 12, 1998).

It is perhaps self-evident and comes as no surprise to anticipate that 'knowing' will be seen in knowledge societies as a domain in urgent need of policing and as a site to study the functioning of power in modern society.[5] Inasmuch as the widespread dissemination of knowledge increases the fragility of modern societies (cf. Stehr 2000) efforts designed to control knowledge may be interpreted as strategic attempts to reduce or manage their fragility. Whether such attempts are likely to be successful is therefore an important issue.

---

[2] The discussion and formulation of the novel moral principle for a "right to ignorance" by Jonas (1974: 161–163) is germane in the context of this discussion.

[3] The new political field I identify as "knowledge politics" is, certainly, not immediately connected with the often-described ambivalent sense of crisis in modern societies, based on the over- and/or mass production of knowledge. The tension between the extent of knowledge production in advanced societies and the limited capability of the individual person to assimilate the huge amount of knowledge available, was already described by Simmel ([1907] 1978) a hundred years ago in a theory of the current age in the final chapter of his *Philosophy of Money.* The tragedy of culture manifests itself in the cleavage between objective culture made independent and the obstinacy of subjective culture. The problem of the policing of knowledge is not related to the production of knowledge in total—even if it is related to overproduction, however that may be defined—but rather to the range of incremental knowledge, which is conceived as being capable of changing reality.

[4] Nelkin (1995: 447–456) has published an informative typological summary of the public controversies in which science has found itself embroiled in the United States in the past.

[5] Fuller (1993: 377) advances a similar assertion, as far as I can see. He indicates that "in the world of tomorrow, breakthroughs in the natural sciences are regarded as triumphs of applied sociology and political economy, rather than of, say theoretical physics, chemistry, or biology". It is better understood and presumed that the implementation of a specific knowledge claim can alter the social fabric of society and the anticipated transformation is no longer seen as mainly beneficial.

But the issue of the control of knowledge becomes significant for another reason as well. Insofar as knowledge, especially 'additional' knowledge, assumes growing importance within the economic system and becomes subject to economic interests, efforts to control, restrict or privatize its use will grow as well. A prominent example comes from genetic research and the Human Genome Project in particular. In light of the intensive competition among hundreds of researchers worldwide in the Human Genome Project, the concern intensifies that findings that might "alter the world economy" will be monopolized, at last temporarily, if they can be protected by patents or other forms of intervention by the state. And since it is not only knowledge about genes that may turn into a valuable raw material, the fear of a progressive privatization of science grows.

Finally, demands to cope with the growth of knowledge refer to the attendant extension in capacities to act. Actors increasingly find themselves in situations in which the need for novel decisions emerges; and with it, of course, new apprehended dangers and risks. The potential openness, and not the self-evident traditional closure, of situations calls for, it seems, regulation and policing of knowledge now that knowledge is seen as the motor of new possibilities to 'manipulate' elements of a situation that in the past had been apprehended as beyond the control of all participants. The role and the prominence of references to fate, nature or the design of some higher being that symbolized the closure of conditions of action lose their relevance. What was seen as forever beyond the control of everyone now becomes—initially in the thought experiments of a few individuals, at least—subject to control and manipulation. And what was in the past seen as an exceptional moral dilemma, or the need to arrive at a decision in an extreme situation or under rare circumstances, now becomes increasingly common.

18.3.1 Regulating Knowledge

Efforts to police knowledge are not new. The notorious and ongoing struggle in some parts of the United States, for example, to ban the teaching of evolution in schools is therefore a relevant case in point. The vote of the Kansas Board of Education to delete virtually any mention of evolution from the state's science curriculum[6] is one of the more recent examples of successful efforts of creationists to ban mention not only of biological evolution but also of the big bang theory from the curricular guidelines of schools in the United States. But most of the efforts to regulate and police the possible ideological and cultural effects of science that have been and continue to be undertaken from time to time in different societies have not

[6] Cf. "Kansas Votes to Delete Evolution from State's Science Curriculum", *New York Times*, National, August 12, 1999.

been overly successful.[7] In addition, there is a distinctive shift in the kinds of concerns and consequences that may prompt efforts directed toward the regulation of knowledge. In the last couple of decades, there is a noticeable shift from concerns that revolve around security, to concerns with risk and now more and more to questions of uncertainty (cf. Bechmann/Stehr 2000).[8]

A transformation in public sentiment in favor of policing knowledge signals a basic change in the legitimacy of science,[9] in particular a shift away from a preoccupation with the 'ideological' or cultural implications of basic knowledge claims generated by science and possible conflicts with established world views, and toward an increasing preoccupation with its practical application and consequences. What I have in mind is perhaps best described as an attempt to directly control or regulate the immediate use or anticipated consequences of incremental knowledge but not the '*secondary*' implications of knowledge.[10] Attempts to police the secondary consequences of knowledge claims could refer, for instance, to action in the form of regulations prompted by the claim that passive smoking increases blood pressure. Efforts to curtail smoking in certain spaces or by certain individuals may be based on and justified by this claim. But in such a context the claim itself is not the subject of any regulation.

The now widely discussed public demystification of experts may be seen not only as a prime example of a fundamental change in the nature of the relations between knowledge-based occupations and clients, consumers, patients, students, trainees, customers, etc., but also as a profound transformation in the public image of scientific knowledge. This change considerably enlarges the number and range of individuals who relinquish their traditional subordinate role in such expert/client relations as recipients of advice that rests on an *a priori* suspension of doubt. Helen

[7] The regulation or the stratification of access to knowledge is nonetheless a constitutive component of everyday life. The world of adults, for example, is differentiated from that of children. These stratified worlds go hand in hand with the ability to impede or even to obstruct children's access to certain forms of knowledge. The quotidian forms of regulating access to knowledge are not under discussion here.

[8] I am grateful to Günther Küppers for this observation.

[9] Whether the public willingness to support the field of knowledge politics will intensify in connection with what some scientists have defined as a "comprehension gap" among the population, or whether this willingness will have any significance at all, remains to be seen. In a lead article, the English Sunday paper *The Observer* (21 February 1999, 28) describes the perceived wide comprehension gap as follows: "Between the scientific upper class, the latter-day Leonardos trekking into the brain or sketching the universe, and the majority of voters and politicians in all Western democracies, there is now a deep comprehension gap". This deficit in comprehension, however, should not be underestimated in the sciences themselves either, given the growing division of labour among the disciplines.

[10] A shift toward concerns with the externalities of science does not mean that contested efforts to regulate the conduct of "scientific inquiry" (cf. Wulff 1979) and, for that matter, attempts to manage or plan scientific research (e.g., Van den Daele et al. 1979) will disappear. On the contrary, issues of ethics, accountability, and conflict, as they relate to the genesis and execution of inquiry, will of course remain highly significant. At the same time, discussions about the conduct of inquiry will be affected by anticipated outcomes of research.

Lopata has described the process I have in mind as the "sophistication and the rebelliousness of the client" in contexts in which expert knowledge is dispensed (Lopata 1976: 437). Lopata notes that several social changes are responsible for the difficulty in monopolizing knowledge, by the professions for instance, and for the refusal of consumers and clients to remain passive and conforming recipients of expert advice. There is, first of all, the very increase in the volume of knowledge-based occupations, which reduces the ability to strictly enforce and control the boundaries and the nature of discourse and increases the fragmentation of fields of expertise. The fragmentation of expertise becomes public knowledge. Secondly, the astuteness and cognitive skills of the public increases. New organizations and pressure groups emerge, reinforcing the decline in the authority of experts.

Efforts to regulate and police knowledge are typically undertaken and/or initiated as well as legitimated outside the boundaries of the scientific community (naturally with repercussions for the production of knowledge within the science system). For the purposes at hand, 'regulating' refers, in the most general sense, to the conscious, strategic use of political and legal power, as well as economic resources and cultural authority, to shape—whatever the specific objective—the utilization of scientific-technical knowledge.[11] It involves a complex set of mainly formal ventures designed to encourage, restrict, shape, or banish knowledge claims and set standards for their use through pressure, the creation of institutions, and the deployment of norms and beliefs to make certain that knowledge evolves along a desired path and has only sanctioned consequences.

The source of the standards chosen to police knowledge, the regulatory procedures put in place, and the intellectual systems legitimizing the cultural dismissal of certain uses of knowledge typically also do not originate in science and technology itself. For example, in the face of demands to preserve and defend the nature of human nature in response to developments in scientific and technical capacities to alter the *status quo* of human reproduction, scientific "notions of nature do not provide us with unambiguous standards of naturalness to which we can appeal for normative orientation" (Van den Daele 1992: 549). Since scientific notions of naturalness allow for the construction of a range of possible natures, regulation efforts advancing the cause of abstaining from practical steps intervening into human nature have to appeal to moral claims and political action that may or may not succeed in arresting human nature. The anchoring of standards and justifications outside of science does not mean that individuals who are scientists may not be found among those who vigorously support attempts to regulate knowledge.

My list of the available measures to control knowledge may at first leave the impression that I include science and technology policies as primary examples of

---

[11] My use of the concept of 'regulation' resonates with the way in which Steinmetz (1993) deploys the term to analyze the regulation of the emergence of the welfare state in Imperial Germany. This concept takes its distance from the economic literature on regulating the practices of capital accumulation (e.g. Jessop 1990) because that approach tends to rely on an overdetermined image of the ultimate efficacy of regulation practices.

such efforts. Strategies designed to regulate knowledge are mostly responses to changed and novel knowledge, not vice versa. Science and technology policies aim to encourage the development of knowledge, but they generally do so in highly ambivalent and open-ended fashion. Many decades of experience demonstrate, furthermore, that it is difficult or even impossible to steer and control the dynamics of developments in science and technology by way of political standards (cf. Van den Daele 1992: 553–555). In contrast to strategic efforts designed to plan and encourage future knowledge, attempts to 'police' knowledge cover a much wider social field than science and technology policies, including more informal control processes. The controls knowledge politics may impose could extend to the ways in which knowledge is disseminated and travels, is dispensed, made accessible, employed and interpreted.

The ideal-typical concepts of research and knowledge policies and their separate strategic functions for the development of knowledge and its societal deployment may increasingly be blurred in knowledge societies as the boundaries of science and society become more fluid and porous. Efforts to regulate knowledge will influence science policies and sciences policies will have an impact on attempts to police knowledge.

Shifting boundaries between science and politics for example may be manifest with respect to the process of the fabrication of knowledge; in particular, the emergence of cognitive closure, consensus formation or the evolution of uncontested facts in scientific fields increasingly may incorporate non-scientific actors and non-systemic groups. The more or less direct intervention into cognitive processes in science perhaps is most evident in the case of problem-oriented research such as environmental research, risk and technology assessment. Some fields of medical research may serve as another example. In France, the involvement and support of patient groups for the treatment of muscular dystrophy has lead to considerable investments by their organisation into molecular biology and the human genome (cf. Latour 1998: 208).

18.3.2 The Social Control of Knowledge Claims in Science

In yet another sense, the social control of knowledge claims *in* knowledge-rich and knowledge-based social systems is not a novel phenomenon. What makes science unique among social systems, for example, is the way in which and the extent to which the social task of maintaining the "quality of the products" of science is accomplished "with so little difficulty that the problem of quality control has received no more than passing mention in any systematic discussion of science" (Ravetz 1971: 273). Assessment of 'quality' is constitutive of much of the work done in science.

For Karl Popper, as is well known, the willingness to submit ideas to critical scrutiny and commitment, and not to accept knowledge claims at face value, constitutes the demarcation criterion between science and other social systems, including systems driven by ideas. Whether or not such a demarcation criterion linked to the motives of individual scientists and the institutional norms allows us to distinguish in an unambivalent manner between science and other increasingly knowledge-based social institutions is not at issue in this context. Nor do I intend to inquire into the functions of quality control, how the standards of the quality control may be elaborated, the precise mechanisms and enforcement of the social control of knowledge in science, whether these processes are effective in weeding out "shoddy science", and how science may be stratified with respect to the policing of knowledge. Much has been written about these matters in recent years. Quality control in present-day science is clearly no longer as invisible and taken-for-granted as in the past. However, a more extensive discussion of the internal control mechanisms of science is accompanied by skepticism about the efficacy of self-policing, and therefore by demands that control within the scientific community must become a strictly formalized undertaking. In a society that is itself knowledge-based, the problem of the social control of knowledge both within and outside of science inevitably becomes a central social and political problem.

The social regulation *of* science-in-progress is a highly difficult and perhaps impossible undertaking that, furthermore, has the unintended consequence of reducing the authority of science as an asset to politics. Perhaps the most significant barrier in the way of extensive external social control mechanisms on science-in-progress is the size and organization of the scientific enterprise today, as well as its competitive and its international texture.[12] The politics of science must not be conflated with the politics of society. The politics of knowledge cannot simply be reduced to political power, and science generates many kinds of knowledge, not only knowledge that is essentially political and therefore of immediate practical use.

18.3.3 The Societal Regulation of Knowledge

It seems highly likely that not only the state and major social institutions, but also social movements and groups of affected 'laypersons', will demand and organize to implement measures to increasingly regulate knowledge. In the past two decades, for example, AIDS research in the United States has been marked "by a sustained lay invasion of the domain of scientific fact-making" (Epstein 1996: 330) breaking down some of the entrenched barriers between science and society.[13] The

[12] The enlargement of the scientific community into an international or even global community is becoming a focus of reflection and research in science studies (e.g. Schott 1988, 1993).

[13] Assessing the impact of the interventions by uncredentialed participants in biomedical research and in AIDS care, Epstein (1996: 346) concludes that "the impact of the AIDS movement on biomedical institutions in the United States has been impressive and conspicuous [and] it has rapidly become something of a cliché to say that the doctor-patient relationship will never be the same in the wake of AIDS".

experience of AIDS research signals that efforts to control the application of knowledge—in this case prominently the aspects of who is to benefit, when and for what 'price'—has repercussions for the development of knowledge in academic science and for research and development in corporations.

It is perhaps self-evident that the growing efforts to police knowledge signal that claims about the inevitability of a self-propelled domination of society by science are simply unsupportable. The specific issue I will therefore discuss is not what I consider almost beyond dispute, namely that the deployment of control and regulation measures will increasingly be aimed at knowledge, but rather the entirely unresolved issue of the likely efficacy of all efforts to police knowledge. There is a yawning gulf between approaches that stress the ease with which knowledge is monopolized and controlled by an elite and the very different perspective advanced here, which emphasizes the extent to which the expanded role of knowledge significantly diminishes the ability of either major societal institutions or small groups to harness without serious challenge the gains that result from the growth of knowledge.

During the evolution of industrial society, liberal democracies successively instituted increasingly elaborate legal frames pertaining to the social status and use of property and labor. Thus the freedom of economic actors to exercise power and authority by virtue of their individual or collective ownership over labor power or the means of production is increasingly constrained and circumscribed by a host of legal norms, as well as organizations and political programs that emerge around these factors. Ownership is restrained not only spontaneously by the market, for example, but also by the state. Deliberate and anticipatory legal constraints on the use of property and labor are not neutral. Legal norms convey, from the point of view of certain actors, especially those who feel impotent in acquiring ownership and in affecting the legal rules pertaining to their disposition, privileges; while they signal (natural) rights to those who control property and labor. Unequal access to ownership, and therefore any stratification of effective influence on the construction of the legal restraints and rights, is in turn typically—but not always exclusively—based on an unequal distribution of labor and property in industrial society, elements that are constitutive for its social and economic existence.

It is almost self-evident that *legal* efforts and legislation in knowledge societies will be increasingly directed toward ways of controlling the employment, and indirectly the development, of knowledge. I emphasize political and legislative efforts to control the implementation of scientific knowledge rather than more tenuous forms of informal or spontaneous social control because the latter are simply part and parcel of the conventional state of affairs of science and its relation to society, namely the standard selectivity with which knowledge develops and is utilized. Vigorous opposition to political ventures to limit the considerable autonomy of the modern scientific community and to control knowledge will be as common as was opposition to efforts to control the use of property or the ways in which labor power might be utilized by the owners of the means of production.

One question that must be examined in the face of demands for the regulation of scientific findings has to do with the problem (which is not merely a new problem) of the extent of the social independence of science, its origins, its foundation and development; as well as the demand, which under certain circumstances opposes such independence, for some kind of control over scientific development, the communication of scientific findings and/or the consequences of scientific knowledge, whether through a kind of voluntary self-control by scientists or by means of externally implemented measures.

The type of control over science that is chiefly of interest here is therefore not related to the (primary) social control of scientific findings, that is to say, to forms of control that arise from the existence of such social constructs as the "scientific community' itself. The system-specific regulation of knowledge has already been mentioned. Robert K. Merton, in one of the most influential treatments of this topic, has attempted to describe the peculiar form of primary or system-immanent social control in the modern scientific community by drawing attention to the existence of a number of special social norms that regulate the social relations among scientists. The presence of a particular social convention, such as for example the demand for unimpeded access of all scientists to all research findings, which also simultaneously means a ban on any form of secrecy or selective communication of scientific results, represents, no matter what attitude one takes to the concrete rules of conduct, a form of social control that influences or regulates, for example, the possible content, extent, goals and methods of communication. In summary, only a limited palette of possibilities from a multitude of other possibilities in the relevant context can be realized. In terms of primary social control, it is therefore a matter of a control taken for granted by scientists, and of a form of constraint on their social and intellectual life that is largely regarded as legitimate and necessary. Whenever the control and/or the freedom of science are under discussion, this taken-for-granted social control cannot be at issue. This control, which certainly must vary in its extent and manner and in the degree to which it is accepted, is, if you like, one of the indispensable resources of the social cohesion or solidarity of any institution, and thus of the scientific community as well.

Against the background of system specific social control within science, therefore, those discussions that lead to a revision or extension of the already existing forms of control in the scientific community are of interest. With mounting efforts outside of science to regulate new knowledge produced by science, the nature of social control within science is bound to be effected and changed. I do not merely mean to refer to what constitutes a kind of anticipatory regulation of research efforts and the informal or formal acceptance of zones that constitute investigatory matters and methods that are off limits, for instance, in the form of ethical certification requirements. In fact, what can and likely may increasingly occur is a convergence or mixture of regulatory practices.

Appended to the United States Human Genome Project is an NIH/DOE Committee to Evaluate the Ethical, Legal, and Social Implications Program of the Human Genome Project (ELSI). The committee has a short but contested history. The National Institute of Health (NIH) has proposed to attach ELSI units to its other institutes and research endeavors (cf. Murray 2000). Such a program, though peer-review based but not in the usual sense since assessments of research proposals are interdisciplinary, represents at least an enlargement of traditional system specific mechanisms of social control in science if not, in this instance, an intrusion of the state and the public concerns into the regulation of the development of knowledge and obviously difficult anticipatory judgments about its social implications. Such committees also raise the general question of the role of democratic order and the influence civic society ought to have on the ways in which the results of scientific research are deployed if at all.

18.3.4 The Public and Science

And in this context, the "loss of contact" (Holton 1986: 92) between science and the larger public is today emerging as a salient attribute of the interrelation between knowledge and society. Large segments of the public have become disenfranchised, at least in the view of the scientific community. This loss of contact is not only the result of a growing cognitive distance between science and everyday knowledge; it is also affected by the ever increasing speed of knowledge expansion and by the deployment of knowledge as a productive capacity. The decreasing cognitive proximity increases the political distance from science, for example by restricting public reflection on both anticipated and unanticipated transformations of knowledge resulting from the application of knowledge.[14] The scientific community shares responsibility for this diminishing intellectual proximity, since the preferred self-image of science as a consensual, even monolithic and monologic, enterprise is increasingly in conflict with both its public role and its own internal struggles about research priorities, as well as the generation of data and their interpretation.

However, on political and moral grounds many groups, constituencies and institutions must be consulted before decisions are made about issues that affect the regulation of knowledge and indirectly the development of science and technology. It would be misleading to think that the distance from and the loss of contact with science, or the considerable scientific illiteracy in modern societies, is somehow a "potentially fatal flaw in the self-conception of the people today" (Holton 1992: 105) and/or signals the possibility of a dramatic collapse in public support for science. It is more accurate, perhaps, to speak of a state of precarious balance affecting the autonomy and dependence of science in modern society. A loss of close intellectual contact between science and the public is perfectly compatible with both a diffuse support for science in modern society and an assent to legal and political efforts to control the impact of science and technology. In another sense,

[14] As late as in the 1970s, confidence in the capacity of 'disinterested' scientists to resolve public issues in the area of space exploration, nuclear power or food additive regulation, etc., was still considerable and significantly exceeded confidence in other groups or agencies (cf. Miller 1983: 90–93; Jasanoff 1990: 12). The general decline in the last two or three decades among the public of developed societies of the trust in science and technology as a problem-solver, a trust that had hitherto been a core element of modernity, has been documented by Inglehart (1995: 391).

however, the loss of cognitive contact is almost irrelevant, and highly controversial; for example, when 'contact' is meant to refer to close cognitive proximity as a prerequisite of public participation in decisions affecting scientific and technological knowledge. Such a claim is practically meaningless because it almost requires public engagement in science-in-progress (cf. Collins 1987: 691).

From the point of view of the scientific community, the lack of cognitive proximity to the general public has advantages and disadvantages. The loss of contact between science and the public can perhaps explain, at least in part, why the scientific community, in view of its attractiveness and usefulness for corporations, the military and the state, has been able to preserve a considerable degree of intellectual autonomy (cf. Gilbert/Mulkay 1984). Such autonomy, however, is contingent on a host of factors within and without the scientific community. The loss of contact is a resource for the scientific community. It signals a symbolic detachment and independence that can be translated into an asset vis-a-vis the state and other societal agencies. Science becomes an authoritative voice in policy matters; or it represents, in ideological and material struggles with other political systems, the openness of society (cf. Mukerji 1989: 190–203). But the cognitive distance also limits the immediate effectiveness of the "voice of science' in policy matters,[15] and extensive autonomy and independence of science may result in an excessive celebration of 'normal' scientific activity and lead to a lack of innovativeness.

From the point of view of the non-scientific institutions, the lack of intellectual proximity of the public to scientific knowledge in general and research fronts in particular also has both advantages and drawbacks. Selected disaffection with science and technology has always accompanied its development; strong demands and efforts to legislate selectivity in the ways in which knowledge is implemented and deployed can lead to even stronger disaffections with science, although such a response may be dismissed as part of an anti-science crusade or movement. But the term 'anti-science' is vague and brings together a broad range of things that typically "have in common only that they tend to annoy or threaten those who regard themselves as more enlightened" (Holton 1992: 104).

18.3.5 The Development of Social Controls

The social control and regulation of scientific knowledge that has moved from the stage of being-in-progress to some form of completion and desires to be

[15] Mukerji (1989: 197) describes the trade-off: "What reassures scientists the most when they face the power of the voice of science and their powerlessness to use the voice in the public arena is the idea of their autonomy. Scientists are not, in the end, politicians, and they suffer political defeats better than the loss of face among their peers. As long as they can conduct research with which they can advance science [both science itself and their positions in it], they can feel potent. But the cost is that scientists cultivate an expertise that empowers someone else".

implemented outside of the scientific community is already quite extensive. In all modern societies, we now find elaborate drug regulations and corresponding agencies that register, test, control or permit pharmaceutical substances to enter the market as legalized drugs. Until a few decades ago, decisions about the production and marketing of chemicals as drugs were typically made by corporations, by individual pharmacists or by physicians (cf. Bodewitz et al. 1987). As scientific knowledge is 'applied', it becomes embedded in social contexts external to science. As a part of such embeddedness, knowledge is subject to the kinds of control mechanisms and social constraints found in these contexts. It simply cannot escape the selectivity that issues from such external contexts, even if only in efforts designed to generate trust toward a certain artifact or solution offered by novel knowledge.

The whole area of national and international intellectual property and copyright protection is another arena in which legislation to control the deployment of scientific and technical knowledge is already extensive. In many ways, such controls date back at least to the 1883 Paris Convention for patents and related industrial matters and to the 1886 Berne Convention for copyrights. The acceleration in the speed with which inventions reach the market, their shortened economic life-span and the extent to which recent inventions, for example in the field of microelectronics, the organization of production, medical treatments and biotechnology, are difficult to protect from copying efforts will increase pressures to enact further protective legislation (cf. Vaitsos 1989).

In social theory, the institution generating knowledge and the institution contemplating and executing political action were once regarded as entirely unrelated domains. At the beginning of the twentieth century, the dilemma of the indispensable separation of science and politics found perhaps its most influential expression in Weber's ([1921] 1948: 77–128, [1922] 1948: 129–156) essays on science and politics as a vocation. Today, the intellectual foundations that allowed Weber to legitimize the fundamental division between the practices of knowledge and politics have fallen into disrepute. Confidence in the neutrality, instrumentality and political neutrality of science has been thoroughly eroded. Reference to the politics of knowledge therefore no longer constitutes a profound break or a violation of the norms of scientific action and the essentially means-like attributes of scientific knowledge. Science is deeply implicated in social action and political agendas hold sway over science. Precisely how dependent or interdependent science and politics are is a matter of ongoing debate and empirical analysis. But the widespread disenchantment with science and the extensive material dependence of the scientific community on the state do not justify the equally unrealistic proposition that the boundaries between politics and science have altogether vanished. Science remains embedded in particular political realities, and as long as it is situated in a form of civil and political society free of totalitarian strains, scientific activity tends to benefit. By the same token, as long as traffic across the boundaries of science remains widely unimpeded and subject to negotiation, both science and society gain.

In as much as knowledge becomes the constitutive principle of modern society, the production, distribution and especially the application of knowledge can avoid political struggles and conflicts less than ever. The distribution and implementation (and with it the fabrication) of knowledge increasingly becomes a domain of explicit legislation and a target of political and economic decisions. Such a development is inevitable, because "as the institutions of knowledge lay claim to public resources, some public claim on these institutions" (Bell 1968: 238) and their results are unavoidable. Even more significant is that, as the importance of knowledge as a central societal resource increases, its social, economic and political consequences for social relations grow rapidly, together with demands to regulate the specific utilization and access to knowledge.

The dissemination and application of knowledge does not occur in the imaginary world of perfect competition and equality of opportunities. As a result, a politics of knowledge must confront the consequences of the social distribution of knowledge, especially the stratified access to and utilization of knowledge. It remains an open question, for example, to what extent dispossession of knowledge generates social conflicts and in what specific ways such struggles manifest themselves. Bell (1964: 49) warned several decades ago that right-wing extremism may 'benefit' from any exclusion of social groups from access to and acquisition of technical expertise.

However, such predictions about the intellectual, social and economic gaps sustained by knowledge overestimate the extent to which knowledge and its use can in fact be controlled. It will be increasingly difficult to control knowledge, in spite of the many efforts that will undoubtedly be made. Efforts to control knowledge encounter contradictions. Sustaining economic growth, for example, requires an expansion of knowledge. And knowledge that expands rapidly is difficult to control. The expansion of knowledge enlarges the segment of knowledge-based occupations. Knowledge expansion and knowledge dissemination rely on conditions that are themselves inimical to control. Nonetheless, as I have observed, the typically expressed fear that an inevitable outcome of such developments is the greater ease with which knowledge (and information) can be monopolized and effectively employed for repressive (even totalitarian) purposes, or even as a tool of maintaining the benign status quo, had been a widely accepted premise of discussion of the social control of knowledge even before Orwell's classic book on the subject. What exactly nourishes this point of view? What is the basis for the widespread conviction that knowledge and technical artifacts are relatively easy to control and that access to knowledge can be easily denied?

18.3.6 Knowledge Hierarchies and Monopolies

One of the ways to understand the various means by which knowledge is seen to be controlled, perhaps even monopolized, and its gains—following the Matthew principle—primarily allocated to the rich and powerful, is to examine the literature that has incessantly informed us that precisely such outcomes are built into the very logic of scientific and technological development. What exactly is it, in the view of these critics, that gives technology and scientific knowledge such potency and discriminatory power? And what kinds of *mundane encounters* with modern science and technology may have prompted or at least reinforced the critics' theoretical conceptions of science and technology? Typical encounters with science and technology in everyday life must have left their mark and strengthened otherwise rather abstract assessments of the technical artifacts and scientific knowledge. I will suggest that these essential and affirming encounters are experiences with 'frozen' or arrested technical artifacts and knowledge forms.

My aim is not an exegesis of the epistemological or theoretical ancestry of such views. I presuppose that the critique of modernity, insofar as it touches upon the rationality of science and technology, represents a form of civilizational critique that has accompanied the emergence of modern societies from the beginning. The critics of modern civilization flatly reject the claim that science and technology, as celebrated by its proponents, are socially and politically neutral. As Marcuse pointedly asserts: "Science, *by virtue of its own method* and concepts, has projected and promoted a universe in which the domination of nature has remained linked to the domination of man" (Marcuse [1964] 1989: 166). For illustrative purposes, I refer in some detail to two representative philosophical and sociological critiques of the interrelations between the social and intellectual fabric of society, knowledge and technology; namely, the analysis of modern science and technology by Herbert Marcuse and Helmut Schelsky.[16]

Marcuse's views of the role of modern science and technology gained considerable public resonance with the publication in 1964 of his *One-Dimensional Man,* subtitled "Studies in the Ideology of Advanced Industrial Society"; but they can be traced back to his writings and those of both Adorno and Horkheimer in the early 1940s. Critical theory, in effect, abandons Marx for Weber on the issue of the emancipatory potential of modern reason. Marcuse observes at the time, "National Socialism is a striking example of the ways in which a highly rationalized and mechanized economy with the utmost efficiency in production can operate in the interest of a totalitarian oppression and continued scarcity. The Third Reich is indeed a form of 'technocracy'" (Marcuse 1941: 414). In the case of National Socialism, politics is still a decisive force; yet technical knowledge is already seen as an indispensable instrument of political control.

A quarter of a century later, Marcuse assails the scientific mind and the transformation of knowledge into a form of scientific-technical rationality that has perverted the project of emancipation and has led to the human domination of nature. Marcuse (1964: 146) argues that such outcomes are inherent in science, that "scientific-technical rationality and manipulation are welded together into new forms of social control" resulting in a kind of epistemic enslavement of modern

[16] A more extensive description and analysis of both Schelsky's and Marcuse's critiques of the excessive power of modern science and technology in society may be found in Stehr (1994: 203–221).

individuals. Modern individuals become incapable of seeing and dealing with the world in any other manner, hence their entrapment.

The technical presumption of science becomes a political presumption and has consequences for human social organization because the transformation of nature, according to the logic of technology, also involves changes in the social relations of individuals. Whatever claims may be made on behalf of the essential political neutrality and potential of technology, Marcuse stresses emphatically, even against Marx, that a technology that has become the *universal* form of material production, "circumscribes an entire culture; it projects a historical totality—a 'world'" (Marcuse 1964: 154). In other words, the relation and respective implication of science and its technical application, and of the nature of the society that is thereby created, can in the final analysis only be viewed as an intimate connection that operates under the same logic. Technological reason and its universals, namely the discipline and control of production resulting in regimentation, the pursuit of narrow goals or specialization and the absolute uniformity of regimented and specialized labor or standardization, are bound to predominate throughout society.[17]

The same inherent force, the rationality of domination, soon propels the universes of scientific and ordinary discourse. All sectors of society, all social activities and all subjectivities are brought under the control of technical forms of discourse. The domination of nature and society go hand in hand. Science and society become reflections of the logic of technical rationality. Marcuse therefore concludes that the "scientific method which led to the ever-more-effective domination of nature thus came to provide the pure concepts as well as the instrumentalities for the ever-more-effective domination of man by man *through* the domination of nature [...] Today domination perpetuates and extends itself not only through technology but *as* technology, and the latter provides the great legitimation of the expanding political power, which absorbs all spheres of culture" (Marcuse 1964: 158).[18] The resulting lack of freedom and autonomy appears neither as irrational nor as the result of political forces but as a 'rational' submission to the technical necessities of existence. In the final instance, therefore, instrumental reason becomes ubiquitous and turns life in society into a 'totalitarian' existence. The sphere of the political becomes, as in Schelsky's scientific civilization, the sphere of the merely technical ["the incessant dynamic of technical progress has become permeated with political content" (Marcuse 1964: 159)] and rationality becomes irrationality. The state becomes merely an expression of the technical base and is depoliticized. Social change will be arrested for the most part, especially by virtue of the power and the primacy of the

[17] The decisive outcome of these developments is that the workers are incapable of acquiring a critical view of the repressive social order. The "masterly enslavement" is pervasive throughout society, affecting all individuals at all levels of production.

[18] Adorno's ([1966] 1973: 320) image of the extension of the rule of nature to a rule over man by man is similar. Adorno warns that the "unity of the control over nature, progressing to rule over man and finally to that over men's inner nature" is one of the enormous dangers of the present age.

society's administrative apparatus, and this containment of social transformations is perhaps the most singular achievement of advanced industrial society.

Marcuse's analysis of scientific rationality is highly abstract and lacks congruence with social reality, especially with the ways in which and the extent to which many modern individuals experience spheres of autonomy and responsibility. He provides no examples of how technological means are turned into mere means of social control and domination; for example, how the telephone or television invariably become instruments of domination. The reluctance of dictators to promote a modern telephone system in the early part of this century would indicate that they feared its subversive possibilities. To this very day, differences in economic and demographic factors do not satisfactorily account for the large disparities in the dissemination of the telephone in state socialist and capitalist societies after the Second World War (cf. Buchner 1988). But even more to the point is Alain Touraine's observation that Marcuse's theory of modern society lacks reality congruence: "The image of a totally unified society, in which there is a perfect correspondence between technology, firms, the State, and the behavior of consumers and even citizens could not be further removed from observable reality" (Touraine [1992] 1995: 159).

Helmut Schelsky's thesis that advanced industrial society is evolving into "scientific civilization" was first expounded by him in a lecture in 1961 entitled "Humans in scientific civilization". For Schelsky, *modern* technology represents not merely an adaptive capacity to the constraints of nature, but a reconstruction of nature by society, and therefore of society. In the context of modern technology, humans no longer confront nature with the assistance of organs aided, improved and developed in their capacity by technology, but on the basis of a 'detour' via the brain, or the application of theoretical knowledge in practical contexts. The outcome is that, using the language Schelsky employs, an 'artificial' nature as well as an 'artificial' change of humankind itself. The result therefore is a "re-construction and re-creation of man [...] in his corporal, psychological and social existence" (Schelsky [1961] 1965: 16). We produce, as Schelsky observes, "the scientific civilization not only as technology but necessarily also in a much broader sense continually as 'society' and as 'soul'" (Schelsky [1961] 1965: 17).

Modern technology changes the relations of humans to nature, to themselves and to others. The result of this dual transformation is the "circulation of self-determined production" (Schelsky [1961] 1965: 16) representing the real foundation of scientific civilization. The self-regulated and self-propelled nature of this process, the constant production and reproduction, evolves into a self-steering process which does not appear to allow for any escape:

> Every technical problem and every technical solution invariably becomes also a social, a psychological issue because the self-propelled nature of this process, created by man, confronts humans as a social and psychological dictate which in turn requires nothing but a technical solution, a solution planned and executed by man since this is the nature of the condition to be tackled (Schelsky [1961] 1965: 16–17).

Modern technology constitutes a particular logic, and this logic necessarily becomes the dominant logic of human life. One of the significant consequences of such a conception of technology is that the traditional 'logic' of technology reverses itself. That is, technology as a producer of mere means of human action becomes a producer of ends or meaning, or in other words, 'means' of action determine its ends and prefigure the direction of social change. Schelsky describes technology as an intellectual process which dissects varied natural objects into their elementary parts in order to re-assemble them according to the principle of the least effort or maximum efficiency. The result of modern technological construction, therefore, is a novel product or process with *artificial* features and, in analogy, an *artificial* human being.

Schelsky's and Marcuse's theories evidently converge. They share the thesis that there is the distinct danger that technology in modern society will increasingly displace spontaneous social and political action and significantly reduce individual spheres of responsibility and autonomy, resulting, in the end, in the "death of the self".

Marcuse and Schelsky are by no means alone in their assessment of the trajectory of the social, political and economic development of advanced industrialized societies. Nor are they alone in attributing the societal changes they describe to intrinsic and enslaving 'laws' of science and technology. On the contrary, their observations and warnings resonate with a broad intellectual trend that actually began to take on its peculiar characteristic in the 1950s, when social theorists first noted distinctive and presumably irreversible trends in industry and production.[19] Social scientists asserted a tendency in industry toward increasing technological progress, manifesting itself in the rapid mechanization or *automation* of production. While the increased automation of production that is, as Marcuse (1964: 35) observes, inherent in technological progress itself enormously enlarges the output of commodities, it does not, as many observers then noted, make work more meaningful, demanding and challenging. The result is summed up by David Riesman and his collaborators in *The Lonely Crowd* (1950): Industry is now producing bored workers through simplified work routines, and the central meaning of life is increasingly shifting away from work toward a search for creative expressions in leisure activities.

---

[19] The genealogy of Schelsky's and Marcuse's fears about the impact of modern science and technology is of course much longer. I will refer to Max Weber but could list many more observers who have expressed concerns about the fateful consequences of science and technology in the age of modernity. Marcuse's and Schelsky's diagnoses resonate closely with Max Weber's analysis of the modern age as a demystification of the world resulting from the growing rationalization of social relations through science and technology. Weber emphasizes the painful tension between rational, empirical knowledge and meaning systems found in the life-world. Moreover, Weber's intellectual 'grandchildren' often share an "Exodus impulse", namely the attempt "to explode the fatalistically closed 'steel-hard casing' of the demystified world" (Bolz 1989: 7). Schelsky and Marcuse therefore also make use, although for the most part implicitly, of a long established radical as well as conservative (romantic) intellectual tradition that launched a highly critical and skeptical analysis of the impact of technology and science on culture and social relations.

Schelsky's and Marcuse's observations resonate with Bell's (1960) thesis about the end of ideology, as well as with the prognosis by Lane (1966) that we are about to enter an age in which scientific knowledge increasingly dislodges the political element from politics. By the same token, the futurists Kahn/Bruce-Briggs (1972: 8–29) in the early 1970s discern multi-trends within modern society that have been widely noticed by 'macro-historians', including the "centralization and concentration of economic and political power" as well as "innovative and manipulative social engineering". The growing rationality that comes with the rapid accumulation of scientific and technical knowledge, according to Kahn and Bruce-Briggs, is increasingly applied to "social, political, cultural, and economic worlds" (Kahn/Bruce-Briggs 1972: 9). Although this trend may not accelerate, the desirability of social engineering is widely supported and an "almost universal belief among the educated" (Kahn/Bruce-Briggs 1972: 29).

The influence of ideological and, more generally, of political factors on scientific and technical developments remains unanalyzed, however. This suggests that the conventional central theoretical categories employed in the analysis of modern society, partly inherited by present-day social science from the past century, such as class or economy but also such notions as capitalist or socialist, have lost their crucial role in social theory. Observers were increasingly convinced that the distinction between capitalist and state socialist economic orders was becoming obsolete. At the same time, however, confidence in the power and the uniqueness of scientific knowledge remained strong. Aron ([1962] 1967: 42) embraced and highlighted these assumptions in his theory of 'progressive' industrial society. At the same time, questions about the motor of "social change" or the centrality of the economic system for societal transformations were raised anew. It is at this time that theorists began to advance the thesis that technology and science, rather than the economy, are the real motor of societal change in modern social systems (cf. Parsons 1970: 619).

More generally, however, Schelsky's and Marcuse's accounts of the social and political force of modern science and technology suffer from an unintended but nonetheless misplaced confidence in the practical efficacy of scientific reasoning and quantification. Knowledge and technology are for the most part treated as a black box. The concern with technical artifacts is primarily functionalist. The major question posed concerns the psychological, social and political consequences of objects in the sphere of social relations. What exactly confers such power on objects is never examined. Marcuse and Schelsky presuppose an image of science and technology that then gives them reason to despair. One perceived consequence of technology and science, the extent to which the world of objects begins to dominate the world of subjects, paradoxically rests on an acceptance by both Marcuse and Schelsky of the positivist image of science as a most efficient, rational enterprise that produces highly useful devices and knowledge claims. As a result, we must return to our initial question: what nourishes such a view of science and technology, in spite of Marcuse's and Schelsky's otherwise deep misgivings about such a science and such efficient technical objects?

At this point, we must take cognizance of some kind of phenomenological analysis of everyday experience and common sense understanding of science, especially regarding technical matters, that are not further investigated by Marcuse and Schelsky, even though they serve as starting point and as affirmation of their observations. The primary experience in everyday contact with technology is the *finished* product. The everyday experience of technology is not rooted in an understanding of the conception and fabrication, in short: The decisions that constitute the nearly always invisible "technical code" of a matter and that co-determine the ways of using such technologies in everyday contexts are not manifest to the user.

Feenberg has provided us with a fruitful explication of the concept of the technical code: The technical code refers to those attributes of an object that reflect the hegemonic values and beliefs that prevail in the design process. Such codes are usually invisible because, like culture itself, they appear self-evident. For example, tools and workplaces are designed for adult hands and heights not because workers are necessarily adults, but because our society expelled children from the work process at a certain point in history with design consequences we now take for granted. Technical codes also include the basic definition of many technical objects insofar as they become universal, culturally accepted features of daily life. The telephone, the automobile, the refrigerator, and a hundred other everyday devices have clear and unambiguous definitions in the dominant culture (Feenberg 1995: 4).

While the technical code of an object originates or is provided in the context of its production, it is thus not yet necessarily decided how ultimately to handle an object—in the context of its use—as if it were natural. For this, the "cultural code" is a further requirement, since it contributes to the decision of which possibilities for use are connected with an object. Technical and cultural codes may overlap, but they can also diverge. Ultimately, the cultural code can also change. In any case, technical and cultural codes more or less definitively limit the imaginative possibilities for use, and have as a consequence the fact that everyday experiences with objects are primarily 'successful' experiences. And this counters the disappointments that naturally also continually occur, nonetheless probably basically confirming the confidence in the predetermined technical and cultural process of the object. The technical and cultural code endow the object with a specific process or even a purpose which will be fulfilled by it. The codes stabilize usage. Objects confer certainty. The degree of security that allows these coded processes to be reproduced again and again is then associated primarily with an image of reliability—although the goals that can be realized with this reliability can be of various different kinds. In any case, in the process an emotional connection with the object takes form. This certainty, security and reliability in principle in everyday dealings with technical objects at the same time induces, according to my thesis, a high degree of confidence in the efficiency of objects. The fact that connected with this efficiency there might at the same time be a feeling of helplessness or of the "power of objects over us" is understandable. The limited technical and cultural code of an object, even if "the radical constraints on possible integration of objects are in the interest of those integrations that serve to satisfy the needs of powerful individuals or groups" (Joerges [1979] 1996: 25) obstructs alternative possibilities for use and confirms one's helplessness in handling objects. This is, to be sure, nothing other than a reification of the dominant code.

A phenomenology of technology underlines some general observations by Alain Touraine about the actual role of technology in a society that is increasingly based on technology:

> We live in a society in which means were completely divorced from ends. Far from determining or absorbing ends, the same means could therefore be used for both good and evil ends, for both reducing inequality and exterminating minorities. The increasingly dense networks of technologies and signs in which we now live, and which orient and govern the ways in which we behave, by no means imprison us in a technological world and by no means destroy social actors. They impose neither a logic of efficacy and production nor a logic of control and reproduction. The image of technocracy triumphant is pathetically inadequate if we contrast it with the increase in consumption, the rise of nationalisms and the might of transnational companies (Touraine [1992] 1995: 148–149).

18.4 Prospects

In my view, efforts to police knowledge and to defend society against some of the anticipated but also uncertain effects of the utilization of recent gains in knowledge ultimately will do little to seriously limit its application, in one way or the other. But this will not keep various societal agents from trying.

One of the most immediate and controversial questions that awaits regulation and resolution as the result of evolving knowledge about the susceptibility to certain health risks in relation to specific genes is the question of how insurance companies (and other organizations and institutions), in particular health insurance companies, will use such information.

Private *health* insurance companies in Germany have announced (*Frankfurter Allgemeine Zeitung*, July 21, 2000: 17) that they plan to continue to use established procedures when it comes to a determination of calculating the risks individual applicants represent (also Murray 2000: 242–245; *Task Force on Genetic Information and Health Insurance* 1993). That is, full disclosure of all relevant information is required. The applicant is under no obligation to disclose information she/he does not happen to have. A genome analysis will not, the insurer's indicate, become a prerequisite in issuing a policy. However, individuals who happen to such information, for example, as the result of taking part in a research study, are expected to divulge the genetic information.

But how is one to insure that insurance companies limit their usage of such information voluntarily? What exactly is genetic information? How broad or narrow can or should one define genetic information? And, how does one treat the interaction between genetic and nongenetic 'causes'? How does one attribute responsibility? Can an insurer acquire genetic information indirectly, for example, on the basis of a family history? Are special legal norms required? Genetic tests are bound to become more common, more accessible, and less and less expensive. Policing knowledge looks like work that Sisyphus might know.

References

Aron, Raymond. [1962] 1967. *18 Lectures on Industrial Society*. London: Weidenfeld & Nicolson.

Adorno, Theodor W. [1966] 1973. *Negative Dialectics*. London: Routledge and Kegan Paul.

Bechmann, Gotthard and Nico Stehr. 2000. "Risikokommunikation und die Risiken der Kommunikation wissenschaftlichen Wissens. Zum gesellschaftlichen Umgang mit Nichtwissen", *Gaia*, 9: 113–201.

Bell, Daniel. 1960. *The End of Ideology*. Glencoe, IL: Free Press.

Bell, Daniel. 1964. "The Post-Industrial Society", in: Eli Ginzberg (ed.), *Technology and Social Change*. New York, NY: Columbia University Press: 44–59.

Bell, Daniel. 1968. "The Measurement of Knowledge and Technology", in: Eleanor B. Sheldon and Wilbert E. Moore (eds.), *Indicators of Social Change. Concepts and Measurements*. Hartford, CT: Russell Sage Foundation: 145–246.

Bodewitz, Henk J.H.W., Henk Buurma and Gerard H. de Vries. 1987. "Regulatory Science and the Social Management of Trust in Medicine", in: Wiebe E. Bijker, Thomas P. Hughes and Trevor Pinch (eds.), *The Social Construction of Technological Systems: New Directions in the Sociology and History of Technology*. Cambridge, MA: MIT Press: 243–259.

Bolz, Norbert. 1989. *Auszug aus der entzauberten Welt. Philosophischer Extremismus zwischen den Weltkriegen*. München: Fink.

Böhme, Gernot and Nico Stehr (eds.). 1986. *The Knowledge Society*. Dordrecht: D. Reidl.

Buchner, Bradley J. 1988. "Social Control and the Diffusion of Modern Telecommunications Technologies: A Cross-National Study", *American Sociological Review*, 53: 446–453.

Collins, Harry M. 1987. "Certainty and the Public Understanding of Science: Science on TV", *Social Studies of Science*, 17: 689–713.

Epstein, Steven. 1996. *Impure Science. Aids, Activism, and the Politics of Knowledge*. Berkeley, CA: University of California Press.

Feenberg, Andrew. 1995. *Alternative Modernity: The Technical Turn in Philosophy and Social Theory*. Berkeley, CA: University of California Press.

Fukuyama, Francis and Caroline S. Wagner. 2000. *Information and Biological Revolutions. Global Governance Challenges—Summary of a Study Group*. Santa Monica, CA: Rand.

Fuller, Steve. 1993. *Philosophy, Rhetoric, & the End of Knowledge: The Coming of Science & Technology Studies*. Madison, WI: University of Wisconsin Press.

Gibbons, Michael et al. 1994. *The New Production of Knowledge. The Dynamics of Science and Research in Contemporary Societies*. London et al.: Sage.

Gilbert, Nigel G. and Michael Mulkay. 1984. *Opening Pandora's Box*. Cambridge, MA: Cambridge University Press.

Haeckel, Ernst. 1878. *Freie Wissenschaft und freie Lehre. Eine Entgegnung auf Rudolf Virchow's Münchener Rede über "Die Freiheit der Wissenschaft im modernen Staat"*. Stuttgart: Schweizerische Verlagbuchhandlung.

Holton, Gerald. 1986. "The Advancement of Science and its Burdens", *Daedalus*, 115: 77–104.

Holton, Gerald. 1992. "How to Think About the 'Anti-Science' Phenomenon", *Public Understanding of Science*, 1: 103–128.

Inglehart, Ronald. 1995. "Changing Values, Economic Development and Political Change", *International Social Science Journal*, 145: 379–403.

Jasanoff, Sheila. 1990. *The Fifth Branch. Science Advisors as Policymakers*. Cambridge, MA: Harvard University Press.

Jessop, Bob. 1990. "Regulation Theories in Retrospect and Prospect", *Economy and Society,* 19: 153–216.

Joerges, Bernward. [1979] 1996. "Die Macht der Sachen über uns", in: Bernward Joerges (ed.), *Technik. Körper der Gesellschaft.* Frankfurt am Main: Suhrkamp: 15–32.

Jonas, Hans. 1974. *Philosophical Essays: From Ancient Creed to Technological Man.* Englewood Cliffs, NJ: Prentice Hall.

Kahn, Hermann and B. Bruce-Briggs. 1972. *Things to Come. Thinking about the Seventies and Eighties.* New York, NY: Macmillan.

Krohn, Wolfgang and Günter Küppers. 1989. *Die Selbstorganisation der Wissenschaft.* Frankfurt am Main: Suhrkamp.

Lane, Robert E. 1966. "The Decline of Politics and Ideology in a Knowledgeable Society", *American Sociological Review,* 31: 649–662.

Latour, Bruno. 1998. "From the World of Science to the World of Research?", *Science,* 280: 208–209.

Lopata, Helen Z. 1976. "Expertization of Everyone and the Revolt of the Client", *Sociological Quarterly,* 17: 435–447.

Marcuse, Herbert. 1941. "Some Social Implications of Modern Technology", *Studies in Philosophy and Social Science,* 9: 414–439.

Marcuse, Herbert. [1964] 1989. *Der eindimensionale Mensch. Studien zur Ideologie der fortgeschrittenen Industriegesellschaft. Schriften* 7. Frankfurt am Main: Suhrkamp.

Miller, Jon D. 1983. *The American People and Science Policy.* New York, NY: Pergamon.

Murray, Thomas H. 2000. "Das Humangenomprojekt, das ELSI-Programm und die Demokratie", in: Matthias Kettner (ed.), *Angewandte Ethik als Politikum.* Frankfurt am Main: Suhrkamp.

Mukerji, Chandra. 1989. *A Fragile Power. Scientists and the State.* Princeton, NJ: Princeton University Press.

Nelkin, Dorothy. 1995. "Science Controversies. The Dynamics of Public Disputes in the United States", in: Sheila Jasanoff, Gerald E. Markle, James C. Petersen and Trevor Pinch (eds.), *Handbook of Science and Technology Studies.* London, Thousand Oaks, CA: Sage Publications: 444–456.

Parsons, Talcott. 1970. "The Impact of Technology on Culture and Emerging New Modes of Behavior", *International Social Science Journal,* 22: 607–627.

Ravetz, Jerome. 1971. *Scientific Knowledge and its Social Problems.* New York, NY: Oxford University Press.

Riesman, David. [1950] 1961. *The Lonely Crowd: A Study of the Changing American Character.* New Haven, CT: Yale University Press.

Schelsky, Helmut. [1961] 1965. *Der Mensch in der wissenschaftlichen Zivilisation.* Köln, Opladen: Westdeutscher Verlag.

Schott, Thomas. 1988. "International Influence in Science: Beyond Center and Periphery", *Social Science Research,* 17: 219–238.

Schott, Thomas. 1993. "World Science: Globalization of Institutions and Participation", *Science, Technology, and Human Values,* 18: 196–208.

Simmel, Georg. [1907] 1978. *Philosophie des Geldes,* 2. Auflage, Leipzig: Duncker & Humblot.

Stehr, Nico. 1994. *Knowledge Societies.* London: Sage.

Stehr, Nico. 2000. *The Fragility of Modern Societies: Knowledge and Risk in the Information Age.* London: Sage.

Steinmetz, George. 1993. *Regulating the Social. The Welfare State and Local Politics in Imperial Germany.* Princeton, NJ: Princeton University Press.

Task Force on Genetic Information and Insurance. 1993. Bethesda, ML: National Institutes of Health, National Center for Genome Research.

Touraine, Alain. [1992] 1995. *Critique of Modernity.* Oxford: Blackwell.

Vaitsos, Constantine V. 1989. "Radical Technological Changes and the New 'Order' in the World-Economy", *Review,* 12: 157–189.

Van den Daele, Wolfgang. 1992. "Concepts of Nature in Modern Societies and Nature as a Theme in Sociology", in: Meinolf Dierkes and Bernd Biervert (eds.), *European Social Science in Transition. Assessment and Outlook*. Frankfurt am Main: Campus: 526–560.

Van den Daele, Wolfgang, Wolfgang Krohn and Peter Weingart (eds.). 1979. *Geplante Forschung. Vergleichende Studien über den Einfluss politischer Programme auf die Wissenschaftsentwicklung*. Frankfurt am Main: Suhrkamp.

Virchow, Rudolf. 1877. *Die Freiheit der Wissenschaft im modernen Staat,* (Rede gehalten in der 3. Allgemeinen Sitzung der 50. Versammlung deutscher Naturforscher und Aerzte zu München am 22. September 1877). Berlin: von Wiegand, Hempel & Perry.

Weber, Max. [1921] 1948. "Politics as a Vocation", in: Hans H. Gerth and C. Wright Mills (eds.), *From Max Weber*. London: Routledge and Kegan Paul: 77–128.

Weber, Max. [1922] 1948. "Science as a Vocation", in: Hans H. Gerth and C. Wright Mills (eds.), *From Max Weber*. London: Routledge and Kegan Paul: 129–156.

Weingart, Peter. 2001. *Stunde der Wahrheit? Wissenschaft im Verhältnis zu Politk, Ökonomie und den Medien in der Wissensgesellschaft*. Weilerswist: Velbrück Wissenschaft.

Willke, Helmut. 1999. "Die Wissensgesellschaft. Wissen ist der Schlüssel zur Gesellschaft", in: Armin Pongs (ed.), *In welcher Gesellschaft leben wir eigentlich? Gesellschaftskonzepte im Vergleich,* Vol. 1: 261–279.

Willke, Helmut. 1998. "Organisierte Wissensarbeit", *Zeitschrift für Soziologie,* 27: 161–177.

Wulff, Keith M. 1979. *Regulation of Scientific Inquiry. Societal Concerns with Research*. Boulder, CO: Westview Press.

Chapter 19
The Social Role of Knowledge

Nico Stehr

The branch of learning that has concerned itself with the nature of knowledge in general has traditionally been philosophy. Questions it investigates are the relationship between knowledge and belief (or ideology), the validity and reliability of knowledge claims pertaining to the external world and based on sense perception, the presuppositions required for the production of knowledge and the use of language in the construction of knowledge claims. Knowing in the philosophical tradition has often been reduced to the relationship between the individual subject (the knower) and the object (the known). The idea that our knowledge is a social construct is of recent origin. Similarly, the growing study of the effects of knowledge on social relations has less of a tradition than the philosophical examination of the secure foundations of knowing. The emphasis in this essay will be on the social role of knowledge in modern society.

That knowledge is of crucial importance in social, political and economic relations in modern society is increasingly accepted in the social sciences (Stehr 1994). But the most serious theoretical deficiency of existing theories that assign a central role to knowledge in society is their rather undifferentiated treatment of knowledge. The role knowledge plays within these theories has not been matched by extended and enlightened discussions of the concept of knowledge. As a result, our knowledge about knowledge is, despite, and for a time because of the sociology of knowledge, still not very sophisticated and comprehensive.

The following exposition of the social role of knowledge in modern societies will not be chronological. I will begin the essay with a discussion of the current and what is a new understanding of the social role of knowledge. I then turn to the

This text was first published as: Stehr, Nico. 2004. "The Social Role of Knowledge", in: Nikolai Genov (ed.), *Advances in Sociological Knowledge*. Wiesbaden: VS Verlag für Sozialwissenschaften: 83–105. The permission to republish it here was granted by Springer.

M. T. Adolf (ed.), *Nico Stehr: Pioneer in the Theory of Society and Knowledge*, Pioneers in Arts, Humanities, Science, Engineering, Practice 16,
https://doi.org/10.1007/978-3-319-76995-0_19

development of these ideas and trace their emergence, first, to the programme of the classical sociology of knowledge, its genesis, and interpretation in the post war era, followed, second, by an examination of the limitations of the classical sociology of knowledge that also are characteristic of the dominant post-war interpretation of the sociology of knowledge.

The prevailing conception of the sociology of knowledge began to be transformed in the late 1960s and early 1970s. I refer to the nature and the reasons for a profound transformation of the sociology of knowledge at this juncture as particularly evident in science studies. The discussion then turns to the present state once more by offering a more detailed discussion of a sociologically viable concept of knowledge that is embedded, in turn, in the contention that modern societies are best interpreted as knowledge societies. Finally, I turn to a discussion of the political economy of knowledge in modern societies.

19.1 The New Understanding of the Social Role of Knowledge

Knowledge is no longer, if it ever was, the key to unlocking the secrets of the world but the becoming of a world. The recognition of this role of knowledge in society has led to a significant change in the knowledge-guiding interests of the sociology of knowledge in recent years. The traditional form of the sociology of knowledge is until this day in its most prominent contemporary form of science studies, primarily interested in the social foundations or dependence of knowledge. As Max Scheler put it, the issue is that the "intimate linkage" (Scheler [1924] 1960: 193) between class and ideology, labor and knowledge, society and truth or interests and knowledge-guiding interests, constitutes and in many ways still represents the primary research foci of inquiries into knowledge.

Interest in the social foundations of knowledge as the research focus of the sociology of knowledge emerges out of epistemology. But in the tradition of epistemological themes, the question of the genesis, its conditions and consequences continue to form the core concern of the sociology of knowledge (Mannheim [1929] 1965: 164–167).

The sociology of knowledge has of course—aside from self-exemplifying questions about its own foundations—always problematised the societal role of knowledge (for example in the sense of *Herrschaft kraft Wissen* [authority based on knowledge] Weber [1922] 1964: 339). The sociology of knowledge also displays an interest in the social role, the carriers of knowledge (for example intellectuals, bureaucrats, experts, and cultural elites) and the civilisational transformation of forms of knowledge (for example in the work of Norbert Elias 1983). Nonetheless, and using economic concepts to describe the core foci of the sociology of knowledge as is still widely practiced, interest centres on the supply and not the demand or the consumption of knowledge. At the same time, the traditional research guiding interest always saw knowledge as something akin to a dependent variable.

The emphasis of the production of knowledge, which is of course an important issue, forms the almost singular interest of science studies as we know it. The many faceted laboratory studies are a part of this focus (e.g. Latour/Woolgar 1986) as well as the controversial finalisation debate some two decades ago (Böhme et al. 1973) and, more recently, the thesis of new forms of the production of knowledge labelled for example "post-normal science" (Funtowicz/Ravetz 1993), "mode 2" (Gibbons et al. 1994) or "postacademic science" (Ziman 1994). In the context of these observations about science, knowledge is not viewed as a closed, autonomous canon but as the *societal orientation of science,* as Böhme (1993: 19) puts it, which assures what kind of knowledge we ultimately have in science and what social processes are responsible for the inner structure of knowledge and its conceptual apparatus.

The transformation of the standard sociology of knowledge approach from questions concerned with the production and supply of knowledge to issues more concerned with the demand for knowledge can perhaps best be traced to various theories of modern society. For example, one may establish the connection to Helmut Schelsky's (1961) notion of the emergence of a scientific-technical civilization or the pessimistic vision of the dawn of the technical state as argued by Herbert Marcuse ([1964] 1989). Another example is the much more upbeat idea of the emergence of post-industrial society (Bell 1964, 1973) and its socialist equivalent of the scientific-technological revolution (Richta 1971; see also Stehr 1994: 99–174). The notable commonalties of the different theories of modern societies just enumerated extend—despite their manifest political and economic differences—to the observation that the primacy of the relations between human and machines in production are replaced by the rise in the economic significance of knowledge.

Initially, however, there were few, if any, explicit connections of these theories of modern society to the sociology of science and knowledge, especially as it concerns the images of the social production and utilisation of scientific and technical knowledge in theories of modernity. Daniel Bell and other authors discovered knowledge as the new axial principle of modern society, but spend little time reflecting on the nature of knowledge, how it was produced or, for that matter, why there appeared to be a growing demand for knowledge in society. To use what has become a cliché, theoretical knowledge was treated as a black box.

Yet, the various theories of society which stressed that the new social role of knowledge, primarily within and for the economic system, did touch upon and reflect societal transformations that force us to shift to knowledge-guiding interests that are concerned with the use or the impact of knowledge on the life world and the kind of society we will live in. What was required then is a shift from a mainly supply-side driven perspectives to a demand-side view of knowledge in modern societies. Nonetheless, the intellectual origins of such a theoretical interest may be found in the classical sociology of knowledge.

19.2 The Classical Sociology of Knowledge

The sociology and knowledge is often regarded as a rather atypical and in some respects unique specialty of sociology (Merton [1946] 1957; Barber 1975). Even though it has been a recognised sociological specialty since the late 1920s, its subject matter has always been a major part of the intellectual domain of sociology itself. Moreover, it has, perhaps to a greater degree than other sociological specialties, also met with considerable interest from scholars in other social and human sciences. This may be the case because sociology of knowledge has at times been regarded as representing a kind of Copernican revolution in the analysis of cultural products.[1]

The recognition that knowledge in the broadest sense is context-dependent and constrained by social factors is of recent origin,[2] as is sociology itself. Sociology could only arise after the dogma of a congruence between natural and social inequality had fallen into disrepute. The French philosophers of the Enlightenment and the Scottish moral philosophers were among the first to recognise that all social differences also had social origins and were thus the result of factors subject to human control (Dahrendorf 1962). They were also aware that a wide range of social, economic, and political factors shape the genesis, structure, and contents of human consciousness. However, these early reflections on the effect of social conditions on consciousness did not crystallise into a more systematic examination of the questions that later became a focus of the sociology of knowledge. Nevertheless, some of these early writings, which either denied the very possibility of a *sociology* of knowledge or which represent first tentative steps in the direction of a sociological examination of knowledge, must be acknowledged as intellectual precursors of the sociology of knowledge.

The histories of science and philosophy, but also of philosophy and epistemology more generally, can be read as an effort to demonstrate that any sociological analysis, especially of the 'higher' forms of knowledge such as scientific knowledge, is inevitably seriously flawed. Philosophers have extended considerable effort in order to demonstrate that a sociology of knowledge is neither possible, necessary, nor desirable. Immanuel Kant, for example, argued that while there cannot be perception without conception, the constitutive components of conception remain *a priori*. Similarly, empiricists of various persuasions have maintained that (scientific) knowledge is warranted by direct experience which is unaffected by social conditions. At best, these philosophies maintain that extra theoretical factors influence the genesis of ideas but not the structure and the content of thought. Otherwise quite

[1] As Robert K. Merton ([1946] 1957: 459) argues, "the 'Copernican revolution' in this area of inquiry consisted in the hypothesis that not only error or illusion or unauthenticated belief but also the discovery of truth was socially (historically) conditioned."

[2] I concur with Berger/Luckmann's (1967: 3) generalised contention that the sociology of knowledge must "concern itself with whatever passes for 'knowledge' in a society, regardless of the ultimate validity or invalidity (by whatever criteria) of such 'knowledge'."

different philosophies of knowledge have often shared an explicit rejection of the kind of doubt and suspicion which sociological approaches to knowledge are believed to invariably generate and perpetuate, and they have instead aimed at overcoming doubt by placing knowledge on a firm foundation, frequently outside the realm of socio-historical existence. But these efforts, in their very concern with the possibility of knowledge about knowledge, have nonetheless provided a significant impetus for a sociology of knowledge.

Among its precursors we must count Karl Marx, whose famous sub- and superstructure scheme, i.e. his assertion that there is, at least under certain historical conditions, a primacy of economic realities, a determination of the "ideological superstructure" by socio-economic processes, has been particularly influential. Perhaps the sociology of knowledge represents even today, as has occasionally been argued, a struggle with this Marxist assertion. But even if this is an exaggeration, Marx' conception of a close affinity between social structure and culture continues to constitute the primary theme of the sociology of knowledge. The distinctly Marxist response to the basic sociology of knowledge assertion has resulted in some outstanding sociological analyses of problems of cultural production, for example in the works of Georg Lukács.

Emile Durkheim is usually regarded as a pioneer of the sociology of knowledge, even if he failed to develop a general model of the classificatory process from his assertions about the basic categories of perception and the ordering of experience (space, time, causality, direction), which he saw as derived from the social structure of simpler societies. Durkheim and Marcel Mauss, but also Lucien Lévy-Bruhl examined the forms of logical classification of 'primitive' societies and concluded that the basic categories of cognition have social origins. They were not prepared, however, to extend this kind of analysis to more complex societies. Their inquiry into the social foundations of logic generated considerable criticism of their guiding assumptions, and much of this criticism has withstood the test of time and continues to be invoked against a sociologically informed inquiry into knowledge. However, attempts are now underway to reconstitute, on a new theoretical basis, Durkheim and Mauss' central proposition that the classification of things reproduces the classification of men (Bloor 1984; Schwartz 1980).

Max Scheler and Karl Mannheim extend the Marxist notion of substructure by claiming that different "real factors" (*Realfaktoren*) condition thought in different historical periods and in different social and cultural systems in specific ways. Mannheim's contribution to the sociology of knowledge represents exemplary research into the social conditions associated with different forms of knowledge.[3]

[3] The formulations chosen here to characterise Mannheim's conception of the nature of the relation between consciousness and society is intentionally ambiguous because Mannheim himself wanted to avoid the impression that he viewed the connection between ideas and social conditions in a simple, straightforwardly *causal* manner. The deliberate ambiguity of many of Mannheim's formulations, which recognise the complexity of the relation in question, has been the object of serious criticism, especially in North America, where assessments of the classical sociology of knowledge tend to prefer a deterministic and mechanistic language (see Kettler et al. 1984).

His analyses of competition as a cultural form, of conservative thought, of the problem of generations, and of economic ambition, for example, remain even today examples of the kind of first-rate analysis which the sociology of knowledge is capable of. For Mannheim, the knowledge generated by the sociology of knowledge, as general social inquiry and in the form of political knowledge, is destined to play a considerable role in the intellectual and political life of society, particularly in an age of dissolution and conflict, by examining sociologically the very conditions which have given rise to competing ideas, political philosophies, ideologies, and diverse cultural products. According to Mannheim, the sociology of knowledge must also be a *diagnosis of its time* and provide practical solutions in an age of disenchantment and disorientation (see Kettler et al. 1984).

The widespread and controversial appeal of the sociology of knowledge at the time of its inception owes much to this ambitious formulation of its aims, which go far beyond anything a sociological specialty has ever claimed for itself. But Mannheim's project, while achieving considerable critical acclaim, nevertheless also foreshadowed the subsequent reception and transformation of the sociology of knowledge into a 'mere' sociological specialty as sociology itself evolved increasingly into a professional activity clearly differentiated from philosophy, history, anthropology, economics, and linguistics, and as it was transplanted into other societies and reflected the commitments of disciplinary traditions considerably different from those found in Germany where it had first formulated its intellectual and political mission.

19.3 The Status of Sociology of Knowledge in Sociology in the Post-War Years

Long after the originally rather ambitious programme of the sociology of knowledge had been moderated or even abandoned, the sociology of knowledge for decades remained a relatively marginal sociological specialty in the last half of the 20th century. This observation can be corroborated, for instance, by the fact that the percentage of sociologists with interests in the sociology of knowledge had actually declined by 1970 (Stehr/Larson 1972), reflecting a perception that it had somehow exhausted its intellectual resources long before its extensive programme was realised. This situation changed rather dramatically, however. Beginning in the late sixties, one can witness a renewal of interest in the sociology of knowledge, perhaps even a kind of renaissance.

Some brief observations may help to explain why sociology of knowledge is experiencing a renewal, a change of direction and in many ways moved into the center of contemporary sociological work. While I have so far referred to significant work in the sociology of knowledge, I will now focus on some issues associated with the unusual history and status of the sociology of knowledge within sociology.[4] The following five themes are especially significant in this context: (1) the origin of the sociology of knowledge and its intellectual development and institutional establishment within sociology; (2) closely related to this, its dogmatic history as presented, for instance, in textbooks, articles, and essays reviewing its development and status; (3) the paradigm of the sociology of knowledge; (4) its limits; and (5) its possibilities.[5] My primary concern here is to critically review the representative attitudes of social scientists with sociology of knowledge interests, since these views have until recently largely determined the characteristic evolution and status of the sociology of knowledge within the social and cultural sciences.

19.4 The Paradigm of the Sociology of Knowledge

Nearly all reconstructions of the paradigm of the sociology of knowledge agree that it is concerned with "existentially connected" thinking (*seinsverbundenes Denken*) or, even more generally, with an "investigation of the relations between knowledge and society" (Lieber 1955: 629). In view of this comprehensive characterisation of the scope of the sociology of knowledge, it is rather surprising that it should ever have been regarded as a mere 'specialty' among so many other sociological specialties. Downplaying the actual and the possible significance as well as the implications of the sociology of knowledge has apparently been one of the strategies of sociologists in attempting to achieve the legitimation of the sociology

---

[4] The prevailing reconstruction of the sociology of knowledge until at least the 1970s can be ascertained by examining selected but nevertheless representative (encyclopedic) essays whose explicit objective it is to provide a comprehensive overview of the genesis and the nature of the sociology of knowledge for a broad audience (e.g. Merton [1946] 1957; Lieber 1955; Rueschemeyer 1981; Coser 1968).

[5] One of the striking features of the reconstructions of the history and the status of the sociology of knowledge in the post war era is the consensus which emerges from these writings undertaken by different authors at different times and in different societies. The consensus extends to a wide range of issues significant in the sociology of knowledge, for example, the boundaries of the specialties, its accomplishments or lack thereof, its limits and pitfalls, its characteristic features and excessive ambition. At this time, part of the same consensus is that the sociology of knowledge, in Hesse's (1980: 30) formulation, is a notorious black spot for accidents both sociological and philosophical. Some observers regarded the classical sociology of knowledge as so clearly subversive of all good order and objectivity as to be beyond the pale of rational discussion, and by others as part and parcel of a variety of non-scientific commitments in ideology, morals and politics. Another common emphasis is on attempts to incorporate sociology of knowledge issues into other sociological specialties, for instance, into role theory, mass communications, sociology of labor, sociology of organisations and social psychology, in order to demonstrate how *sociological* its issues really happen to be. And this of course means that there is agreement in these reconstructions that a number of themes which were once an integral part of the classical sociology of knowledge in the pre-war years are now outside of the proper domain of a *sociology* of knowledge. For example, questions which pertain to the validity of knowledge-claims or the problem of relativism are generally seen as outside the domain of the sociology of sociology (cf. Stehr/Meja 1982).

of knowledge *as sociology*. If, just like other sociological specialties, the sociology of knowledge is legitimated by its specialised topic of inquiry, any far-reaching claims are precluded from the outset. Mannheim's ([1929] 1982: 427–437) attempt to transcend specialised sociology in the direction of a philosophically oriented sociology is, by contrast, an example of a more comprehensive understanding of both sociology and the sociology of knowledge.[6] Mannheim (in Wolff 1971: 263) writes: "But what is decisive is that in Germany this possibility, which exists for almost all men now living—namely, to enlarge one's view of the world and to this end avail oneself of the method of sociology—eventually exceeds the problem area of this special discipline. The sociological problem constellation in the narrower sense transcends itself in two directions—in the direction of philosophy and in the direction of a politically active world orientation". And in regard to the epistemological relevance of the sociology of knowledge Mannheim (in Wolff 1971: 263) claims: "The sociology of knowledge, however, is in a position to provide a peculiar kind of factual information concerning the various truth concepts and epistemologies—factual information which itself has epistemological implications that no future epistemology may overlook."

The reconstruction of the sociology of knowledge by sociology of knowledge specialists in the post-Mannheim period, by contrast, is characterised by an increasing renunciation of the *general* claims of the sociology of knowledge. This has led to difficulties. It is, for instance, hardly justified to maintain, on the one hand, that the sociology of knowledge is merely another sociological specialty, and to insist, on the other, that its general objective is the analysis of the relations between knowledge and society, and that the relationship between thinking or culture and social processes in the broadest sense is constitutive of human thought and action. Questions about such issues as social causes of deviant behaviour or the evolution of bureaucratic organisations can therefore hardly be granted similar status and have comparable implications for sociological theory and research generally.

The real success of the sociology of knowledge might lie with the incorporation of its concerns into sociology in general. This is precisely what Coser (1968: 432) observes for the situation in the United States: "As the sociology of knowledge has been incorporated into general sociological theory both in America and in Europe,

[6] A similar programmatic statement by Mannheim may be found in his *Announcement* written in 1929 on the occasion of taking over the editorship of the series *Schriften zur Philosophie und Soziologie* originally founded by Max Scheler and published by Friedrich Cohen, Bonn. In the *Announcement* Mannheim writes: "Collaboration between the two disciplines philosophy and sociology should not obliterate boundaries, yet mutual stimulation should be one aspiration. Neither displacing the initially philosophical context of inquiry with sociology, nor burying empirical methods under empty speculation in the social sciences can be desirable. Co-operation can have but one purpose, namely that philosophical questions can become a part of this newest level of world orientation in science and life, and that sociology in the attempt to permeate reality empirically keeps its investigative impulses always uniformly centered. Philosophy gives up on itself if it does not deal with contemporary problems, sociology if it loses the center of its inquiry."

it has often merged with other areas of research and is frequently no longer explicitly referred to as sociology of knowledge. Its diffusion through partial incorporation has tended to make it lose some of its distinctive characteristics. Thus, the works of Robert K. Merton and Bernard Barber in the sociology of science, the works of E. C. Hughes, T. H. Marshall, Theodore Caplow, Oswald Hall, Talcott Parsons, and others in the sociology of the professions and occupations, and—even more generally—much of the research concerned with social roles, may be related to, and in part derived from, the orientation of the sociology of knowledge. Many practitioners of what is in fact sociology of knowledge may at times be rather surprised when it is pointed out that, like Monsieur Jourdain, they have been 'talking prose' all along." (see also Rueschemeyer 1981: 354–359)

The obliteration by incorporation of the issues of the sociology of knowledge into other sociological specialties also indicates that the sociology of knowledge cannot easily be seen as merely one sociological specialty among many others. The incorporation of sociology of knowledge issues by other areas, in particular by specialties with a history of extensive empirical research, is associated with a neglect of foundational issues initially considered extremely important, in particular the neglect of epistemological, ethical, philosophical or political issues.

Among the central issues of the sociology of knowledge are—if one follows Merton's ([1946] 1957: 372) delineation of its core problematic—the following questions:

1. Where is the existential basis of cognitive products located?
2. Which cognitive products are subjected to sociological analysis?
3. What is the correlation between cognitive products and existential basis?
4. What are the manifest and latent implications of such correlations?
5. Under what conditions *or* at what point in time can these assumed correlations be observed?

It is hardly surprising that the numerous attempts to define existentially connected knowledge in concrete terms are highly contested issues. The sociology of knowledge is, at least in this respect, self-exemplifying.

19.5 Limitations of the Sociology of Knowledge

Most reconstructions of the sociology of knowledge in the post-war era are in nearly unanimous agreement about the necessity that the examination of human thought and ideas, firstly has to be limited to the (evidently) shrinking domain of knowledge in modern societies which might be called "pathological knowledge" (Hesse 1980: 32:), i.e. knowledge that does *not* deserve the distinction of being called scientific knowledge. As Lieber (1955: 633) puts it, in the sociology of knowledge the subject of inquiry ought to be limited to "ideological knowledge".

Secondly, as has been pointed out by Rueschemeyer (1981: 352) for example, the investigation of the relationship between ideas about reality and the existential processes toward which the sociology of knowledge strives, "does not question the validity of these ideas, although the beginnings of the sociology of knowledge may indicate otherwise, since its interest was focused on the study of the social conditions of distorted and false knowledge and on a critique of ideology."

Lewis Coser (1968: 430) refers to a third basic restriction the sociology of knowledge should observe. The sociology of knowledge as sociology should not engage in discussions of relativism and the logical contradiction into which a more 'radical' sociology of knowledge would inevitably become entangled. Referring to Mannheim's programme for the sociology of knowledge, Coser explains that to "him all knowledge and ideas, although to different degrees, are 'bound to a location' with the social structure and the historical process [...]. From its inception, Mannheim's thesis encountered a great deal of criticism, especially on the grounds that it led to universal relativism [...]. If it is assumed that all thought is existentially determined and hence all truth but relative, Mannheim's own thought cannot claim privileged exemption." In spite of Mannheim's own repeated attempts to protect his approach from the charge of relativism, the very charge has usually also been levelled against Mannheim himself.[7]

A notable characteristic of the normalisation of the sociology of knowledge in the first decades after the war is therefore the expulsion of these issues from its domain. The separation from those concerns that were seen as liabilities of the classical sociology of knowledge resonated closely with intellectual developments in sociology in general and closely related fields of the sociology of knowledge such as the sociology of science and its internalist programme of inquiry that abstained from asking questions about the social nature of scientific knowledge.

However, the traditional division of labor in science—epistemology here, substantive scientific disciplines and specialties there—thus sanctioned and implied, at the same time, that the majority of *sociologists* of knowledge fully agreed with the critique levelled against the classical sociology of knowledge by philosophers, epistemologists and others. Niklas Luhmann (1977: 9, 1984: 103–148) has emphasized the sterility that can result from such a turning away from philosophical reflections. Screening out philosophical considerations "[becomes], in turn, too confining. Introduced as a safeguard for unrestricted research against an overwhelming tradition, the impermeable boundary between science and philosophy creates, now that the power of tradition is abating, thought barriers and provincialism, and frequently a too narrow interpretation of that which, indeed, is already being thought."

The openly neutral reconstructions of the sociology of knowledge are, of course, not the result of an unbiased attitude, but rather an acknowledgement of widespread

[7] The sociology of knowledge dispute centering on Karl Mannheim's work at the end of the 1920s and the beginning of the 1930s has only been documented some sixty years later in Meja/Stehr (1982; for an abridged English edition see Meja/Stehr 1990).

criticism of the sociology of knowledge as conceived by its founders. However, it became increasingly obvious that the separation of important issues from the sociology of knowledge contributes to the undesirable consequence of a stagnant, unreceptive attitude toward significant intellectual developments in other disciplines.

19.6 The Transformation of the Sociology of Knowledge

The initial, new developments in the sociology of knowledge in the late sixties and early seventies were stimulated, at least within the scientific community, by revisions to then standard view of science (Rudolf Carnap, Carl Gustav Hempel, Ernest Nagel, Karl Popper) and reconstructions of the history of the cognitive dynamics within the scientific community. Developments in epistemology and the history of science that began to question the traditional concept of science and of the development of scientific knowledge were a crucial contributor to these transformations and ultimately a renewal of the sociology of knowledge (cf. Kuhn 1962; Feyerabend 1975; Lakatos 1978; Toulmin 1972).

The new phase is characterised, in particular, by a gradual lifting of certain intellectual taboos which had been part of the standard view of science (cf. Barnes 1974, 1977; Bloor 1976; Mulkay 1979; Knorr-Cetina 1981; Whitley 1984). One of the results of the emerging concept of scientific knowledge as socially constructed knowledge is that the analysis of human knowledge in the sociology of knowledge is no longer necessarily restricted to a limited and perhaps shrinking area of ('erroneous') human thinking. A sociology of scientific and formal knowledge is now considered a real possibility as well a necessity. As a result, the problems of relativism, of self-refutation and of the desirability for a strict separation of the context of discovery from the context of the justification of human knowledge is now part of social inquiries into all forms of knowledge.

The renewed interest in the sociology of knowledge transcends the classical sociology of knowledge programme in other respects as well. Importantly among these new interests figures a concern with what had been a shortcoming of the classical sociology of knowledge, frequently pointed to by its critics (Merton [1946] 1957), namely the absence of a satisfactory theoretical solution to the question of how the relation between the structure of human groups and consciousness arises, is maintained or evolves, and therefore changes in the course of social evolution.[8] Barry Schwartz, for example, has proposed an ingenious solution to the "missing link' in the tradition of a Durkheimian sociology of knowledge by advancing a

[8] Norbert Elias (1971), in particular, has referred to the gap in theory in the sociology of knowledge concerning the long-term development of knowledge and therefore of empirical or theoretical inquiry into the possible evolutionary changes in the relationship between consciousness and society. However, Elias not only identifies such a gap but actually proposes an interesting theoretical solution.

theory that accounts for the linkage between social conditions and classificatory systems (Schwartz 1981).

These developments in turn coincide with analogous attempts in other disciplines or related intellectual traditions, for example with the attempts by Mary Douglas (1975) in anthropology, by Mary Hesse (1980) in the philosophy of science and in science studies (Knorr-Cetina 1999; Latour 1999), to specify, often on the basis of empirical research in a variety of social science disciplines, what the nature of the relationship between perception and social structure exactly is.

But aside from these efforts, which often occur at the level of a microsociology of knowledge and tend to refrain from an analysis of the societal role of scientific knowledge, another set of genuinely new themes emerged as important issues within the contemporary sociology of knowledge. Significant among these are the assertions concerning the growing power of knowledge in industrial or post-industrial society (Lane 1966; Bell 1973; Elias 1971), about the emergence of a new class (Gouldner 1979), about the increasing influence of a new caste of priests (Schelsky 1975; Bahro 1978; Richta 1971), the growing segment of knowledge workers among the labor force (Machlup 1962) and the emergence of the knowledge-based economy (Drucker 1986). All these writings share a common emphasis on the growing importance of specialised forms of knowledge for modern society and on the corresponding power the carriers of this knowledge are able to exercise in it.

19.7 Possibilities of the Sociology of Knowledge

Lewis Coser (1968: 433) is one of the few commentators on the sociology of knowledge who considers the ambitious original claims of its founders a challenge rather than an obsolete burden. In his survey of the history of the sociology of knowledge he comes to the following conclusions:

> The sociology of knowledge was marked in its early history by a tendency to set up grandiose hypothetical schemes. These contributed a number of extremely suggestive leads. Recently its practitioners have tended to withdraw from such ambitious undertakings and to restrict themselves to somewhat more manageable investigations. Although this tendency has been an antidote to earlier types of premature generalisations, it also carries with it the danger of trivializations. Perhaps the sociology of knowledge of the future will return to the more daring concerns of its founders, thus building upon the accumulation of careful and detailed investigations by preceding generations of researchers. (Coser 1968: 433)

The very question, however, whether such a return to issues that may have already been raised by the founders of the sociology of knowledge and the related question of whether a transformation of the sociology of knowledge is in fact possible or fruitful, is not only the result of purely intellectual efforts, but also, according to the theory of the sociology of knowledge, itself a development influenced by societal conditions. Merton (1957: 368), for example, recognises that an interest in the sociology of knowledge is determined by certain cultural and social conditions of society. One factor among these conditions is that "with increasing social conflict, differences in the values, attitudes and modes of thought of groups develop to the point where the orientation which these groups previously had in common is overshadowed by incompatible differences. Not only do these develop different universes of discourse, but the existence of any one universe challenges the validity and legitimacy of the others [...]. Thought becomes functionalized."

It should at least be indicated, therefore, that the reconstruction of the classical sociology of knowledge by the generation of sociologists succeeding the generation of the founders coincided with economic, political, and social conditions as well as academic and disciplinary circumstances in the first decades of the post-war era, which in their relative tranquillity stood in marked contrast to the crisis conditions of the 1920s. The renewed interest in certain issues first articulated by the classical sociology of knowledge as well as questions that transcend the classical programme of sociology of knowledge, in contrast, 'reflects', as in the earlier period when it was first developed, the experience of a crisis in contemporary society, most notable in the late 1960s. Moreover, the range and forms of essentially contested knowledge that science makes available, the fact that science becomes increasingly the only source of additional knowledge in modern society and that the change in the available knowledge itself (Stehr 2003) dramatically enlarges the available options of social action suggests that the investment in the production, the distribution (supply) and the reproduction of scientific knowledge changes as well and is bound to acquire greater social significance. These questions will form the core of the considerations in the final parts of this essay.

19.8 Toward a Sociological Concept of Knowledge

Scientific discourse developed a kind of natural attitude toward its own knowledge. And for this reason, but not merely for this reason, the number of well explicated categories of knowledge in social science has been fairly limited. We have not moved much beyond the proposals of different forms of knowledge found in Max Scheler's early contributions to the sociology of knowledge. Later explications tend to be quite similar to Scheler's ([1924] 1960: 13–49) categories of knowledge, namely (1) knowledge of salvation, (2) cultural knowledge, or knowledge of pure essences and (3) knowledge that produces effects, although that ancestry is not often spelled out or recognised. The most widely employed conceptions of different forms of knowledge are dichotomies. Dominant is the distinction between scientific and non-scientific knowledge. This distinction was taken for granted for such a long time that it has not really been elaborated for decades, except in the sense that non-scientific knowledge became a residual form of knowledge. In addition, the dichotomy between specialised and everyday knowledge has been widely used in sociological discourse, whereby specialised knowledge was often identical with scientific and technical knowledge.

For the purpose of some further explication of the concept of knowledge, one must distinguish between what is known, the content of knowledge, and knowing. Knowing is a relation to things and facts, but also to rules, laws and programmes. Some sort of participation is therefore constitutive for knowing: knowing things, rules, programmes, facts is 'appropriating' them in some sense, including them into our field of orientation and competence. The intellectual appropriation of things can be made independent or objective. That is, symbolic representation of the content of knowledge eliminates the necessity to get into direct contact with the things themselves. One is able, in other words, to acquire knowledge from books. The social significance of language, writing, printing, data storage etc. is that they represent knowledge symbolically or provide the possibility of objectified knowledge. Thus, most of what we today call knowledge and learning is not direct knowledge of facts, rules, and things but objectified knowledge. Objectified knowledge is the highly differentiated stock of intellectually appropriated nature and society which may also be seen to constitute the cultural resource of a society. Knowing is, then, *grosso modo* participation in the cultural resources of society. However, such participation is of course subject to stratification; life chances, life style and social influence of individuals depend on their access to the stock of knowledge at hand.

Knowledge, ideas, and information are most peculiar entities with properties unlike those of commodities or secrets, for example. If sold, they enter other domains and yet remain within the domain of their producer. Knowledge does not have zero-sum qualities. Knowledge is a public good. When revealed, knowledge does not lose its influence. While it has been understood for some time that the 'creation' of knowledge is fraught with uncertainties, the conviction that its application is without risks and that its acquisition reduces uncertainty has only recently been debunked. While it is very reasonable and in some sense urgent to speak of the limits to growth in many spheres and resources of life, the same does not appear to hold for knowledge. Knowledge has virtually no limits to its growth.

Modern communication technologies ensure that access becomes easier and may even subvert remaining proprietary restrictions, although concentration rather than dissemination is also possible and feared by some. But one could just as easily surmise that the increased social importance of knowledge, and not so much its distinctiveness, may in fact undermine the exclusiveness of knowledge. Yet the opposite appears to be the case and therefore raises anew the question of the persisting basis for the power of knowledge.

I would like to define knowledge as a *capacity for action*. Knowledge is a condition for the possibility of social action. The notion of knowledge as a capacity for social action has the advantage that it enables one to stress not merely one-sided but multifaceted consequences of knowledge for action. The term capacity for action signals that knowledge may be left unused or may be employed for irrational ends and leaves room, therefore, for a 'dialectical' theory of the use of knowledge. The thesis that knowledge invariably is pushed to its limit, that is, is realised and implemented almost without regard for its consequences constitutes of course a view which is quite common among observers, for example, of the nature of technological development. However, the notion that science and technology inherently and inevitably force their own realisation in practice fails to give proper recognition to the context of implementation by assuming such automaticity in the realisation of technical and scientific knowledge. The definition of knowledge as a capacity for action indicates strongly that the material realisation and implementation of knowledge is dependent on, or embedded within the context of specific social and intellectual conditions.

It is important to realise that knowledge, as an element of power relations, generates not merely coercive, distorting and repressive consequences, as many traditional conceptions of power would imply, but also has productive and enabling features. And a more developed theoretical notion of power requires the ability to incorporate both attributes of knowledge as a capacity for action. The use of a more differentiated concept of power, conscious of the constraining and enabling features of power, becomes all the more relevant the greater the actual amorphousness of power and the greater the malleability of social structures and conditions become for the realisation of knowledge. Knowledge has many enabling features which allow individuals and groups to organize resistance, avoidance and opposition. Not only such an emphasis becomes important because the degree and the effectiveness to which knowledge operates as a capacity for action increases, but also the dissemination and access to knowledge among strata, which may have been cut off in the past. Since knowledge and access to knowledge is not evenly distributed, such a world is not a world without power and inequality.

Obviously, *scientific and technical knowledge* represent such "capacities for action" and maybe even quite a special capacity for action in modern society. However, this does not mean that scientific knowledge should be seen as a resource which lacks contestability, is not subject to interpretation and can be reproduced at will. The special importance of scientific and technical knowledge in any modern societies derives not so much from the fact that it is at times treated as if it is essentially uncontested (or, objective), but that it constitutes, more than any other form of modern knowledge, an incremental capacity for social action or an increase in the ability of "how-to-do-it" which, moreover, may be "privately appropriated", if only temporarily. In economic settings, incremental knowledge has particular importance as a source of added value. The strategic importance of incremental knowledge as an immediately productive force in economic contexts specifically may affect the ways in which production and the delivery of services are organized as well as the types of commodities and services that are produced.

The transfer of knowledge does not necessarily include the transfer of the cognitive ability to generate such knowledge; for example, the theoretical apparatus or the technological regime which yields such knowledge claims in the first place and on the basis of which it is calibrated and validated. Cognitive skills of this kind, therefore, are scarce. But they are not the only skills in demand since knowledge constantly has to be made available, interpreted and linked to emerging local circumstances. This is the job performed by *knowledge-based occupations* or by experts, counsellors and advisors. This group of professions is needed to mediate between the complex distribution of changing knowledge and those who search for knowledge enabling them to act because "ideas travel" as "baggage" of people, whereas skills are embodied in people. A chain of interpretations must come to an "end" in order to become relevant in practice and effective as a capacity of action. This function of ending reflection for the purpose of action is largely performed by experts in modern society.

19.9 Modern Societies as Knowledge Societies

The foundation for the transformation of modern societies into knowledge societies is to a significant extent also based, as was the case for industrial society, on changes in the structure of the economies of advanced societies. Economic capital —or, more precisely, the source of economic growth and value-adding activities— increasingly relies on knowledge. The transformation of the structures of the modern economy by knowledge as a productive force constitutes the "material" basis and justification for designating advanced modern society as a "knowledge society". The significance of knowledge grows in all spheres of life and in all social institutions of modern society.

What distinguishes a knowledge society above all else from its historical predecessors is that it is a society which is to an unprecedented degree the product of its own action—keeping in mind of course that this enlarged capacity to act is not necessarily evenly distributed throughout society—or a society in which our secondary nature far outpaces and outgrows our primary nature. Interventions by nature are increasingly the result of prior human interventions into nature. The balance of nature and society or of facts beyond the control of humans and those subject to their control has shifted strikingly. In as much as the greater ability and range of social action is a constitutive feature of the knowledge society, the idea that modern society necessarily is a "technocratic" enterprise, that is, a society which witnesses almost helplessly the widespread inversion of technical means into social ends as sketched, for instance, by Schelsky (1961) or Marcuse (1964) has proven to be an erroneous portrait of modern society.

The historical emergence of "knowledge societies" does not occur suddenly. It represents not a revolutionary development, but rather a gradual process during which the defining characteristics of society change and new traits emerge. Even today, the demise of societies is typically as gradual as was their beginning, even if some social transformations do occur in spectacular leaps. But most major social changes continue to evolve gradually, at an uneven pace, and they become clearly visible only after the transition is already over. The proximity of our time to significant social, economic and cultural changes, however, makes it highly likely that what is now beginning to come into view is of extraordinary significance in the present and in the future.

Moreover, knowledge societies do not come about as the result of some straightforward uni-modal unfolding. They are not a one-dimensional social figuration. Knowledge societies become similar by remaining or even becoming dissimilar. New technological modes of communication break down the distance between groups and individuals, while the isolation of particular regions, cities and villages remains. The world opens up and creeds, styles and commodities mingle, yet the walls between incompatible convictions about what is sacred do not come tumbling down. The meaning of time and place erodes, even while boundaries are celebrated.

Until recently, modern society was conceived primarily in terms of property and labor. Labor and property (capital) have had a long association in social, economic and political theory. Work is seen as property and as a source of emerging property. In the Marxist tradition, capital is objectified, encapsulated labor. On the basis of these attributes, individuals and groups were able to, or constrained to define their membership in society. In the wake of their declining importance in the productive process, especially in the sense of their conventional economic attributes and manifestations, for example as "corporeal" property such as land and manual work, the social constructs of labor and property themselves are changing. While the traditional attributes of labor and property certainly have not disappeared entirely, a new principle ("knowledge") has been added which to some extent challenges as well as transforms property and labor as the constitutive mechanisms of society.

19.10 The Political Economy of Knowledge

The nature of the changes described as transformations that lead to a knowledge society are above all changes in the intellectual appropriation of nature and society. There is now an immense stock of objectified knowledge which mediates our relation to nature and to ourselves. Nature is and can scarcely be experienced other than as a human product or within human products, and social relations are mediated by an increasing stock of arrangements of an administrative, legal, or technical kind.

From the point of view of the production, distribution and reproduction of knowledge, the first characterisation is able to give of our contemporary societal condition a mere quantitative diagnosis: the superstructure of society and nature is currently so immense that the greater part of the overall social activity is not production but reproduction in particular of knowledge itself, and that implies the conditions which make specific effects and processes possible in the first place.

The kinds of constraints which operate in a knowledge society differ from those analysed in theories dealing with more traditional forms of power relations in general, and political power in particular. In the case of traditional notions of power, its possession and use is deliberate and intentional. Responsibilities can be traced to its sources and the benefits or costs associated with its employment are in many instances assigned quite unambiguously. The kinds of emerging power in knowledge societies are not merely the constraints which issue from the exclusive possession and control over specific resources or persons which compel those subjected to the exercise of power because they lack the same means of power and have, therefore, but limited choices at their disposal in the face of overwhelming odds against them in carrying out certain actions or tasks. In knowledge societies, constraints circulate throughout our relations with one another and our dealings with things and pervade the smallest and most ordinary of our doings. They shape the practical configuration within which our action makes sense, both to ourselves and to one another. This shaping occurs most directly through their effects upon the kinds of equipment available to us, the skills and procedures required to use that equipment, the related tasks and equipment that use imposes upon us and the social roles available to us in performing these tasks. Through this process, they change our understanding of ourselves and our lives.

The social and material contexts of social action are altered in fundamental ways. The choice of alternatives, especially sensible alternatives in a given situation, are affected and redesigned. The changes are also related to changes in the fundamental characteristics of knowledge itself and the way knowledge is shaped by the means of its production and dissemination. The differences at issue concern the rise in the need to interpret knowledge and therefore the extent to which knowledge has lost, or appears to have lost, the characteristic of being secure, certain, definitive and even truthful. The progressive elimination of time and space as relevant elements in the production of knowledge has paradoxically injected the importance of time and location in the interpretation or use of knowledge. Since the validation process of knowledge cannot refer back to the original author of the claim, except in rare circumstances, the interpretative task once again becomes crucial.

In the view of some observers, part of the political economy of knowledge in modern societies concerns the possibility that knowledge becomes the basis for class formation in general or the emergence of a "knowledge class" in particular. Peter Berger (1987: 66) has recently argued that the modern middle class increasingly is divided into the old middle class consisting of the "business community and its professionals as well as clerical affiliates" and the newer part, namely the "knowledge class". In other words, the former segment of the middle class consists of individuals who derive their "livelihood from the production and distribution of material goods or services" while the knowledge class includes people whose occupations "deal with the production and distribution of symbolic knowledge". For the purposes of specifying this modern social stratum in greater detail, Berger refers affirmatively to Helmut Schelsky's description of the emerging group of knowledge-producers and knowledge-disseminators in modern society. Schelsky (1975: 14) labels this group as the class of the "distributors and mediators of meaning and purposes" *(Sinn- und Heilsvermittler).* Given Berger's inclusive definition of social class, he maintains that the knowledge class has common material and political interests and a common consciousness. For Berger, the knowledge class tends to be, at least in the United States, on the left of the political spectrum while their economic interests include firm support for an extension of the welfare state—which happens to be the employer of the majority of the knowledge class.

I believe that these claims about a knowledge class are exaggerated but also too restrictive. Individuals who are in the business of distributing and disseminating knowledge are by no means confined to the membership in the "middle class". The probability that this stratum develops a "class consciousness", or in fact has an incipient class consciousness, is remote. As a matter of fact, Schelsky maintained that the group of knowledge-producers and knowledge-disseminators has every interest in hiding its rule. Since the new class Schelsky discerns tends to dominate education, the media, and the public relations field—in short, all those spheres in modern society which somehow have to do with the formation of the public's consciousness and identity—it is relatively easy for this class to deceive the public about its inordinate influence and its interests. One of the most effective strategies employed in this deceptive process is the maintenance of the old myth about the continued existence of the old class conflict between proletariat and capitalist.

To suggest that knowledge plays an increasingly important role in shaping the nature and the structure of inequality of modern society is not to maintain that knowledge in the sense of a resource for action is a novel phenomenon for the analysis of social inequality. On the contrary, knowledge representing a variety of cultural competencies and abilities has throughout history played a significant role determining aspects of inequality and its evaluation in society.

The important question which needs to be addressed therefore is not only how the generally growing importance of knowledge in society affects patterns of social inequality, but also why knowledge is 'capable' of eroding or even replacing what has been for centuries and continues to be for many, the solid foundation of social inequality and whether knowledge, as a result, will lead to the formation of a knowledge class.

Theories of social inequality for the most part continue to be linked in a rather close fashion to the production process, its organisation and outcomes. Moreover, the lingua franca of inequality theories remains tied to an imagery which displays agents as inflexible creatures entangled in single-purpose structures of inequality which imposes its rhythm, often over generations, on passive individuals and groups. This includes the idea that social classes continue to be relevant social formations in modern society.

As long as inequality regimes are reconstructed as inflexible and largely one-dimensional, the gradients of stratification will be steep and the effects of material inequalities long lasting. Discourse on inequality will be concerned with the coercion, that is, the constraints, the vulnerability and effective impotence of individuals and groups in the face of power exercised by the powerful. But new realities require a new language of inequality. The new language should, in contrast, stress agency, malleability, flexibility, multipurpose resources, volatility, heterogeneity etc. of social structures and the extent to which individuals and groups have the capacity to employ and transform these structures, once the generalised vulnerability to the forces of inequality has been reduced to a significant degree. The condition for the possibility of greater and more broadly based agency is linked to knowledge. Knowledge acts as a bundle of more widely accessible social competencies and their impact on social structures of inequality accelerate chances for actors to re-fashion social constructions. In such contexts, inflexible social classes to which actors are tied in an almost fate-like fashion are largely irrelevant social institutions (cf. Stehr 1997).

19.11 Concluding Observations

New realities require a new perspective. In the last fifty years, the social role of knowledge has changed in the developing world. One may witness in the social sciences a significant change in the perception of the nature of knowledge, its production and dissemination. The natural attitude toward scientific knowledge in science that assigned a special status to such knowledge claims has lost its dominance. Outside the boundaries of the scientific community, one still encounters echoes of the once standard image of scientific knowledge. However, the view that knowledge is socially constructed, essentially contested and that its use entails risks has gained in general acceptance. Science increasingly co-determines the political agenda in modern societies. Science discovers the problems to which society has to attend. As a result, a new field of political activity is emerging: knowledge politics. Some of the new questions about the social role of knowledge that will be raised in this context are: what knowledge needs to be regulated and whose responsibility is it? And what knowledge do we require in the context of regulative knowledge politics? Do we need to regulate new scientific knowledge as closely or even more strictly as one regulates, say, the traffic? If not, on what grounds? If yes, for what reasons protecting what interests? If strict restrictions apply, could absolute prohibition (for example, outlawing human germ line gene therapy) be enforced? Is knowledge about to become a private good (again)? What will be the identity of some of the major actors involved and how will knowledge politics be organized? These questions signal that the major future challenge will require sociological work that is concerned with the ways in which society deploys new knowledge.

References

Bahro, Rudolf. 1978. *The Alternative in Eastern Europe*. London: New Left Books.

Barber, Bernard. 1975. "Toward a New View of the Sociology of Knowledge", in: Lewis A. Coser (ed.), *The Idea of Social Structure*. New York: Harcourt Brace Jovanovich: 103–116.

Barnes, Barry. 1974. *Scientific Knowledge and Sociological Theory*. London: Routledge & Kegan Paul.

Barnes, Barry. 1977. *Interests and the Growth of Knowledge*. London: Routledge & Kegan Paul.

Bell, Daniel. 1964. "The Post-industrial Society", in: Eli Ginzberg (ed.), *Technology and Social Change*. New York: Columbia University Press: 44–59.

Bell, Daniel. 1973. *The Coming of Post-Industrial Society*. New York: Basic Books.

Berger, Peter. 1987. *The Capitalist Revolution: Fifty Propositions About Prosperity, Equality, and Liberty*. New York: Basic Books.

Berger, Peter and Luckmann, Thomas. 1967. *The Social Construction of Reality: A Treatise in the Sociology of Knowledge*. New York: Doubleday.

Bloor, David. 1976. *Knowledge and Social Imagery*. London: Routledge & Kegan Paul.

Bloor, David. 1984. "Durkheim and Mauss Revisited: Classification and the Sociology of Knowledge", in: Nico Stehr and Volker Meja (eds.), *Society and Knowledge. Contemporary Perspectives on the Sociology of Knowledge*. New Brunswick: Transaction Books: 51–76.

Böhme, Gernot. 1993. *Am Ende des Baconschen Zeitalters. Studien zur Wissenschaftsentwicklung*. Frankfurt am Main: Suhrkamp.

Böhme, Gernot, Wolfgang van den Daele and Wolfgang Krohn. 1973. "Die Finalisierung der Wissenschaft", *Zeitschrift fur Soziologie*, 2: 128–144.

Coser, Lewis A. 1968. "Sociology of Knowledge", in: *International Encyclopedia of the Social Sciences*. Volume 8. New York: Macmillan: 428–435.

Dahrendorf, Ralf. 1962. "On the Origin of Social Inequality", in: Peter Laslett and W.G. Runciman (eds.), *Politics and Society*. Oxford: Blackwell.

Douglas, Mary. 1975. *Implicit Meanings*. London: Routledge & Kegan Paul.

Drucker, Peter F. 1986. "The Changed World Economy", *Foreign Affairs*, 64: 768–791.

Elias, Norbert. 1971. "Sociology of Knowledge: New Perspectives", *Sociology*, 5(2): 149–168, 335–370.

Elias, Norbert. 1983. *Engagement und Distanzierung. Arbeiten zur Wissenssoziologie* I. Frankfurt am Main: Suhrkamp.

Feyerabend, Paul K. 1975. *Against Method*. London: NLB.

Funtowicz, Silvio O. and Jeffrey R. Ravetz. 1993. "The Emergence of Post-normal Science", in: Rene von Schomberg (ed.), *Science, Politics and Morality. Scientific Uncertainty and Decision Making*. Dordrecht: Kluwer.

Gibbons, Michael et al. 1994. *The New Production of Knowledge. The Dynamics of Science and Research in Contemporary Societies*. London: Sage.

Gouldner, Alvin W. 1979. *The Future of Intellectuals and the Rise of the New Class*. New York: Continuum.

Hesse, Mary B. 1980. *Revolutions and Reconstructions in the Philosophy of Science*. Brighton: Harvester Press.

Kettler, David, Volker Meja and Nico Stehr. 1984. *Karl Mannheim*. London: Tavistock.

Knorr-Cetina, Karin. 1981. *The Manufacture of Knowledge. An Essay on the Constructivist and Contextual Nature of Science*. Oxford: Pergamon Press.

Knorr-Cetina, Karin. 1999. *Epistemic Cultures. How the Sciences Make Knowledge*. Cambridge, MA: Harvard University Press.

Kuhn, Thomas. 1962. *The Structure of Scientific Revolutions*. Chicago: University of Chicago Press.

Lakatos, Imre. 1978. *The Methodology of Scientific Research Programmes*. Cambridge: Cambridge University Press.

Lane, Robert E. 1966. "The Decline of Politics and Ideology in a Knowledgeable Society", *American Sociological Review*, 31: 649–662.

Latour, Bruno. 1999. *Pandora's Hope. Essays on the Reality of Science Studies*. Cambridge, MA: Harvard University Press.

Latour, Bruno and Steve Woolgar. 1986. *Laboratory Life. The Construction of Scientific Facts. With a New Postscript*. Princeton, NJ: Princeton University Press.

Lieber, Hans-Joachim. 1955. "Wissenssoziologie", in: Wilhelm Bernsdorf and Friedrich Bülow (eds.), *Wörterbuch der Soziologie*. Stuttgart: Enke: 629–633.

Luhmann, Niklas. 1977. *Zweckbegriff und Systemrationalität*. Frankfurt am Main: Suhrkamp.

Luhmann, Niklas. 1984. "The Differentiation of Advances in Knowledge: The Genesis of Science", in: Nico Stehr and Volker Meja (eds.), *Society and Knowledge. Contemporary Perspectives on the Sociology of Knowledge*. New Brunswick: Transaction Books: 103–148.

Machlup, Fritz. 1962. *Function and Distribution of Knowledge in the United States*. Princeton: Princeton University Press.

Mannheim, Karl. [1929] 1982. "Zur Problematik der Soziologie in Deutschland", in: Volker Meja and Nico Stehr (eds.), *Der Streit um die Wissenssoziologie*. Volume 2. Frankfurt am Main: Suhrkamp: 427–437.

Mannheim, Karl. [1929] 1965. *Ideologie und Utopie*. Frankfurt am Main: G. Schulte- Bulmke.

Marcuse, Herbert. [1964] 1989. *Der eindimensionale Mensch. Studien zur Ideologie der fortgeschrittenen Industriegesellschaft*. Schriften 7. Frankfurt am Main: Suhrkamp.

Meja, Volker and Nico Stehr (eds.). 1982. *Der Streit um die Wissenssoziologie*. Two Volumes. Frankfurt am Main: Suhrkamp.

Meja, Volker and Nico Stehr (eds.). 1990. *Knowledge and Politics*. London: Routledge.

Meja, Volker and Nico Stehr (eds.). 1999. *The Sociology of Knowledge*. International Library of Critical Writings in Sociology. Two Volumes. Cheltenham: Edward Elgar.

Merton, Robert K. [1946] 1957. "The Sociology of Knowledge", in: Robert K. Merton (ed.), *Social Theory and Social Structure*. New York: The Free Press: 457–60.

Merton, Robert K., Marjorie Fiske, and Alberta Curtis. [1946] 1991. *Mass Persuasion. The Social Psychology of a War Bond Drive*. Stamford, CT: Greenwood Press.

Mulkay, Michael. 1979. *Science and the Sociology of Knowledge*. London: George Allen & Unwin.

Richta, Radovan. 1971. *Politische Ökonomie des 20. Jahrhunderts*. Berlin: Makol.

Rueschemeyer, Dietrich. 1981. "Wissen", in: René König (ed.), *Soziologie*. Frankfurt am Main: Fischer: 352–359.

Scheler, Max. [1924] 1960. *Die Wissensformen und die Gesellschaft*. Bern und München: Francke.

Schelsky, Helmut. 1961. *Der Mensch in der wissenschaftlichen Zivilisation*. Köln und Opladen: Westdeutscher Verlag.

Schelsky, Helmut. 1975. *Die Arbeit tun die anderen*. Opladen: Westdeutscher Verlag.

Schwartz, Barry. 1981. *Vertical Classification*. Chicago: University of Chicago Press.

Stehr, Nico. 1994. *Knowledge Societies*. London: Sage.

Stehr, Nico. 1997. "Social Inequality and Knowledge", in: David Sciulli (ed.), *Normative Social Action. Cross-National and Historical Approaches*. Greenwich, CT: JAI Press: 41–51.

Stehr, Nico. 2003. *Wissenspolitik. Die Überwachung des Wissens*. Frankfurt am Main: Suhrkamp.

Stehr, Nico and Lyle Larson. 1972. "The Rise and Decline of Areas of Specialization", *The American Sociologist*, 3: 5–6.

Stehr, Nico and Volker Meja. 1982. "Classical Sociology of Knowledge Revisited", *Knowledge: Creation, Diffusion, Utilization*, 4: 33–50.

Toulmin, Stephen. 1972. *Human Understanding*. Princeton: Princeton University Press.

Weber, Max. [1922] 1964. *The Theory of Social and Economic Organization*. Edited with an Introduction by Talcott Parsons. New York: Free Press.

Whitley, Richard. 1984. *The Intellectual and Social Organization of the Sciences*. Oxford: Clarendon Press.

Wolff, Kurt H. (ed.). 1971. *From Karl Mannheim*. New York: Oxford University Press.

Ziman, John. 1994. *Prometheus Bound: Science in a Dynamic State*. Cambridge: Cambridge University Press.

Chapter 20
Modern Societies as Knowledge Societies

Nico Stehr

New social realities require a new perspective. In advanced societies, the capacity of the individual to say no has increased considerably. At the same time, the ability of the large social institutions that have significantly shaped the nature of the twentieth century to get things done has diminished in the past couple of decades. Or, appropriating Lowe's (1971: 563) astute insights, we are witnessing a change from social realities in which "things", at least from the point of view of most individuals simple "happened" to a social world in which more and more things are "made" to happen. In this contribution, these new realities are described as representing the emergence of advanced societies as knowledge societies.

I will describe some of these transformations that constitute a real and unprecedented gain from the perspective of the individual and small groups but also what may be described as a rise in the fragility of society. The stress on rights and the growing ability to assert and claim such rights is one of the salient manifestations of the transformations I examine. The same developments are responsible for a crisis in mastering, planning and managing common problems and for a decline in the sense of individual responsibilities. However, there is a trade-off; the decline in the steering capacity of large social institutions and their growing difficulty in imposing their will on society leads to a rise of the importance and efficacy of civil society.

I propose to define knowledge as a capacity to act. I will describe the reasons for the importance of scientific knowledge as one among various forms of knowledge

---

This text was first published as: Stehr, Nico. 2003. "Modern Societies as Knowledge Societies", in: George Ritzer and Barry Smart (eds.), *Handbook of Social Theory*. London: Sage Publications: 494–508. The permission to reprint this text was granted on 24 July 2017 by Mr. Craig Myles on behalf of Sage.

M. T. Adolf (ed.), *Nico Stehr: Pioneer in the Theory of Society and Knowledge*,
Pioneers in Arts, Humanities, Science, Engineering, Practice 16,
https://doi.org/10.1007/978-3-319-76995-0_20

in advanced societies. I also examine the limits to the power of scientific knowledge as well as the emergence of the fastest growing segment of the labor force, namely knowledge-based occupations. The transformation of modern societies into knowledge societies manifests itself most importantly in the sphere of economic activities. I therefore describe some of the features of the changing economy before turning to those consequences of the advancing "knowledgeability" of actors in modern society that give rise to the growing fragility of modern society.

20.1 Introduction

John Stuart Mill, in *The Spirit of the Age* (1831), published after his return to England from France, where he had encountered the political thinking of the Saint-Simonians and of the early Comte, affirms his conviction that the intellectual accomplishments of his own age make social progress somehow inevitable (cf. Cowen/Shenton 1996: 35–41). But progress in the improvement of social conditions is not, Mill argues, the outcome of an "increase in wisdom" or of the collective accomplishments of science. It is rather linked to a general diffusion of knowledge:

> Men may not reason better, concerning the great questions in which human nature is interested, but they reason more. Large subjects are discussed more, and longer, and by more minds. Discussion has penetrated deeper into society; and if greater numbers than before have attained the higher degree of intelligence, fewer grovel in the state of stupidity, which can only co-exist with utter apathy and sluggishness. (Mill [1831] 1942: 13)

Mill's observations in the mid-nineteenth century, a period he regarded as an age of moral and political transition, and in particular his expectation that increased individual choice (and hence emancipation from "custom") will result from a broad diffusion of knowledge and education, strongly resonate with the notion of present-day society—the social structure that is emerging as industrial society gives way—as a "knowledge society".

The foundation for the transformation of modern societies into knowledge societies is to a significant extent also based, as was the case for industrial society, on changes in the structure of the economies of advanced societies. Economic capital—or, more precisely, the source of economic growth and value-adding activities—increasingly relies on knowledge. The transformation of the structures of the modern economy by knowledge as a productive force constitutes the "material" basis and justification for designating advanced modern society as a "knowledge society". The significance of knowledge grows in all spheres of life and in all social institutions of modern society.

20.2 Knowledge Societies

Both the greatly enhanced social, political and economic significance of science and technology and the often narrow, even scientistic conception of knowledge generated by modern science call for a careful sociological analysis of knowledge itself. Knowledge has, of course, always had a major function in social life.[1] That human action is knowledge-based might even be regarded as an anthropological constant. Social *groups*, social *situations*, social *interaction* and social *roles* all depend on, and are mediated by, knowledge. Relations among *individuals* are based on knowledge of each other.[2] Indeed, if (as in the interactionist tradition in sociology) such a general notion of knowledge is regarded as the foundation of social interaction and social order, we find that the very possibility of social interaction requires situation-transcendent knowledge that is deployed by the individuals engaging in social action.[3] Power too has frequently been based on knowledge advantages, not merely on physical strength. Societal reproduction, furthermore, is not just physical reproduction but has always also been cultural, that is, it involves reproduction of knowledge.

The historical emergence of "knowledge societies" does not occur suddenly; it represents not a revolutionary development, but rather a gradual process during which the defining characteristics of society change and new traits emerge. Even today, the demise of societies is typically as gradual as was their beginning, even if some social transformations do occur in spectacular leaps. But most major social

---

[1] A more extended discussion of the sociological meaning of the term "knowledge" as used in this context may be found in Stehr (1994: 5–17). Lane (1966: 650) first employed the term "knowledgeable society", but it was Drucker (1969) who first specifically referred to "knowledge society", a term used later also by Bell (1973) and, more recently, by Böhme (1993). There have been various other attempts to find a term suited to describe the new type of social structure, including "science society" (Kreibich 1986), "information society" (e.g. Nora/Minc 1980), "postindustrial society" (Bell 1973), "postmodernization" (e.g. Inglehart 1995), "technological civilization" (Schelsky 1961), and "network society" (Castells 1996).

[2] Cf. Simmel's ([1908] 1992: 383–455) analysis of the secret and the secret society in his *Soziologie*.

[3] Cf. Barnes' (1995: 85–93) sympathetic account of the interactionist perspective in sociology and the prominent role of shared knowledge in its thick description of elementary social interaction. However, Barnes (1995: 111) dismisses any designation of modern society as knowledge-based, precisely because he is convinced that shared knowledge is an anthropological constant. In this sense, Barnes echoes Znaniecki's (1940: 23) observation "that every individual who performs any social role is supposed by his social circle to possess and believes himself to possess the knowledge indispensable for its normal performance". "It is sometimes said", Barnes writes, that "we are living in an increasingly knowledge-based society, which is a profoundly misleading way of describing the proliferation of technical knowledge, the extraordinary division of mental labor and the ever-growing dependence on specialized expertise, all of which are undoubtedly in evidence." Barnes limits his enumeration of the possible reasons for designating modern society a knowledge-based society to cognitive transformations and he therefore misses most changes that result from the increased penetration of social structure and culture by knowledge as well as the impact this may have on individuals and major social institutions.

changes continue to evolve gradually, at an uneven pace, and they become clearly visible only after the transition is already over. The proximity of our time to significant social, economic and cultural changes, however, makes it highly likely that what is now beginning to come into view is of extraordinary present and future significance.

Moreover, knowledge societies do not come about as the result of some straightforward common pattern of development. They are not a one-dimensional social figuration. Knowledge societies become similar by remaining or even becoming dissimilar. New technological modes of communication break down the distance between groups and individuals, while the isolation of particular regions, cities and villages remains. The world opens up and creeds, styles and commodities mingle, yet the walls between incompatible convictions about what is sacred do not come tumbling down. The meaning of time and place erodes even while boundaries are celebrated.

Until recently, modern society was conceived primarily in terms of property and labor. Labor and property (capital) have had a long association in social, economic and political theory. Work is seen as property and as a source of emerging property. In the Marxist tradition, capital is objectified, encapsulated labor. On the basis of these attributes, individuals and groups were able or constrained to define their membership in society. In the wake of their declining importance in the productive process, especially in the sense of their conventional economic attributes and manifestations, for example as "corporeal" property such as land and manual work, the social constructs of labor and property themselves are changing. While the traditional attributes of labor and property certainly have not disappeared entirely, a new principle, "knowledge", has been added which, to an extent, challenges as well as transforms property and labor as the constitutive mechanisms of society.

Theories of societies, depending on their constitutive principles, mirror these quintessential social mechanisms in the chosen shorthand for the historical era they claim to describe and represent. Thus, bourgeois or capitalist society was originally viewed as a society of owners. Later it became a "laboring society" (*Arbeitsgesellschaft*), and it is now evolving into a knowledge society.

In retrospect, even some ancient societies (Rome, China, the Aztec Empire), that gained and maintained power in part as a result of their superior knowledge and information technology, may be described as knowledge societies of sorts. Ancient Israel was founded upon its lawlike Torah-knowledge, and in ancient Egypt religious, astronomical and agrarian knowledge served as the organizing principle and basis of authority. In this sense knowledge has had an important function throughout history, and humans have always lived in "knowledge societies". But in present-day society knowledge has clearly become much more fundamental and even strategic for all spheres of life, greatly modifying and in some cases replacing factors that until recently had been constitutive of social action.

Thus, and despite the fact that there also have been societies in the past that were based on knowledge-intensive action, the idea that modern society increasingly is a knowledge society is meaningful and has practical relevance. It is as meaningful to refer to modern society as a knowledge society as it made sense to refer to industrial societies even though there had been past social systems that were based on the work of "machines".

20.3 Knowledge About Knowledge

The focus of sociological analysis must therefore increasingly be the peculiar nature and function of knowledge in social relations as well as the carriers of such knowledge together with the resulting changes in power relations and sources of social conflict.[4] In sociology, however, virtually all classical theorists are proponents and even architects of scientism. This also applies to the ways in which knowledge is conceptualized in theories of society designed to capture the unique features of present-day society.

Marxist theories of society have assigned decisive importance to the (cultural) forces or means of production for societal development since "man's understanding of nature and his mastery over it by virtue of his presence as a social body [...] appears as the great foundation-stone *(Grundpfeiler)* of production and of wealth", so that general knowledge becomes a direct force of production (Marx [1939–1941] 1973: 705). Max Weber's seminal enquiry into the unique features of Western civilization stresses the pervasive use of reason to secure the methodical efficiency of social action. The source of rational action and, therefore, of rationalization is located in particular intellectual devices. The theory of industrial society, as developed by Raymond Aron, which encompasses both socialist and capitalist forms of economic organization as a single social reality of industrial civilization, accentuates first and foremost the extent to which science and technology shape the social organization of productive activities.

More recent theories of postindustrial society, in particular those of Daniel Bell, have elevated theoretical knowledge to an axial principle of society.[5] Scholars like Bell, for whom knowledge is an axial principle, nevertheless treat knowledge as a kind of black box. In often polemically charged circumstances, they have tended to

---

[4] Touraine ([1984] 1988: 111) captures well some of the long-term changes in social relations and goals. In mercantile societies, the "central locus of protest was called *liberty* since it was a matter of defending oneself against the legal and political power of the merchants and, at the same time, of counter-posing to their power an order defined in legal terms. In the industrial epoch, this central locus was called *justice* since it was a question of returning to the workers the fruit of their labor and of industrialization. In programmed [or, post-industrial] society, the central place of protest and claims is *happiness,* that is, the global image of the organization of social life on the basis of the needs expressed by the most diverse individuals and groups." Touraine ([1969] 1971: 3) employs the term "programmed" society for the new, emerging type of society in order to refer to the "nature of their production methods and economic organization".

[5] An appreciation as well as a more extensive critique of the theory of postindustrial society is contained in Stehr (1994: 42–90). Bell (1971) discusses the more distant as well as immediate intellectual antecedents of his theory of modern society a postindustrial society in an essay published in 1971. Bell's (1999) book has now been re-issued with a new foreword by the author.

defend positive knowledge as non-problematic, inherently practical, efficient, powerful and even ethical. That "rational knowledge", fabricated in one system, apparently travels with great ease and without loss across the boundaries of social systems, for instance, from science into the economy or state institutions, is hardly ever questioned.

The knowledge referred to in virtually all theories of society that elevate knowledge to prominence, and the groups of individuals that are seen as acquiring influence and control by means of this knowledge, tend to be conceptualized narrowly. This does not mean, however, that such a concept lacks cultural centrality and public or political influence. On the contrary, the narrower notion of knowledge that attributes enormous efficacy to scientific and technical knowledge resonates strongly with the dominant public conception of knowledge and its tasks. This concept of knowledge is a testimonial of the success of the scientific community in installing a particular conception of knowledge as the dominant public concept of knowledge. Whatever the limitations of this "scientistic" conception of knowledge, its centrality clearly reflects the diminishing role of the non-scientific conception of knowing. The scientization of everyday life, for example in the fields of health or the assessments of risks, manifests the cultural centrality of a particular conception of knowledge that has been assimilated by the theories of modern society described above. There exists, then, a perhaps paradoxical tendency to overestimate the efficacy of "objective" technical-scientific or formal knowledge. Theories of modern society generally lack sufficient detail and scope in their conceptualizations of "knowledge" in order to provide explanations for the causes of the increasingly greater demand for ever more knowledge, the ways in which knowledge travels, for the rapidly expanding groups of individuals in society who in some way or another live off knowledge, for the many forms of knowledge considered pragmatically useful and the various effects knowledge may have on social relations. Since the constitutive mechanism of "knowledge" is defined in a restrictive objectivist manner, the social, political and economic consequences to which these theories allude tend to be confined to rather straightforward effects that include the hope for (or the fear of) highly rationalized forms of social action.

20.4 Knowing the Known

The changes that should be examined are developments that occur with respect to the forms and dominance of knowledge itself. The focus should be on the relationships between scientific knowledge and everyday knowledge, declarative and procedural knowledge, knowledge and non-knowledge, and on knowledge as a capacity for social action. In order to demonstrate the significance of knowledge for social action, particularly in advanced societies, it is necessary to formulate a sociological concept of knowledge. What is it that we do know? Knowing represents a relation to things and facts, but also to laws and rules. Knowing involves participation: knowing things, facts, rules means to "appropriate" them, to include them in our field of orientation and competence.

Knowledge can of course be objectified, that is, the intellectual appropriation of things, facts and rules can be established symbolically. In order to know it is not necessary to get into intimate contact with the things themselves, but only with their symbolic representations. This is precisely the social significance of language, of writing, printing and data storage. Most of what is called knowledge and learning today is not direct knowledge of facts, rules and things, but objectified knowledge. Objectified knowledge is the highly differentiated stock of intellectually appropriated nature and society that constitutes society's cultural resource.

However, such participation is subject to stratification; the life chances, the life style and the social influence of individuals depend on access to the stock of knowledge at hand. Modern societies have made dramatic advances in the intellectual appropriation of nature and society. There exists an immense stock of objectified knowledge that mediates our relation with nature and with ourselves. In a general sense, this advancement used to be seen, in earlier contexts, as a form of modernization and rationalization that would lead to a "unity of civilization".[6] This second nature now overshadows the primary nature of humans. The real and the fictional merge and become indistinguishable. Theories become facts, yet facts cannot police theories.

It is only after the societal significance of such opposites and oppositions has been understood that the full sociological significance of knowledge can become clear. Such a perspective leads to the realization that knowledge is increasingly the foundation of authority, that access to knowledge becomes a major societal resource as well as the occasion for political and social struggles.

Although knowledge has always had a social function, it is only recently that scholars have begun to examine the structure of society and its development from the point of view of the production, distribution and reproduction of knowledge.[7] Applied to present-day society, the question arises if knowledge can provide a foundation for social hierarchies and stratification, for the formation of class structure, for the distribution of chances of social and political influence and also for personal life and, finally, whether knowledge may prove to be a normative principle of social cohesion and integration, even though the variations and alterations in the reproduction of knowledge appear to be enormous. Paradoxically, efforts to entrench necessity in history or eliminate the role of chance from it has produced, at least at the collective level, the very opposite tendency. The role of chance, ambiguity and "fragility" at the collective level, continues to be an increasingly important part of the way society is organized.

[6] Cf. the critique of the "unity of civilization" thesis by Toynbee (1946: 36–41).

[7] E.g. Malinowski (1955) and Machlup (1962, 1981, 1984).

20.5 Knowledge as a Capacity for Action

Knowledge may be defined as a *capacity for action.* The use of the term "knowledge" as a capacity for action is derived from Francis Bacon's famous observation that knowledge is power (a somewhat misleading translation of Bacon's Latin phrase: "*scientia est potentia*"). Bacon suggests that knowledge derives its utility from its capacity to set something in motion. The term potentia, that is *capacity,* is employed to describe the power of knowing. More specifically, Bacon asserts at the outset of his *Novum Organum* that "human knowledge and human power meet in one; for where the cause is not known the effect cannot be produced. Nature to be commanded must be obeyed; and that which in contemplation is the cause is in operation the rule."

The definition of knowledge as capacity for action has multifaceted implications and consequences. *Capacity* for action signals that knowledge may in fact be left unused, or that it may be employed for "irrational" ends. The thesis that knowledge invariably is pushed to its limit, that it is often translated into action without regard for its possible consequences (as argued, for instance, by C. P. Snow; cf. Sibley 1973), represents a typical view among observers of technological development. However, the claim that science and technology invariably push for the practical implementation of scientific and technical knowledge does not give proper recognition to the context of implementation of such knowledge. Such a conception of the immediate practical efficacy of scientific and technological knowledge, furthermore, vastly overestimates the inherent practicality of the knowledge claims fabricated in science.

The definition of knowledge as capacity for action strongly indicates that the material realization and implementation of knowledge is open, that it is dependent on or embedded within the context of specific social, economic and intellectual conditions. Knowledge, as a capacity for action, does not signal that *specific* knowledge claims always possess a fixed "value" or even a distinct practical dimension. We cannot, as a result, stipulate a priori that some knowledge claims, for example, those that issue from disciplines in the humanities, are less practical than knowledge that originates in the natural sciences.[8] Inasmuch as the realization

[8] It is a widely shared assumption that social science knowledge and knowledge from the humanities is somehow less useful than natural science knowledge, and perhaps increasingly so as 'modernization' advances: "The more post-industrial society becomes intellectualized, the more it tends to displace traditional value-oriented intellectual disciplines to the benefit of action-oriented ones, that is, those disciplines that can play a direct role in policy-making" (Crozier 1975: 32). For Crozier, the societal debasement of knowledge from the humanities produces a widespread sense of alienation among its carriers, namely intellectuals, and a general drift toward protest and even revolutionary posture.

of knowledge is dependent on the active elaboration of knowledge[9] within specific networks and social conditions, a definite link between knowledge and social power becomes evident because the control of conditions and circumstances requires social power. The larger the scale of a project, the greater the need for social power to control the actual realization of knowledge as capacity for action.

Knowledge is a peculiar entity with properties unlike those of commodities or of secrets, for example. Knowledge exists in objectified and embodied forms. If sold, it enters other domains—and yet it remains within the domain of its producer. Knowledge does not have zero-sum qualities. Knowledge is a public as well as private good. When revealed, knowledge does not lose its influence. While it has been understood for some time that the "creation" of knowledge is fraught with uncertainties, the conviction that its application is without risks and that its acquisition reduces uncertainty has only recently been debunked. Unlike money, property rights and symbolic attributes such as titles, knowledge cannot be transmitted instantaneously. Its acquisition takes time and often is based on intermediary cognitive capacities and skills. But acquisition can be unintended and occur almost unconsciously. Neither the acquisition nor the transmission of knowledge is always easily visualized. The development, mobility and reproduction of knowledge are difficult to regulate. It is "troublesome" to censor and control knowledge. It is reasonable to speak of limits to growth in many spheres and resources of life, but the same does not appear to hold for knowledge. Knowledge has virtually no limits to its growth, but it takes time to accumulate.

Knowledge is often seen as a collective commodity par excellence; for example, the ethos of science demands that it be made universally available, at least in principle. But is the "same" knowledge available to all? Is scientific knowledge, once transformed into technology, still subject to the same normative conventions? The answer provided by one economist is that technology must be considered a "private capital good". In the case of technology, disclosure is uncommon, and rents for its use can be privately appropriated (cf. Dasgupta 1987: 10). But the potentially unrestricted universal availability of knowledge makes it, in peculiar and unusual ways, resistant to private ownership (Simmel [1907] 1978: 438). Modern communication technologies ensure that access becomes easier, and may even subvert remaining proprietary restrictions; however, concentration rather than dissemination is also possible and certainly feared by many, including the late Marshall McLuhan. But it is equally possible to surmise that the increased social importance of knowledge in the end undermines its exclusiveness. Yet the opposite appears to be the case and therefore raises anew the question of the persisting basis for the power of knowledge. Despite its reputation, knowledge is virtually never uncontested. In science, its contestability is seen as one of its foremost virtues. In practical circumstances, the contested character of knowledge is often repressed and/or conflicts with the exigencies of social action.

[9] Compare Lazega's (1992) essay on the "information elaboration" in work groups and the relations between information and decision-making in and dependent on "local" contexts.

Scientific and technical knowledge, while clearly representing such "capacities for action", does not thereby become uncontestable, no longer subject to challenge and interpretation.[10] Scientific and technical knowledge is uniquely important because it produces *incremental* capacities for social and economic action or an *increase* in the ability of "how-to-do-it" that may be "privately appropriated", at least temporarily.[11] And contrary to neoclassical assumptions, the unit price for knowledge-intensive commodities and services decreases with increased production, reflecting "progress down the learning curve" (cf. Schwartz 1981).

Knowledge constitutes a basis for power. As Galbraith (1967: 67) stresses, power "goes to the factor which is hardest to obtain or hardest to replace [...] it adheres to the one that has greatest inelasticity of supply at the margin". But knowledge as such is not a scarce commodity, though two features of certain knowledge claims may well transform knowledge from a plentiful into a scarce resource. First, what is scarce and difficult to obtain is not access to knowledge per se but to *incremental knowledge*, to a "marginal unit" of knowledge. The greater the *tempo* with which incremental knowledge ages or decays, the greater the potential influence of those who manufacture or augment knowledge, and correspondingly, of those who transmit such increments. Secondly, if sold, knowledge enters the domain of others, yet remains within the domain of the producer, and can be spun off once again. This signals that the transfer of knowledge does not necessarily include the transfer of the cognitive ability to generate such knowledge, for example the theoretical apparatus or the technological regime that yields such knowledge-claims in the first place and on the basis of which it is calibrated and validated. Cognitive *skills* of this kind, therefore, are scarce.

20.6 Knowledge as Capital

Among knowledge-based approaches and concepts in social theory, *cultural capital* and *human capital* theories stand out. Pierre Bourdieu distinguishes between different forms of cultural capital—its embodied or symbolic form as internalized culture, its objectified form in material objects and media, and its institutionalized

[10] If knowledge indeed "travelled" almost without impediments and could be reproduced largely at will, the idea would make considerable sense that scientists and engineers, as the creators of the "new" knowledge in modern society, should be located at the apex of power.

[11] Peter Drucker (1993: 184) observes, however, that initial economic advantages gained by the application of (new) knowledge become permanent and irreversible. What this implies, according to Drucker, is that imperfect competition becomes a constitutive element of the economy. Knowledge can be disseminated or sold without leaving the context from which it is disseminated or sold. The edge that remains is perhaps best described as an advantage based on cumulative learning.

form (for example, as academic certificates).[12] These distinctions signal the ways in which cultural capital is stored and passed on by way of becoming an integral habitus of the individual. Bourdieu identifies two additional forms of capital—economic and social capital. These two forms of capital refer to the gains individuals may derive from their network of social relations. I will focus on Bourdieu's concept of cultural capital since it resonates more closely with the concept of knowledge.

In Bourdieu's sense, cultural capital[13] as a form of *symbolic capital* is much broader than the concept of human capital as developed in economic discourse. Modern human capital theory relates deliberate and measurable educational investments (and achievements) in the acquisition of useful skills and knowledge to their *monetary* gains or losses. Skills and knowledge have grown in Western societies at a much faster rate than nonhuman capital, as one of the originators of this idea in economics, Schultz (1961) contends. Investment in human capital (that is, capital embodied in human beings), Schultz argues, has driven much of the growth in real earnings per worker in recent decades.

In strong contrast, cultural capital theory does not proceed from the assumption of a kind of *tabula rasa* that allows every individual to participate in the competitive market where human capital is allocated and where success or failure is at most affected by unequal natural aptitudes.

Cultural capital theory acknowledges not only pre-existing unequal access to the distributional channels for its accumulation, but also the different ways in which the "market" from the beginning favors the chances of particular players. In a largely undifferentiated society or community, of course, culture does not function as a vehicle for the emergence of cultural capital. There the social conditions of its transmission tend to be much more disguised than those that govern economic

[12] Bourdieu's discussion of cultural capital resonates with Simmel's observations ([1907] 1978: 439–40) in *The Philosophy of Money* about the role of the "intellect". Simmel notes that "the apparent equality with which educational materials are available to everyone interested in them is, in reality, a sheer mockery. The same is true of the other freedoms accorded by liberal doctrines which, though they certainly do not hamper the individual from gaining goods of any kind, do however disregard the fact that only those already privileged in some way or another have the possibility of acquiring them. For just as the substance of education—in spite of, or because of its general availability—can ultimately be acquired only through individual activity, so it gives rise to the most intangible and thus the most unassailable aristocracy, to a distinction between high and low which can be abolished neither (as can socioeconomic differences) by a decree or a revolution, nor by the good will of those concerned […]. There is no advantage that appears to those in inferior positions to be so despised, and before which they feel so deprived and helpless, as the advantage of education."

[13] Bourdieu ([1983] 1986: 243) first encountered the usefulness of the notion of "cultural capital" in social inequality research designed to explain the unequal scholastic achievement of children from different social classes; unequal academic success or the "specific profits" (failures) students are able acquire in the academic market, are related to the stratified distribution of cultural capital among social classes and the unequal chances of acquiring such capital domestically. One might say that the benefits that derive from the unequal distribution of cultural capital represent a form of unearned income.

capital. The portion of individual lives that can be afforded for the acquisition of cultural capital is regarded as highly significant. Cultural capital yields benefits of distinction for its owner.

The most evident drawback of Bourdieu's explication of cultural capital theory is, first, its strong individualistic bias, that is, the extent to which Bourdieu stresses the fusion of cultural capital and the personality of the individual owner. The emphasis remains for the most part on cultural capital as an inherent attribute of the *individual* carrier. Cultural capital declines and dies with its bearer since both have identical biological limits. Bourdieu's individualistic conception of cultural capital appears to be linked to his determination not to dispossess cultural capital theory of the ability to calculate and attribute investment gains that derive from cultural capital. And such returns of investment are seen to accrue primarily to the investor. In this sense, cultural capital theory resonates with human capital theory. It contains crucial residues of economic discourse.[14]

It is important to recognize that cultural capital is embodied in collective processes and structures; hence the benefits often do not accrue only to those who have invested resources. The production as well as consumption of such capital is not charged to the individual. It is borne by the collectivity. At one extreme, such capital can even be seen to be entirely free, in that its use by certain individuals does not diminish its utility or availability to others. Cultural capital is human-made capital and as such subject to limits applicable to all human products and creations. Secondly, Bourdieu discovers and utilizes the concept of cultural capital in the context of social inequality research. The concept derives its coherence from this context in which distinction, processes of inclusion and exclusion, cultural frames and meaning production are the hallmark of the work that cultural capital accomplishes for individuals. Bourdieu thereby implies the continued social, economic and political relevance of social class in modern society.[15] But it must be asked if class divisions are not undermined by virtue of the transformation of economic realities. Distinctions linked to cultural processes are not merely derivative but foundational. Thirdly, although the notion of human cultural capital is not employed in a fully ahistoric manner, it is for the most part devoid of historical specificity. Bourdieu ([1983] 1986: 255) refers to relatively undifferentiated societies in which embodied culture, since not stratified, does not function as cultural capital; however, this does not permit differentiation between various forms of society beyond a straightforward dichotomy of "simple" and "complex" societies. New "structures of consciousness" (to use Benjamin Nelson's phrase) cannot be

---

[14] It needs to be recognized, however, that the actual acquisition of capital, even if the quantity of capital acquired depends on capital previously accumulated by the family of an individual, is—as Simmel ([1907] 1978: 439) observed—ultimately an individual activity.

[15] Sympathetic critics of Bourdieu's capital theory have pointed to other attributes of his approach as problematic, for example, the holistic presupposition as a general theoretical assumption. Bourdieu tends to postulate cultural capital as a generalized medium of accumulation and distinction ill-suited for the analysis of a society with multiple cleavages and divisions (see Hall 1992; Lamont/Lareau 1988).

captured by this term. In many ways, the structure of consciousness of knowledge societies is not novel. It resonates with the consciousness of modernity that dates—although this too is a highly contested question—at least from the sociohistorical context of the French Revolution. In other respects, the *conscience collective* in knowledge societies is at variance with the belief systems and mental sets that are usually identified as uniquely modern and therefore warrants the designation of a new structure of consciousness. In any event, the notion of cultural capital is not well designed to capture such transformations.

20.7 The Limits of the Power of (Scientific) Knowledge

A critical analysis of the limits of the social power of scientific knowledge requires an understanding of the special nature as well as the similarities of scientific and non-scientific knowledge and action. Classifying scientific knowledge as a unique form of human knowledge is of little value. Such a classification is too closely linked to now obsolete epistemological conceptions of science—to such notions and ideals as universality, experience, rationality, necessity and practicality. Conceptions of scientific knowledge that adhere to such notions tend to deny that scientific knowledge is socially based and a collective as well as historical enterprise.

Merton ([1942] 1973) has suggested that for most people everyday knowledge provides greater plausibility and more useful means of comprehension than scientific knowledge, as well as considerable substantive affinity to existing cultural "prejudices", thereby constituting a potential source of competition for scientific knowledge claims. Merton's is an early theoretical conception of the limits of scientific knowledge and goes far beyond considerations primarily driven by epistemological concerns.

Toulmin's (1972: 378) useful characterization of organized human activities generally may be taken as a description of the social conditions within which the social production of scientific knowledge occurs: "[H]uman activities and enterprises [...] in which decisions are made, procedures followed, considerations taken into account, conclusions arrived at, new possibilities entertained, and 'reasons' given for the resulting conclusions or actions." The *special* social and intellectual circumstances that prevail in the manufacture of scientific knowledge affect the structure and the possibility of reproducing such knowledge.

Among the special conditions that should be noted here are that knowledge claims or knowledge effects produced under special conditions in scientific laboratories can, first, only be reproduced outside the laboratory if the special conditions that allowed such outcomes are also reproduced outside the laboratory. That is, the special circumstances that led to the original observation of the effect must be extended to the context in which a successful transfer is to be made (see Rouse 1987: 227). The notion that scientific knowledge, unlike other forms of knowledge, is not bound or limited institutionally has to be questioned in light of the conditions necessary for the reproduction of scientific knowledge claims outside the circumstances of their initial discovery.

Secondly, knowledge claims not only take on features derived from the material conditions of their production, but also reflect institutionally bound cognitive attributes. These attributes include, importantly, a suspension of the pressure to act as constitutive of scientific discourse. Knowledge produced within the scientific community is released from the tasks it must perform outside of science.

One of the most salient attributes of everyday life situations is, in contrast, the persistent pressure to reach a decision, to observe a specific rule, to follow a particular course of action by discarding alternative possibilities, or to provide an account of completed action *ex post facto*. This suspension of the constraint to act within scientific discourse may be described, on the one hand, as a virtue of intellectual activity taking place under privileged conditions that moderate the effect of the pressing interests, rapidly passing opportunities and ambiguous dependencies of everyday contexts on the production of scientific knowledge claims. On the other hand, the result of this suspension of the pressure to act is that scientific knowledge takes on qualities of incompleteness, provisionality, fragmentariness or expansiveness that reduce its effectiveness as knowledge in circumstances in which action is the foremost requirement. As Durkheim ([1912] 1965: 479) observed so well: "Life cannot wait" (cf. also Gehlen [1940] 1988: 296–7).[16] Finally, knowledge must be made available and interpreted, and also linked to local, contingent circumstances. The complexity of the linkages and the volume of resources required delineate further limits of the power of scientific and technical knowledge.

The set of limits to the social power of scientific knowledge constitutes an inevitable part of the fabrication and the utilization of scientific knowledge. But they also explain why the knowledge work performed by the stratum of experts of knowledge-based occupations, generally speaking, attains greater and greater centrality in advanced society. The knowledge work performed by knowledge-based occupations or by experts, counselors and advisors, is crucial in that their work "heals" some of the practical deficiencies of scientific knowledge. For example, a chain of interpretations must come to an "end" in order for knowledge to become relevant in practice and effective as a capacity for action. This function of ending reflection or remedying the lack of immediate practicability of scientific and technical knowledge—as it emerges from the scientific community—for the purpose of action is largely performed by various groups of experts in modern society.

[16] Incompleteness or the lack of any impetus to action is constitutive for scientific knowledge: "Faith is before all else an impetus to action, while science, no matter how far it may be pushed, always remains at a distance from this. Science is fragmentary and incomplete; it advances but slowly and is never finished" (cf. also Luhmann [1986] 1989: 154–9). The probability that myths and half-truths are employed by large segments of the population in "advanced" societies may well be even more characteristic of crisis situations in which various dangers appear to be imminent, as Elias ([1989] 1996: 500–1) argues, for example. In this respect, present-day societies do not differ from "primitive" societies in which similar responses were elicited by the dangers brought about by illness, draught, thunderstorms or floods. However, Elias is convinced that this state of affairs can be corrected in principle as adequate knowledge is diffused more widely throughout society.

The centrality of knowledge-based occupations or, using a narrower term, experts in knowledge societies, does not mean that we are on the way, as social theorists have feared in the past, to a technocratic society. A technocratic model of society and its major social institutions which "sees technicians dominating officials and management, and which sees the modern technologically developed bureaucracies as governed by an exclusive reliance on a standard of efficiency" (Golden 1976: 257) is but a nightmare, an ideal type or a utopia. Quite a number of arguments can be deployed to demystify the threat of technocracy and a new ruling class made up of faceless experts. The most persuasive argument is social reality itself, which has failed to support the transformation of society in this direction. The emergence of technocratic regimes long predicted has not materialized.

20.8 The Knowledge-Based Economy

The emergence of knowledge societies signals first and foremost a radical transformation in the *structure of the economy*. Productive processes in *industrial society* are governed by factors that—relative to the increasing importance of the exchange of symbolic goods—have greatly changed and for the most part declined in significance as preconditions for economic growth: the dynamics of the supply and demand for primary products or raw materials; the dependence of employment on production; the importance of the manufacturing sector that processes primary products; the role of manual labor and the social organization of work; the role of international trade in manufactured goods and services; the function of time and place in production and of the nature of the limits to economic growth.

The most common denominator of the changing economic structure is a shift away from an economy driven and governed by "material" inputs into the productive process and its organization, towards an economy in which the transformations of productive and distributive processes are increasingly determined by "symbolic" or knowledge-based inputs. The development and impact of modern information technology exemplifies these transformations (and not just in the sphere of economic activities). They include the dematerialization of production that represents diminished constraints on supply, lower and still declining cost, and a redefinition of the social functions of time, place and the increasing acceleration of change (cf. Miles et al. 1988; Perez 1985).

The economy of industrial society, in short, is primarily a *material economy* on the way of becoming a monetary economy. Keynes' economic theory, particularly his *General Theory* (1936), reflects this transformation of the economy of industrial society into an economy substantially affected by monetary matters. But, as more recent evidence indicates, the economy described by Keynes is best understood as a *symbolic economy*. The structural changes of the economy and its dynamics increasingly reflect the fact that *knowledge* is emerging as the leading dimension in the productive process, the primary condition for its expansion and for a change in the limits to economic growth in the developed world. In the knowledge society, most of the wealth of a company is embodied in its creativity and information. In short, for the production of goods and services, with the exception of the most standardized commodities and services, factors other than "the amount of labor time or the amount of physical capital become increasingly central" (Block 1985: 95) to the economy of advanced societies.[17]

20.9 Individual and Collective Social Conduct in Knowledge Societies

The transformation of modern societies into knowledge has profound consequences aside from those that pertain to its economic structure. One of the more remarkable consequences is the extent to which modern societies become fragile societies. This observation has to be qualified.

Modern societies tend to be fragile from the viewpoint of those large and once dominant social institutions that find it increasingly difficult to impose their will on all of society, to give direction and determine the fate of its individual components. From the perspective of small groups and social movements more and uncoupled from the influence of the traditional large-scale social institutions, however, modern societies are not particularly fragile at all. For such groups and social movements, the social transformations under way mean a distinct gain in their relative influence and participation, even if typically mainly in their ability to resist, delay and alter the objectives of the larger institutions. I regard precisely the growing importance of such knowledge in modern society as the prime and immediate reason for the enlargement of the capacity of individuals and social movements to assert themselves in traditional as well as new contentious circumstances. The increase in the "knowledgeability" of actors and the decrease or static capacity to act of large collectivities have to be seen as complementary developments since the decline in the ability of large institutions to impose their will is linked to the enlargement of the capacity to act by individuals and small groups in society, for instance, in their capacity to say no or mobilize effective strategies of contention.

Knowledge societies are (to adopt a phrase from Adam Ferguson) the results of human action, but not of deliberate human design. They emerge as adaptations to persistent but evolving needs and changing circumstances of human conduct. Among the most significant transformations in circumstances that face human conduct is the continuous "enlargement" of human action, including an extension of its "limits to growth". Modern societies as knowledge societies are becoming more fragile. But this does not mean that they are disintegrating. Increased individualism, for example, does imply an uncoupling from certain collective obligations and constraints and the distinct possibility that the role of the stranger becomes less and less strange for more and more individuals. But it does not suggest a complete

[17] See especially Drucker (1986) and Lipsey (1992).

uncoupling from collective consciousness and action restraints. In much the same way, while knowledge societies become more fragile, they do not lead to an arrest of social action. On the contrary, they lead to an enlargement and extension of forms of conduct, forms of life, chains of social interaction and channels of communication.

The enlargement in capacities to act occurs at an uneven pace and to an uneven degree. The outcome is a hitherto unknown contradiction: An increasingly larger proportion of the public acquires and exercises political skills, for example—including the choice of non-participation (cf. Stehr/Meja 1996), or the denial that political activities are indeed political (cf. Magnussson 1996: 29–32)—while the ability of the state and its agencies to "impose its will" or to exercise sovereignty is arrested, and typically even decreases.[18] This leads to more fragile and volatile forms of legitimate authority and more fragile powers of the state and of other major social institutions. In that sense the growth and broader dissemination of knowledge paradoxically produces greater uncertainty and contingency rather than providing a resolution of disagreements or the basis for a more effective domination by central societal institutions.

Modern societies are also increasingly *vulnerable* entities. More specifically, the economy, the communication or traffic systems are vulnerable to malfunctions of self-imposed practices typically designed to avoid breakdowns. Modern infrastructures and technological regimes are subject to accidents as the result of fortuitous, unanticipated human action,[19] to non-marginal or extreme natural events that may dramatically undermine the taken-for-granted routines of everyday life in modern societies or to deliberate sabotage.[20] That societies appear to be assailable and sometimes even defenseless in the face of damaging or murderous attacks launched by dedicated individuals represents a fear as well as a now taken-for-granted risk. However, my analysis of modern societies as fragile societies does not extend to its vulnerability in the face of attacks launched by "rebel" groups, revolutionary dissidents, extremists, assassins, terrorists bent on destroying the institutions they choose to assault, accidents or extreme natural events. It may indeed be difficult to clearly separate the profound susceptibility and vulnerability

[18] As Crozier ([1979] 1982: 5) observes, "the average citizen has never been so free in the range of choices as he is now and has never been able to exert so much influence when grouped together with others as he currently can". Inglehart (1990: 335–70) examines the enlargement of political skills of the public in Western societies in terms of a shift from "elite-directed" to "elite-challenging" politics.

[19] The *Globe and Mail* (17 July 1999, National News) describes the breakdown of much of the communications system in the City of Toronto on 16 July 1999 as the result of an accidentally dropped tool that was the beginning of a chain-reaction disaster affecting not only Canada's largest city as "a series of failures that revealed the fragility of the complex communications society takes for granted".

[20] My conception of fragility therefore excludes what are clearly *illegal* activities that could hamper and interfere with establish patterns of social conduct, for example, the consequences that follow on the fabrication and at times fast spread of computer viruses (such as the one dubbed Melissa in early 1999; cf. "Melissa virus suspect caught", *New York Times*, 3 April 1999).

of modern society to such assaults and forms of aggression from what I am describing and analysing here as the essential fragility of modern society. However, the two refer to entirely different sets of processes, motives and consequences. A society is vulnerable because—prompted by profound disagreements about its very fabric and legitimacy—large or small groups of individuals are determined to negate it. "Extraordinary" events[21] that occur as the result of such a constellation of motives may be anticipated in principle; at least many large social institutions act and plan as if such events can be anticipated. The state for one prepares itself for events of this kind. "Revolutionary" activities are not new. In short, we cannot say with any confidence that modernity equals stability as Huntington (1968: 47), for example, proposed.

Present-day social systems may be seen to be fragile and vulnerable entities in yet another sense. Such fragility results from conduct as well as the deployment of artifacts designed to stabilize, routinize and delimit social action. I am referring specifically to what Rochlin (1997) has called the "computer trap" or the unintended outcomes and secondary effects of computerization. In the process of even more deeply embedding computers into the social fabric of society, that is, re-designing and re-engineering large-scale social and socio-technical systems in order to manage the complexities of modern society, novel risks and vulnerabilities are created. Computerization becomes more and more invisible but the potential consequences as the result of a breakdown are enlarged. The long-term secondary and destabilizing consequences range from "the loss of the basis from which such skills can be constructed to the creation of a socio-technical dependency on operating at such high levels of efficiency and complexity that human beings will not be able to manage or control the effects, intended or unintended, whether the computers break down or not"; the increased tightness of linkages, lack of back-up systems, and the speed of response of such systems will make "human intervention or control difficult at best when (and not if) something goes wrong" (Rochlin 1997: 217). Hence, one might argue, a basic fragility is inscribed into social systems via the deployment of technical regimes designed to achieve the opposite, namely to stabilize, constrain, routinize and even control conduct. Such an outcome of computerization might be particularly unexpected, cataclysmic and far-reaching but does not differ in principle from the unintended and unanticipated consequences of the widespread deployment of other technical devices in the past.

The fragility of modern societies as described here, however, is a unique condition. Societies are fragile because—propelled by a marked enlargement of their capacities to act—individuals are capable, within certain established rules, to assert their own interests by opposing or resisting the—not too long ago—almost

[21] Euchner (1996) in his study *Extraordinary Politics* analyses protest movements of recent decades and stipulates that their common denominator is that their members reject or violate the rules of conventional politics. Aside from the distinct possibility that modern protest movements change the rule of politics (cf. Clark/Hoffmann-Martinot 1998) and are themselves transformed in the course of their struggles, the list of movements Euchner develops shows that it is very difficult to clearly distinguish between "ordinary" and "extraordinary" political events.

unassailable monopoly of truth of major societal institutions. That is to say, legitimate cultural practices based on the enlargement and diffusion of knowledge enable a much larger segment of society to effectively oppose power configurations that turned out or are apprehended to be tenuous and brittle.[22]

Among the major but widely invisible social innovations in modern society is the immense growth of the "civil society" sector. The civil society sector recognizes the "plethora of private, nonprofit, and nongovernmental organizations" (Salomon/Anheier 1997: 60) that have emerged and grown considerably both in volume and in public influence in recent years in many countries of the developed world. This sector provides an organized basis through "which citizens can exercise individual initiative in the private pursuit of public purposes" (Salomon/Anheier 1997: 60).[23]

I also interpret the considerable enlargement of the informal economy, crime, corruption and the growth of wealth in modern society as well as increasing but typically unsuccessful efforts to police these spheres as evidence of the diverse as well as expanded capacity of individuals, households and small groups to take advantage of and benefit from contexts in which the degree of social control exercised by larger (legitimate) social institutions has diminished considerably.[24]

The enlargement of the various social activities known as the informal economy or the growth of wealth, despite the ambiguity of its connotations, constitutes a major structural feature of advanced societies. Among the consequences is a distinctive shift in value-orientations in advanced societies. In political terms, this represents a displacement of the prominence of leftist by more centrist and

---

[22] My emphasis on the individual ability to oppose and contest established power resonates with recent research that shows power relations to be multidimensional configurations. Such a perspective stresses, for example, that the notion of resistance must be redefined so "that it can be applied to a much wider range of sociocultural practices and take into account the ways in which the subjectivity of the dominated is constrained, modified and conditioned by power relations" (Haynes/Prakash 1992: 2).

[23] Salomon/Anheier (1997: 62) have attempted to quantify the growth and presence of the civil society sector in different countries: "In France, over 60.000 associations were created in 1990 alone, compared to less than 18,000 in 1961. Similarly, in Germany the number of associations per 100.000 population nearly tripled from 160 in 1960 to 475 in 1990. Even Hungary, within two years of the fall of communist rule, boasted over 13.000 associations. And Sweden, often regarded as the prototypical welfare state, displays some of the highest participation rates in civil society worldwide." The growth of international non-governmental organizations is sketched in Boli/Thomas (1997).

[24] The point is made very well in a study of street vendors and the state administration in Mexico City or the political economy of informality as the author calls it. Cross (1998: 228) observes that street vending in Mexico City has "experienced vigorous growth despite state policies designed to control it and even, at times, to reduce or eliminate it". The growth of street vending activity the author describes certainly has economic reasons but what is important as well, and perhaps more significant in this context, is the "ability of street vendors collectively to thwart or reverse administrative attempts to control them that would, if successful, have prevented such an explosive growth" (Cross 1998: 228).

conservative agendas in all political parties and no longer a trend to the left, as may have been the case in the decades of the 1960s and 1970s.[25]

However, much of social science discourse has been preoccupied with the opposite phenomenon,[26] namely the probable and dangerous enlargement of the ability of modern social institutions, especially various state institutions but also the economy, to more ruthlessly impose its will on its citizens. Thus, the classical social theorists as well as many of their more recent successors were concerned with discovering the conditions that produce and reproduce domination and repression rather than greater autonomy, freedom and independence. Modern science and technology typically were viewed, in the context of such analyses, as the handmaidens of regressive civilizational developments.

But whether the kinds of societal developments we are sketching constitute, as John Stuart Mill anticipated one hundred and fifty years ago, a reconciliation of order and progress remains in doubt. Today, in fact, order and progress are essentially contested concepts and objectives. What is reconciliation to some invariably represents an unsustainable agenda for others. We are living in an age in which the expansion of individual choices is in conflict with traditional sentiments as well as with objectives that favor their restriction.

20.10 Prospects

History has by no means ended, but it certainly has changed. The old rules, certainties and trajectories no longer apply. Of course, there are few opportunities of fresh starts in history. None the less, the future of modern society no longer mimics the past to the extent to which this has been the case. That is to say, the future is made from fewer fragments of the past. As a result, sentiments with respect to history that are becoming more pervasive are those of fragility and dislocation. History will increasingly be full of unanticipated incertitudes, peculiar reversals, proliferating surprises, and we will have to cope with the ever-greater speed of significantly compressed events. The changing agendas of social, political and economic life as the result of our growing capacity to make history will also place

[25] Bialer (1977: 36–9) tries to account for the "resurgence and changing nature of the left" in the 1960s and 1970s. He argues that he is not convinced that either "the stress on the structural societal changes and on the growth of the 'knowledge industry'" which various observers (for example, Peter Berger) invoked at the time should count as adequate explanations nor should the reference to the cyclical nature of the interest in and attraction to socialism. For Bialer, the key development that accounts for the shift toward the political left is ideological, namely the dissatisfaction with the political status quo which in turn reflected a fundamental discontinuity in the economic, social and political developmental base of industrialized democracies.

[26] A growing number of studies are investigating these structural changes in industrialized and less developed countries. The informal economy, for example, is examined in the form of a number of case studies in Portes et al. (1989).

inordinate demands on our mental capacities. The fit or lack of fit between our knowledgeability and what society, the economy and culture mentally demands is one of the major challenges of knowledge societies.

References

Bahro, Rudolf. 1978. *The Alternative in Eastern Europe*. London: New Left Books.

Barber, Bernard. 1975. "Toward a New View of the Sociology of Knowledge", in: Lewis A. Coser (ed.), *The Idea of Social Structure*. New York: Harcourt Brace Jovanovich: 103–116.

Barnes, Barry. 1974. *Scientific Knowledge and Sociological Theory*. London: Routledge & Kegan Paul.

Barnes, Barry. 1977. *Interests and the Growth of Knowledge*. London: Routledge & Kegan Paul.

Bell, Daniel. 1964. "The Post-industrial Society", in: Eli Ginzberg (ed.), *Technology and Social Change*. New York: Columbia University Press: 44–59.

Bell, Daniel. 1973. *The Coming of Post-Industrial Society*. New York: Basic Books.

Berger, Peter. 1987. *The Capitalist Revolution: Fifty Propositions about Prosperity, Equality, and Liberty*. New York: Basic Books.

Berger, Peter and Luckmann, Thomas. 1967. *The Social Construction of Reality: A Treatise in the Sociology of Knowledge*. New York: Doubleday.

Bloor, David. 1976. *Knowledge and Social Imagery*. London: Routledge & Kegan Paul.

Bloor, David. 1984. "Durkheim and Mauss Revisited: Classification and the Sociology of Knowledge", in: Nico Stehr and Volker Meja (eds.), *Society and Knowledge. Contemporary Perspectives on the Sociology of Knowledge*. New Brunswick: Transaction Books: 51–76.

Böhme, Gernot. 1993. *Am Ende des Baconschen Zeitalters. Studien zur Wissenschaftsentwicklung*. Frankfurt am Main: Suhrkamp.

Böhme, Gernot, Wolfgang van den Daele and Wolfgang Krohn. 1973. "Die Finalisierung der Wissenschaft", *Zeitschrift fur Soziologie*, 2: 128–144.

Coser, Lewis A. 1968. "Sociology of Knowledge", in: *International Encyclopedia of the Social Sciences*. Volume 8. New York: Macmillan: 428–435.

Dahrendorf, Ralf. 1962. "On the Origin of Social Inequality", in: Peter Laslett and W.G. Runciman (eds.), *Politics and Society*. Oxford: Blackwell.

Douglas, Mary. 1975. *Implicit Meanings*. London: Routledge & Kegan Paul.

Drucker, Peter F. 1986. "The Changed World Economy", *Foreign Affairs*, 64: 768–791.

Elias, Norbert. 1971. "Sociology of Knowledge: New Perspectives", *Sociology*, 5(2): 149–168, 335–370.

Elias, Norbert. 1983. *Engagement und Distanzierung. Arbeiten zur Wissenssoziologie* I. Frankfurt am Main: Suhrkamp.

Feyerabend, Paul K. 1975. *Against Method*. London: NLB.

Funtowicz, Silvio O. and Jeffrey R. Ravetz. 1993. "The Emergence of Post-normal Science", in: Rene von Schomberg (ed.), *Science, Politics and Morality. Scientific Uncertainty and Decision Making*. Dordrecht: Kluwer.

Gibbons, Michael et al. 1994. *The New Production of Knowledge. The Dynamics of Science and Research in Contemporary Societies*. London: Sage.

Gouldner, Alvin W. 1979. *The Future of Intellectuals and the Rise of the New Class*. New York: Continuum.

Hesse, Mary B. 1980. *Revolutions and Reconstructions in the Philosophy of Science*. Brighton: Harvester Press.

Kettler, David, Volker Meja and Nico Stehr. 1984. *Karl Mannheim*. London: Tavistock.

Knorr-Cetina, Karin. 1981. *The Manufacture of Knowledge. An Essay on the Constructivist and Contextual Nature of Science*. Oxford: Pergamon Press.

Knorr-Cetina, Karin. 1999. *Epistemic Cultures. How the Sciences Make Knowledge*. Cambridge, MA: Harvard University Press.
Kuhn, Thomas. 1962. *The Structure of Scientific Revolutions*. Chicago: University of Chicago Press.
Lakatos, Imre. 1978. *The Methodology of Scientific Research Programmes*. Cambridge: Cambridge University Press.
Lane, Robert E. 1966. "The Decline of Politics and Ideology in a Knowledgeable Society", *American Sociological Review*, 31: 649–662.
Latour, Bruno. 1999. *Pandora's Hope. Essays on the Reality of Science Studies*. Cambridge, MA: Harvard University Press.
Latour, Bruno and Steve Woolgar. 1986. *Laboratory Life. The Construction of Scientific Facts. With a New Postscript*. Princeton, NJ: Princeton University Press.
Lieber, Hans-Joachim. 1955. "Wissenssoziologie", in: Wilhelm Bemsdorf and Friedrich Bülow (eds.), *Wörterbuch der Soziologie*. Stuttgart: Enke: 629–633.
Luhmann, Niklas. 1977. *Zweckbegriff und Systemrationalität*. Frankfurt am Main: Suhrkamp.
Luhmann, Niklas. 1984. "The Differentiation of Advances in Knowledge: The Genesis of Science", in: Nico Stehr and Volker Meja (eds.), *Society and Knowledge. Contemporary Perspectives on the Sociology of Knowledge*. New Brunswick: Transaction Books: 103–148.
Machlup, Fritz. 1962. *Function and Distribution of Knowledge in the United States*. Princeton: Princeton University Press.
Mannheim, Karl. [1929] 1982. "Zur Problematik der Soziologie in Deutschland", in: Volker Meja and Nico Stehr (eds.), *Der Streit um die Wissenssoziologie*. Volume 2. Frankfurt am Main: Suhrkamp: 427–437.
Mannheim, Karl. [1929] 1965. *Ideologie und Utopie*. Frankfurt am Main: G. Schulte- Bulmke.
Marcuse, Herbert. [1964] 1989. *Der eindimensionale Mensch. Studien zur Ideologie der fortgeschrittenen Industriegesellschaft*. Schriften 7. Frankfurt am Main: Suhrkamp.
Meja, Volker and Nico Stehr (eds.). 1982. *Der Streit um die Wissenssoziologie*. Two Volumes. Frankfurt am Main: Suhrkamp.
Meja, Volker and Nico Stehr (eds.). 1990. *Knowledge and Politics*. London: Routledge.
Meja, Volker and Nico Stehr (eds.). 1999. *The Sociology of Knowledge*. International Library of Critical Writings in Sociology. Two Volumes. Cheltenham: Edward Elgar.
Merton, Robert K. [1946] 1957. "The Sociology of Knowledge", in: Robert K. Merton (ed.), *Social Theory and Social Structure*. New York: The Free Press: 457–460.
Merton, Robert K., Marjorie Fiske, and Alberta Curtis. [1946] 1991. *Mass Persuasion. The Social Psychology of a War Bond Drive*. Stamford, CT: Greenwood Press.
Mulkay, Michael. 1979. *Science and the Sociology of Knowledge*. London: George Allen & Unwin.
Richta, Radovan. 1971. *Politische Ökonomie des 20. Jahrhunderts*. Berlin: Makol.
Rueschemeyer, Dietrich. 1981. "Wissen", in: René König (ed.), *Soziologie*. Frankfurt am Main: Fischer: 352–359.
Scheler, Max. [1924] 1960. *Die Wissensformen und die Gesellschaft*. Bern und München: Francke.
Schelsky, Helmut. 1961. *Der Mensch in der wissenschaftlichen Zivilisation*. Köln und Opladen: Westdeutscher Verlag.
Schelsky, Helmut. 1975. *Die Arbeit tun die anderen*. Opladen: Westdeutscher Verlag.
Schwartz, Barry. 1981. *Vertical Classification*. Chicago: University of Chicago Press.
Stehr, Nico. 1994. *Knowledge Societies*. London: Sage.
Stehr, Nico. 1997. "Social Inequality and Knowledge", in: David Sciulli (ed.), *Normative Social Action. Cross-National and Historical Approaches*. Greenwich, CT: JAI Press: 41–51.
Stehr, Nico. 2003. *Wissenspolitik. Die Überwachung des Wissens*. Frankfurt am Main: Suhrkamp.
Stehr, Nico and Lyle Larson. 1972. "The Rise and Decline of Areas of Specialization", *The American Sociologist*, 3: 5–6.
Stehr, Nico and Volker Meja. 1982. "Classical Sociology of Knowledge Revisited", *Knowledge: Creation, Diffusion, Utilization*, 4: 33–50.
Toulmin, Stephen. 1972. *Human Understanding*. Princeton: Princeton University Press.

Weber, Max. [1922] 1964. *The Theory of Social and Economic Organization*. Edited with an Introduction by Talcott Parsons. New York: Free Press.
Whitley, Richard. 1984. *The Intellectual and Social Organization of the Sciences*. Oxford: Clarendon Press.
Wolff, Kurt H. (ed.). 1971. *From Karl Mannheim*. New York: Oxford University Press.
Ziman, John. 1994. *Prometheus Bound: Science in a Dynamic State*. Cambridge: Cambridge University Press.

Chapter 21
The Moralization of the Markets in Europe

Nico Stehr

The subject of my essay is about products and services, about the places where such commodities and services are exchanged, and how products and services are transformed as a result of new preferences and motives that affect not only production but first and foremost decisions that consumers make in the market place. Decisions are no longer solely fixed on the price of the product and the service being purchased; for example, the homeowner who sells her house to a couple for less than a competitor because she is impressed with the way they will use the home. My essay is about people who purchase large quantities of candles even though their illumination needs can easily be served by electricity. My focus is on consumers who refuse to buy milk that may come from animals that have been fed genetically modified feed. My essay is about families who purchase cars with hybrid engines even though the price may be dearer than a competitor featuring a more conventional motor. Finally, my essay is about production chains and products that acquire moral value because of the nature of the resources that go into their manufacture or the way in which they have been put together.

I will advance my argument about a moralization of the markets in Europe in a number of steps: First, I will offer general observations about the ways in which 'markets' and their major groups of actors have been treated primarily in the economic literature in the last century or more. This includes a brief discussion of what I mean by referring to morally coded markets. This will be done in form of a set of questions that pertain to market conduct in contemporary advanced societies. Second, I will advance some ideas that speak for a fundamental, yet gradual transformation of markets in modern societies. Third, I will discuss the major social processes accounting for a distinctive trend toward a moralization of the markets in

---

This text was first published as: Stehr, Nico. 2008. "The Moralization of the Markets in Europe", *Society*, 45(1): 62–67. The republication was granted as the publisher's own imprint.

M. T. Adolf (ed.), *Nico Stehr: Pioneer in the Theory of Society and Knowledge*, Pioneers in Arts, Humanities, Science, Engineering, Practice 16,
https://doi.org/10.1007/978-3-319-76995-0_21

Europe. Fourth, I will examine the impact that globalization processes have on the moralization of the markets. I will try to make the case that the moralization of the markets manifests itself in changes in global value and commodity chains. In a brief conclusion, I will point to some of the objections that may be raised to my thesis about a moralization of the markets.

A century ago, the typical family household in one of today's OECD countries spent four-fifths of its income on essential necessities, such as food, clothing and shelter. Today, spending on these commodities accounts for less than one-third of consumption. In the developed world in the past 50 years, average *real* incomes have risen four or fivefold.

Even spending toward satisfaction of "absolute needs" has changed significantly. Our clothing, shelter and food hardly resemble that of a century or 50 years ago. Not only has the volume changed radically, but also what, where and how we buy food, clothing and shelter and other commodities and services.

Most of the important ideas about markets were conceived in a world that did not know general affluence or broad-based education—only poverty, hunger and illiteracy. Wage-earner poverty was seen as an essential condition for the expansion of production. By the same token, the wealth of a nation was seen to be a function of the poverty of its working class. Poverty also had strong moral attributes because it induced discipline in the laboring population and nourished the status quo. From the eighteenth century on, it was common to conclude that prosperity came with serious demoralizing effects.

The paradox is that the theory of the market still widely in use today emerged in a society that no longer exists in terms of its system of communication, its forms of social differentiation and social and political inequality, and the role of the state and its economic system. Consumers were hardly in evidence, if at all, in early theories of the market. Consumption did not create wealth. In fact, most families did not 'consume'. Most families subsisted.

The interest of political economists was focused on the production, distribution and exchange of goods. It was a world in which control of both the production process and the workplace was privileged. The consumer was a powerless laborer and nature was a resource to be exploited by forces of production. In many present-day accounts of the consumer, the picture is equally bleak. The masses are suffering from either material deprivation or from the debilitating effects of over consumption.

21.1 What Is the Moralization of the Markets?

In contrast to the knowledge-guiding focus of the social sciences for much of the last century, are we now witnessing a departure from the rationalization process driven by purely self-centered, mechanical acquisition of money and decisions independent of its effect on others? Is it possible that "substantive rationality" (Max Weber)—and therefore consumer decisions and producers' judgments not strictly driven by monetary considerations and the rational pursuit of self-interest are on the ascendancy in the economic system? Along with changing power relations in the marketplace, is there a rise in the significance of consumer politics as well as an increased capacity of consumers to resist? This would mean that norms, worldviews, and varied interests would now play a key role in the modern economy with repercussions for the conduct of other social institutions as well.

A trend toward a moralization of the markets implies that regulation of market relations in response to major societal changes makes reference to, and is based on, constraining notions. These include fairness, authenticity, compassion, goodwill, fear, guilt, solidarity, sustainability, revenge, age, exclusivity, reciprocity, duty, benevolence, and other moral concerns and political considerations that not only negate persistent short-term pursuit of the best buy but also take a long-term perspective.

Importantly, such social goals, individual apprehensions, and modes of conduct are becoming *inscribed* in products and services offered in the marketplace as well as in the rules and regulations that govern market relations. One is therefore able to speak of the *moral,* rather than the mere economic, *content* of commodities, as in the case for solar energy and services or the *lack* of moral status in the case for nuclear energy.

It is also necessary to examine the interaction between the norms of consumption, production and procedure because social desires, values and beliefs manifest in modern societies are implanted not only in commodities, but also in the production and consumption *process.* Value becomes attached to the reputation of the producer or the supplier of commodities and services. As a result, the morally coded market becomes, at least to some extent, a self-actualizing and self-amplifying process.

This self-amplifying process is also moved forward by those whose conduct does not explicitly support a trend toward a moralization of the markets. In the final instance, moralizing markets are extending across the world with the globalization process, especially through the rapid enlargement of trade since the 1970s, global marketing campaigns, and the diffusion of modern communication media, such as the Internet.

In short, I will argue that a strong case can be made for a trend toward a moralization of the markets. In practice, moralization of the markets is transforming what many economists still praise as a social sphere of cultural indifference. My thesis of a morally coded market extends the basic theoretical postulate and advice of Emile Durkheim's work on modern societies, namely that the morality of society changes with transformations in the 'physical' and social milieu of men and women.

Moralization of the markets does not signal that morally 'superior,' more 'civil' or 'humane', or even 'sustainable' norms are on the ascendant, although it is possible that such conventions are gaining. Nor can it imply that economic conduct now consensually follows and is governed by norms that are widely shared. As a matter of fact, some of the norms beginning to influence economic decisions are still those of minorities, while others are more widely shared by both producers and consumers in different countries.

Nonetheless, it is important to discover what orientations and experiences, perhaps even what common experiences and orientations, prompt people in societies with different cultures and histories to fashion a new, moral approach to economic conduct. Whether these experiences and orientations produce moral commonalities or a minimal morality that is more widely shared across and within societies, is an empirical issue.

21.2 What Is Responsible for Morally Coded Markets?

It seems to me that there are two major processes under way in modern societies that are responsible for moralization of the markets. Reference will be to social processes that have been largely neglected in their significance by social scientists. However, emphasizing these processes requires that any firm distinction between non-market norms and market norms does turn out to be helpful. A strong social differentiation as found, for example, in social systems theory and elsewhere, implies that the work of market norms is restricted to a specific social form.

My premise is that not only do markets shape culture but also that cultural processes impact markets. A vantage point that emphasizes the *interrelations* among social systems or the traffic across the boundaries of social systems, as it were, represents more interesting and challenging issues than the view that the economy is largely autonomous and self-contained.

The two social transformations mainly responsible for the possibility of morally coded markets are:

1. Western societies are characterized by a historically unique level *of private wealth* supported by a highly effective social safety net. During the early 1950s, fear that one might not be able to secure one's daily food was still a main concern of many households in some of the most developed societies. Today it is not only the elite that is doing well. In contemporary societies of the developed world, the majority of households are materially well off. The capacity to act and the liberty to decide among consumers has increased significantly.

Thus, nothing in the history of the industrialized countries in Western Europe and North America resembles their experience between 1950 and 2000. By the end of this period, the possibility of serious economic hardship that had hovered over the lives of three-quarters of the population now menaced only about one-fifth of it. The considerable increase in the wealth of average households has enhanced their discretionary economic power. Although absolute poverty still exists in even the richest countries, the material standard of living for most people has improved almost without interruption, and often very rapidly, for 35 years. The farmhand or the farmer in the nineteenth or early twentieth century who was forced to leave the rural countryside could have only dreamt of the kind of social security an unemployed person is today eligible to receive in many developed societies. Above all else, these are the marks of the uniqueness of the experience and perhaps even a basic transformation in the ecology of the human species.

The societal importance of the unprecedented level of average wealth is further heightened by the fact that most Western societies are 'aging.' In the next few decades, most Western societies will not only have a larger proportion of older citizens, but also a larger proportion of persons who have acquired their household wealth during the sustained level of economic growth in the postwar era or have benefited through inheritance from the same economic forces.

Keynes ([1930] 1984: 331) anticipated these developments in 1930, "I look forward," he observed that "in days not so very remote, to the greatest change which has ever occurred in the material environment of life for human beings in the aggregate. The course of affairs will simply be that there will be ever larger and larger classes and groups of people from whom the problems of economic necessity have been practically removed."

Moreover, it is possible to define affluence and wealth other than merely in terms of changes in monetary income and the volume of money-equivalent assets that households command—but rather with a much broader definition of wealth that would include aspects of human capital (knowledge, skills), social capital (networks of social relations, civil society linkages) and cultural capital. With this new definition, the growth in average wealth is even more remarkable in recent decades in the developed world.

Lack of attention to the immense rise in average wealth and to a broader concept of wealth is the result of a variety of factors. These range from established economic traditions that measure wealth by focusing exclusively on GNP data to our present inability to advance more than tentative and contentious quantitative information about the exact value of specific components of human, social and cultural capital.

2. Although my broad definition of the significant growth of human prosperity already includes a general reference to the significant increase of "human capital" (as economists prefer to call it), or, more precisely, the growth of cultural capital in the developed world in the last few decades and of skills and knowledge in particular, I will refer to the *growth of knowledge* separately.

A separate discussion of knowledge as a capacity to act (as I would prefer to define it) is warranted, since it permits explication of its function for groups of actors, both within the economic system and elsewhere in society, as an anchor of the trend toward moralization of the markets in modern societies. In addition, a discussion of the enhanced societal role of knowledge permits reference to the changing commodities and services typical of a knowledge-based economy.

The growth of knowledge refers to significant changes in the potential capacity of larger numbers of individuals to change what has appeared to be immovable in the past. *Knowledge as a capacity to act* not only opens up new technical possibilities and vistas, but also enhances the ability of many actors to act, that is, to obtain and be open to relevant information and knowledge, to make informed judgments, to utter an opinion, and to develop horizontal social ties, etc.

Greater knowledgeability is coupled with changes in ethical norms. Individuals are more prepared to assume personal responsibility, where in the past culpabilities were explained away by reference to phenomena beyond the control of individuals, such as fate, powerful social forces or even supernatural powers.

The enhanced *knowledgeability*—in the sense of an increasingly self-determined and self-confident capacity to conduct one's life generally and as an economic actor in particular—allows, or even demands, a view of consumers as agents rather than as docile creatures trapped in a culture of dependency, perhaps duped by the media, uninformed about production processes and products, and vulnerable targets of advertising. More recently, the dependent, helpless consumer has been confronted with an overload of choice. Knowledgeability is now linked, for example, with the capacity to forecast future movements in actual inflation and therefore better protect one's assets from undue depreciation.

The growing knowledgeability of actors generally refurbishes their relations to the major social institutions in society. Major social institutions are no longer seen as excessively powerful, simply beyond the reach of influence of most individuals or small groups of actors. In contemporary societies, the consumer's agency is nurtured, enhanced and sustained at low cost by the multi-directional information medium of the Internet.

In short, on the basis of an enhanced knowledgeability, the relations of market participants change, for example, what once were non-transparent production processes and realities have potentially become more transparent and understandable. The more the world of markets that we experience fills up with new services and products, the greater the importance of the knowledgeability of the actors involved.

Finally, it is important to refer to the fact that consumption, or the refusal to consume specific products and services (including so-called 'cultural' products or services), conveys knowledge and information that instructs us about our social environment and the groups that populate the marketplace.

Economic actors not only bring knowledge to the marketplace, as Friedrich Hayek, for example, stresses, but the market also confers knowledge on individuals and groups as participants. The origins, organization of production, composition, appearance, texture, color and other attributes of commodities or different forms of service—all these communicate knowledge *about* society and the world. Knowledge mediated by the consumption process has repercussions in turn for future consumption patterns and the production of commodities and services. Knowledge is always a form of participation.

21.3 Markets in an Age of Ecological and Global Modernization

Given the strong emphasis on the contentious phenomenon of 'globalization' in recent discussions of the future shape of modern societies, one cannot but ask what added theoretical value, if any, is gained by an examination of the moralization of the markets in light of powerful forces that are supposedly accelerating globalization, such as the creation of global advertising agencies and marketing campaigns. Are the forces of globalization in fact strengthening or weakening trends toward a moralization of the markets?

What is the result of the emergence of globalizing markets in which decisions are not merely driven by a highly rational calculus? What are the consequences of the growing emphasis on goals in consumption and production that attempt to optimize the aesthetic qualities of goods, to reduce the environmental impact of services and improve sustainability, that stress fair treatment of workers and their employment conditions, the health and safety of products, prices paid to producers, and the extent to which the product may or may not be fashionable?

Indeed, the political promises that "discriminating consumers" or "political consumers" may represent can be revolutionary. This theory is advanced by some Marxist thinkers who have recently discovered the 'class' of modern consumers as the vanguard of major social, economic and political change in advanced and developing societies.

Moreover, many commodities and services sold and purchased as brands that appear to originate in a certain company and country are in many instances commodities composed of parts from numerous regions of the world. Not only is the moral content of commodities and services exported via an intensifying trade around the world, but parts that then make up a commodity assembled in a specific location are also importing moral content from abroad. It is in this sense that the moralization of the markets is spreading as a result of both growing imports and exports.

Substantiation of the coupling of economic and moral trends on a global scale comes from examining global trade and manufacturing, particularly with certain commodities. Recent studies devoted to *global value or commodity chains* are a case in point. Commodity chains are best defined as global links and successive phases of raw material, production, distribution, sale and consumption of products. The advantage of studies that concentrate on the governance, coordination and control of global industries is that the unit of analysis is not the nation-state, as was and is common in internationally comparative studies. Another advantage of the global commodity chain perspective is an analysis of the moralization of the markets—in contrast to the world-systems or world society theories, for example—in that its focus is not only on the present age, but also that the starting points of the value chain perspectives are products, services, producers, distributors and consumers.

As the very notion of globalization suggests, globalization processes are attached to and become the motor of a multitude of changes, including changes below the level of the nation-state, such as transformation in the nature of markets. The most dynamic economic units today are organized in a transnational manner. Global industries, global economic corporate strategies and rivalries—manifest, for example, in global campaign and takeover battles, as well as in social and cultural networks—are among the focuses of the global commodity chain perspective.

As Ponte/Gibbon (2005) point out, the attributes of *global value or commodity chains* have become increasingly dependent on consumer tastes and preferences and not on preferences that producers may have. In the case of foodstuffs, for example, observers speak of an increasing "demand oriented food system" that poses special challenges and vulnerabilities to agriculture, the "final foods industry" and, of course, retailers.

Assertions of the increased power of consumers in the food markets are invariably met with skepticism that points, for example, to large retailers that continue to disempower consumers in today's marketplace. There are undoubtedly many producer-driven global commodity chains in existence today. They are characteristic of capital-intensive industries, such as aircraft and heavy machinery manufacturing. But even in the aircraft industry, producer-driven control is by no means absolute.

The focus of empirical analysis of the nature of global commodity chains is on the control processes and control agencies that can be detected in such global networks. Questions of interest concerning social control are: what products should be delivered, when and in what quantities, how should the products be produced, and, of course, at what price? Ponte/Gibbon (2005) argue that decisions along the chain, and therefore the answers to the questions about generating commodities, are more and more driven by consumer decisions and not by decisions that could be attributed to the group of producers active in a commodity chain. Separating consumer and producer decisions and motives and the control exercised by each group, of course, raises the issue of the reasons for the decline of the power of producers and the gain in influence by consumers.

The reasons to which Ponte/Gibbon (2005: 6) refer concern the importance of *quality* attributes and judgments (not in an objective sense but in a subjective one, which attributes quality content to products) in the genealogy of commodity chains made by consumer groups. Decisive in the shift toward the relative influence of consumers versus producers in global commodity chains are the 'preferences' of consumers—more to the point, societal and necessarily contested conventions that influence market conduct. In their theory of the "economy of quality", Ponte and Gibbon suggest that the quality standards held by consumers are of decisive importance in who exercises control in global value chains. Whether consumers are indeed capable of imposing their quality standards in a sovereign manner in the marketplace is an empirical issue and cannot be answered in the affirmative a priori.

The products that Ponte and Gibbon examine in their empirical analysis of global commodity chains are textiles and coffee. In the case of the market for roasted coffee, traditional, well-established firms dominate. After raw coffee beans are roasted and a brand name is attached, the coffee is sold to the consumers. Companies in the coffee roasting business use the *reputation* of their firm as a substitute for separate coffee quality standards. They usually do not make available to the consumer additional information, which they in fact control, about quality attributes or the nature of the production process. Loss of control over quality standards in the global value chain by coffee roasting companies has resulted from rapidly changing notions of what constitutes quality, in the case of coffee, and of conventions surrounding the consumption of coffee by consumers. These changes particularly apply to the type of consumers who not only "require more information on (and higher levels of) 'intrinsic' coffee quality (product attributes), but also include environmental and socio-economic aspects in their consideration of quality (attributes on production and process methods)" (Ponte/Gibbon 2005: 12).

Companies roasting coffee are pressured not only to develop indicators, control processes and certification procedures of quality standards for these special types of coffee, but also to impose the same standards on producers and distributors both earlier and later in the value chain. Finally, a kind of control of quality standards evolves, which is monitored and enacted by the roasting companies.

The same changes can also be observed with respect to quality standards linked to sustainability criteria. Changing quality standards thereby are becoming the power lever exercised by consumers. Consumers impose their orientations, which are often moral standards, on the production process and thereby assure that the quality criteria relevant in their life world becomes an integral part of the global value chains. Global commodity chains continue to be organized in a hierarchical manner, but relative control of different collective actors involved in the network is changing. Inasmuch as the quality standards of the consuming public becomes attached to commodities or to rules of the game governing transactions, including codified indicators or embedded in technical devices, the globalization of moral contents is enhanced, multiplied, and diffused into different regions of the world. In the case of food items, of course, demand-driven production weakens the structural position of large, global players in the food industry. As Wilkinson (2002: 342–343) therefore concludes, the market position of such food operations has become much less secure in recent years, as the "food system has shifted to a sharper focus on nutrition and health foods and as the biotechnology revolution has increased its grip on the driving forces of innovation".

The globalization of environmental problems and, in response, the globalization of environmental politics and policies can now be taken for granted. The drive for the globalization of environmental politics is not necessarily linked to the fact that environmental problems ignore national boundaries. The motor for the globalization of environmental politics and policies is closely associated with the sciences in which experts for environmental issues can be found operating international non-governmental organizations, with the international presence of the media and the Internet, and with the decisions by producers and consumers, which, of course, have an impact far beyond the boundaries of a nation-state. However, we have few studies that focus on the international role and impact of consumer decisions in environmental politics (see Spaargaren/Martens 2005). As a rule and insofar as such issues are discussed at all, the debates are prompted and dominated by political agendas. This applies, for example, to reflections on the notion of *ecological citizenship* (e.g. Urry 2000), which may be driving the moralization of the markets.

21.4 Possible Objections to a Theory of Morally Coded Markets

In place of a conclusion, I will simply name some objections that my thesis about morally coded markets may encounter. Critics may point to the difficulties that the social sciences already have in projecting future societal transformations, let alone anticipating social trends, such as the moralization of the markets, in what is today a rather fragile social world. In theoretical discourse, the "rock of positivist solidity" (Hirschman) and therefore the dichotomous opposition of phenomena, such as 'rational' and 'irrational', 'efficient' and 'inefficient', or 'useful' and 'moral,' stand in the way. Markets, much like viruses, know no morality. Economists, therefore, prefer a formal conception of preference structures of market participants. Professional economists urge us to concentrate our attention on the consequences of economic action and not to waste scarce attention on questions that deal with the kind of preferences market actors may pursue. The ends that guide economic action are random in any case.

Finally, there may be strident voices arguing firmly that the thesis of a moralization of the markets is not *new*, be it as a theoretical construct or as an empirical statement about the practice of market conduct. The objection may be that morally coded markets are just a means of presenting an old idea dressed up in perhaps more fashionable garb. Well, my point is, of course, that the moralization of the markets is not the eternal return of the same.

References

Hirschman, Albert O. [1980] 1981. "Morality and the Social Sciences: A Durable Tension", in: Albert O. Hirschman (ed.), *Essays in Trespassing. Economics to Politics and Beyond*. Cambridge: Cambridge University Press: 294–306.

Keynes, John Maynard. [1930] 1984. "Economic Possibilities for Our Grandchildren", in: John Maynard Keynes (ed.), *Collected Writings. Volume IX: Essays in Persuasion*. Cambridge: Cambridge University Press: 321–332.

Ponte, Stefano and Peter Gibbon. 2005. "Quality Standards, Conventions and the Governance of Global Value Chains", *Economy and Society*, 34: 1–31.

Spaargaren, Gert and Susan Martens. 2005. "Globalisation and the Role of Citizen-Consumers in Environmental Politics", in: M. Boström, A. Føllesdal, M. Klintman, M. Micheletti, and M. P. Sorenson (eds.), *Political Consumerism: Its Motivations, Power, and Conditions in the Nordic Countries and Elsewhere*. Copenhagen: Nodisk Ministerrad: 25–58.

Urry, John. 2000. *Sociology Beyond Societies*. London: Routledge.

Wilkinson, John. 2002. "The Final Foods Industry and the Changing Face of the Global Agro-food System", *Sociologia Ruralis*, 42: 329–346.

Chapter 22
Climate Change: What Role for Sociology?

Nico Stehr (with Reiner Grundmann)

Lever-Tracy (2008) has called for sociologists to become more involved in the debates about climate change and the future of planet Earth. We support some of the points of her argument wholeheartedly. Nonetheless, we want to query four tenets of her article that seem particularly problematic to us. First, she argues that sociology has long been silent on climate change because sociologists are suspicious of naturalistic explanations and because they are sceptical of teleological arguments, or discourses that identify plans for the future. While this is partly true, we see other, more important processes at play that need examining. For instance, we argue that sociologists have been cautious due to the overly political nature of the debates thus far, due to methodological differences with the natural sciences, and because they have learned to be sensitive about locating the phenomenon in the *longue durée*. Second, she argues that natural and social change are now in lockstep with each other, operating on the same time scales. This requires effective multidisciplinary cooperation between natural and social sciences. Here we argue that the true challenge is interdisciplinarity (as opposed to multidisciplinarity). Third, Lever-Tracy argues that the natural sciences have made a convincing case of the reality of anthropogenic climate change. In her view, sociologists should not cast doubt on the robustness of the scientific consensus or on the findings of the scientific community, as expressed by the Intergovernmental Panel on Climate Change (IPCC). In contrast, we emphasize the virtues of constructivism. Fourth, Lever-Tracy holds that we need to place climate change firmly at the heart of the discipline. In our view, if this is going to more than wishful thinking, we need to carefully consider the prospects of such an enterprise.

This text was first published as: Grundmann, Reiner and Nico Stehr. 2010. "Climate Change: What Role for Sociology? A Response to Constance Lever-Tracy", *Current Sociology*, 58(6): 897–910. The permission was granted on 31 July 2017 by Craig Myles on behalf of SAGE Ltd. Permissions Team in London.

M. T. Adolf (ed.), *Nico Stehr: Pioneer in the Theory of Society and Knowledge*, Pioneers in Arts, Humanities, Science, Engineering, Practice 16,
https://doi.org/10.1007/978-3-319-76995-0_22

22.1 "Time is Running Out"

At the 2008 Annual Conference of the British Sociological Association, John Urry gave a plenary talk on climate change in which he made the strong case that sociologists so far had not engaged very much with the topic of climate change and that it was high time to get involved. In his concluding words he spoke of a "call to arms" for sociology. This mixing of political urgency and need for profound rethinking of sociological traditions and conceptual frameworks is what we also encounter in Lever-Tracy's piece in the pages of this journal [*Current Sociology*]. The urgency and the political nature of climate change should lead us to think about our faults, omissions and options carefully. However, sociologists should not rush into the discursive arena without asking some critical questions in advance, questions such as: What exactly could sociology contribute to the debate? And, is there something we urgently need that is not addressed by other disciplines or by political proposals?

The urgency of the matter to which Lever-Tracy points is becoming more apparent as evidence mounts that efforts to date to mitigate greenhouse gas (GHG) emissions are falling short of what is required to *arrest* or *slow* climate change (see Matthews/Caldeira 2008), that global emissions are likely to rise significantly in the next years (cf. Sheehan 2008) and that transnational policies and treaties put in place to reduce emissions fail to reach their targets (let alone achieving the objectives needed to put a brake on climate change; see Prins/Rayner (2007).

The inherent alarmism in many social science contributions on climate change merely repeats the central message provided by mainstream media. It is curious that little guidance is provided in terms of what could be done to deal with climate change, or at least pointing out what options we have. Can sociology deliver on this? One would expect substantial proposals besides the exhortation to listen more to what the natural sciences have to tell us.

If the political urgency of the matter is one thing, sociology's contribution is another. While we share Lever-Tracy's concern and agree with her general diagnosis that sociologists have been fairly invisible in professional circles as well as in public and policy discussions about global warming, we are less clear what the consequence would be. Should sociology's primary task be to help sound the alarm assuming this would force politicians into action, thereby lowering citizens' and businesses' carbon dependency? We are not convinced it should. But neither is it clear that Lever-Tracy wants to make this point. The essence of her article seems to waver between a call upon sociology to make theoretical innovations that are taken up by the core of the discipline, on the one hand, and a call to make a political contribution to mitigate against climate catastrophe, on the other.

While this tension is symptomatic of many social scientific attempts to get involved, we propose to study these very underlying assumptions. For example, we should ask: Does the dramatization of events lead to effective political responses? Do we need a politics of fear? Is scientific consensus instrumental for sound policies? And more generally, what are the relations between a changing technological infrastructure, social shifts and belief systems? What contribution can bottom-up initiatives have in fighting climate change? What roles are there for markets, hierarchies and voluntary action? How was it possible that the "fight against climate change" rose from a marginal discourse to a hegemonic one (from heresy to dogma)? And will the discourse remain hegemonic or will too much public debate about climate change lead to "climate change fatigue"?

22.2 Reasons for Sociological Abstinence

While we agree with much of the diagnosis of the scarce engagement of sociology with climate change, we identify different reasons at the root. We argue that as anthropogenic climate change has evolved from a science-based issue to a top global policy issue, social scientists have developed a mixed but largely uneasy relationship with the topic. This is due to the politicization of the debate. Science studies scholars, for example, who were in principle well placed to contribute to the debates, felt uneasy in a polarized debate where academic research might be seen as politically counterproductive.[1]

Lever-Tracy argues that global warming issues have been, at best, of marginal interest to sociologists. Her pertinent encouragement, addressed to mainstream sociologists, to engage more vigorously with the issue of global warming and its possible societal consequences is therefore warranted. For example, she takes Beck and Giddens to task for not paying enough attention to climate change. However, it has to be said that Giddens has just published a book on the matter (without much application of sociological theory; see later).

This leads us to the crucial question of sociology's core contribution to the problem. It has been said many times that sociology as an academic discipline came into being at a time when western societies were making a transition from agricultural societies to industrial societies. This change led to enormous shifts in social structure and cultural values. Both of these core elements were seen as influencing each other and are still today sociology's core programme: the study of social change together with technological change, change in social relations and cultural systems. We are now witnessing global change that is at least of a similar order of magnitude as the earlier change. Some societies in developing countries make the

[1] Brian Wynne put it this way: "a sociological deconstruction of knowledge may find itself in unwelcome company, politically speaking. [...] These sociological observations about the scientific knowledge of global warming could of course contribute to a deconstruction of the intellectual case for the environmental threat, and thus also to a political demolition of the 'environmentalist' case for internationally effective greenhouse gas controls" (Wynne 1996: 363, 372).

transition to a less agricultural, more industrialized (and *potentially* de-carbonized) type while a part of its workforce is migrating around the globe. The workforce in developed countries is encouraged, even forced to participate in higher education and to adopt a lifestyle that is flexible enough to manage transitions in the labor market. From this follows that we have a more educated population, sometimes eager and ready to participate in complex decision-making, while at the same time trying to cope with cultural value systems they have inherited from an earlier epoch. Sociology has been one of the leading disciplines in the study of globalization. The question is whether it can do the same in the study of climate change.

Posing the question this way already suggests an answer. While sociology was extremely well placed to detect the process of globalization as a social change issue, climate change was from the beginning the domain of natural scientists who addressed and framed the issue according to the parameters of their specific disciplines. Dominated by the modelling community, the study of climate change largely, though not exclusively, amounted (and still amounts) to the development and refinement of global circulation models that depend crucially (among other things) on sound input data. Climate change modellers are keen to get useable data from other academic communities but only if these are in the "right format", so to speak. They want cost of life estimates from economists and predictions about mitigation and adaptation costs. They would like from sociologists some guidance about the relation between climate-induced stress on social communities and the likelihood of wars or revolutions. Such estimates would be valuable for predictions of further GHG emissions and their abatement. Few sociologists would subscribe to the notion that social processes can be 'modelled' in this way (or even that it should be a prime task of sociologists to do so).[2] Since few sociologists are doing so, let alone engaging in interdisciplinary cooperation with climate scientists, the latter are literally encouraged to become social scientists themselves. The outcome of such cognitive shifts is very likely technocratic perspectives and policy advice. Scientists and engineers have acted as "lay sociologists" on many an occasion and have—based on their personal and professional beliefs and approaches—built several assumptions about human behaviour into their theories.

There are other reasons for the neglect of addressing global warming issues in sociology, not mentioned extensively in Lever-Tracy's discussion. These reasons, especially the deep-seated conviction within the sociological imagination of the firm boundary between nature and society, have to be dissected in greater detail in order to gain insights into the degree of resistance to any efforts to transcend the

[2] The same cannot be said for members of other social science disciplines, for example for economists, anthropologists, or geographers. The significant role of economists in discussions of the consequences of climate change and climate policies (e.g. Stern 2006) may also be one of the reasons why sociologists are at best marginal voices in climate change discussions. Economists have dominated debates on the appropriate responses to the climate change by successfully defining such responses as economic issues, for example, how costly will it be to mitigate greenhouse gas emissions and what economic measures may be put in place to encourage energy efficiency (carbon taxes, trading).

reluctance of sociologists to engage in discussions that assign a role to 'natural' processes within sociological discourse. The extent to which the dichotomy is firmly entrenched in the consciousness of professional sociologists also explains why the *mere appeal* to the profession to treat global warming issues as part of the core of sociological concerns will in all likelihood fall on deaf ears.

The main issue that remains therefore is the question: why has *mainstream sociology* and its major journals failed to incorporate global warming within its horizon? Aside from the fact that anthropogenic global warming as a largely uncontested issue constitutes a recent consensus within climate science, the lack of uptake in core sociological reasoning and research has other reasons than those listed by Lever-Tracy. While she identifies a drift of sociologists into postmodern ways of thinking, thereby favouring a short-term orientation, and a rejection of teleology, we would like to point to two different reasons. The first is sociology's legacy of social constructivism and its scepticism of ecological determinism, including climatic determinism. We find this largely unproblematic. The second is the politicization of the climate change debate and a problematic reaction from the sociological community. We address both in turn.

The grand narrative of climate refers to stories (found in scholarly literature and in public discourse) in which descriptive and prescriptive elements are liberally mixed. They inform us about the rise of civilizations, the fate of nations and the character of humans under favourable or adverse climatic conditions. 'Permissive' or 'harsh' environments of a geographical area determine not only the material but the immaterial social and cultural realities. And, as the French historian Edmond Demolins, at the turn of the last century, confidently asserts, "if history of mankind were to begin over, without any change in the world's surface, it would broadly repeat itself" (as cited in Landes 1998: 533). Unfortunately for the grand narrative of climate, different cultures emerge in similar natural environments and similar cultures in different natural environments. What is more, anthropogenic climate change alters the climatic conditions in historical times and global warming may change climates within the time span of a generation. This should dampen the enthusiasm towards determinist accounts. Lever-Tracy, for example, appraises the work of Jared Diamond in such overwhelming terms.

22.3 Prospects for Multidisciplinarity

Lever-Tracy argues that society and nature are now in lockstep and influencing each other on the same time scale. She uses this argument to urge a closer cooperation between the sciences (although in practice this amounts to little more than admonishing sociologists to give up their urge of deconstructing scientific knowledge claims since they are counterproductive, politically). This is, for her, a reason to call for more multidisciplinary cooperation. But is it true that nature and society are in lockstep? We would agree with a notion of a dialectical relationship between society and nature *à la* Marx (see Grundmann 1991). However, the time scales of the mutual transformations are not always similar. We may be able to pollute a river within a few years and clean it up in about the same time or we may drastically improve air quality in specific areas of the globe (inner cities) through technological and political measures. However, there are other examples such as nuclear waste, or even the very conditions of climate change, that operate on huge time scales. These issues have implications of scale (both temporal and spatial) that are new to political decision-making. We have accumulated GHGs in the atmosphere for over 200 years and we are only now starting to debate the consequences of their presence. We need similar time scales for reversing the process because of the long lifetime of some of the main GHGs (such as CO_2). In a way, climate change exemplifies the opposite of what Lever-Tracy wants to show: we cannot reverse or halt climate change within the current or next generation. Our mitigation policies will, at best, benefit our great-grandchildren. In sum, while the lockstep argument is not convincing, neither it is necessary for advocating greater multidisciplinarity.

Despite calls for multidisciplinarity,[3] Lever-Tracy reaffirms the traditional division of labor between natural and social scientists. Writing against the dangers of falling prey to climate sceptics she writes: "it seems to me that a respectful division of labor is essential now that natural and social change are operating in tandem, on the same time scales. Since we are not ourselves competent to evaluate the debate between climatologists and sceptics, we have no option but to accept the professional authority and integrity of the accredited experts" (Lever-Tracy 2008: 457).

There are two flaws in this argument: first, what follows if we grant the validity of the results of mainstream climate science? Does anything follow in terms of us addressing climate change on a practical level? We suspect that if the purpose of Lever-Tracy's argument is to galvanize sociologists, nothing specifically practical follows from it. Second, if we do not approach this arena with a more refined sociological apparatus, then we will not be able to participate in an interdisciplinary debate at all. We would, instead, simply be enrolled by the dominant discourse. This is a kind of 'irrelevance' that does not figure in her discussion.

Upon realizing that the discipline has neglected 'nature' for so long, some sociologists have now adopted opposite approaches, from showing no concern for environmental issues, to actively engaging in environmental causes (Elias 1956). Making up for the intellectual neglect that has prevailed for so long, now the time seems right to embrace a political programme associated with climate change, "to prevent the worst", or "to save the planet". While this is the terminology of some enviornmentalists and politicians, we doubt that it represents a sound response from a sociological viewpoint. The severity of a problem does not mean that we as sociologists should forget about our analytical apparatus.

[3] It may be significant that she uses the term multidisciplinarity, not interdisciplinarity. While the latter would require a thorough mutual learning process, the former could content itself with a mutual supplementarity (see Klein 1990; Weingart/Stehr 2000).

Political decisions about climate change are made on the basis of scientific research and a host of other (economic, political, cultural) considerations. Regarding the scientific dimension, it is a common perception (one that Lever-Tracy seems to share) that the more knowledge we have, the better the political response will be. This is the assumption of the linear model of policy-making that has been dominant in the past but debunked time and again (Godin 2006). What we increasingly realize is that knowledge creation leads to an excess of information and 'objectivity' (Sarewitz 2000). Even the consensual mechanisms of the IPCC lead to an increase in options because knowledge about climate change increases.

22.4 The Virtues of Constructivism

It is an often repeated myth that constructivism ultimately means playing into the hands of climate change deniers. One of the key sociological insights when dealing with policy-relevant information is the notion of framing. This is a social constructivist concept that rejects the assumption that there is anything self-evident to be found "out there". The framing of climate change as a social and political issue is therefore likely to be among the prime concerns for sociologists researching this area. Frames are used in the sense of organizing devices that allow the selection and emphasis of topics to decide "what matters" (Gitlin 1980). Gamson/Modigliani (1989) consider frames as being embedded within "media packages" that can be seen as "giving meaning" to an issue. According to Entman (1993), framing comprises a combination of different activities such as: problem definition, causal interpretation, moral evaluation and/or treatment recommendation for the item described (see Grundmann/Krishnamurthy 2010 for an application to media discourse).

It is odd that Lever-Tracy is highly dismissive of 'constructivist' accounts. She writes: "Preoccupied with analysing [...] social facts, sociologists are unwilling to be disturbed by the voices of natural scientists, reporting from inaccessible upper atmospheres, ancient ice cores or deep oceans, where no social fact exists" (Lever-Tracy 2008: 454). This misses the point that sociologists might, *qua* citizens, be very aware of such developments and act accordingly. Professionally, sociologists have indeed studied the emergence (or, to take a term she finds suspicious, *construction)* of scientific claims as facts (see, for example, Demeritt 2001; Lahsen 2005; Skodvin 2000; Van der Sluijs et al. 1998). To define global climate change as a social construction is not to diminish its importance, relevance, or reality. It simply means that sociologists study the process whereby something (like anthropogenic climate change) is transformed from a conjecture into an accepted fact. With regard to policy, we observe a near exclusive focus on carbon dioxide emissions. This framing has proven counter productive, as the Hartwell paper and other sources demonstrate (see Eastin et al. 2010; Prins et al. 2010). Reducing carbon emissions in the short term is among the most difficult tasks. More progress could be made by a re-framing of the issue, not as an issue of human sinfulness, but of human dignity.

Many social scientists and scientific experts more generally are aware of the dangers that arise from a politicized environment. If it is true that controversies abound in risk debates, this leads to a dilemma for researchers in the field. On the one hand, they may want to see the problem solved according to their value preference, no matter how strong the scientific evidence is. On the other hand, they may be very reluctant to be drawn into a controversy which could harm the integrity of scientific research. Many scientists believe that it is imperative to "stop global warming" and therefore critical voices within the academic world (or in public discourse) are seen as a hindrance. Because the political project of CO_2 mitigation represented by the United Nations Framework Convention on Climate Change (UNFCCC), supported by the IPCC, and resulting in the Kyoto Protocol, is for many the only game in town, other options have been pushed off the agenda. The perception was, and still is, that critical discussions of the IPCC and the Kyoto framework would lend support to the contrarian forces who wanted to derail any political project addressing climate change. It may be that this constellation had an impact on science studies, and there has been very little research on this topic in recent years.[4] No doubt, many scholars working in this field are sympathetic to environmental values and thus uncomfortable being seen as part of the sceptical bandwagon. One of the methodological principles of science studies is to study controversy from a standpoint of symmetry, i.e. not privileging one account over the other. The so-called Strong Programme (Bloor 1976) holds that science studies scholars should remain neutral with respect to the truth claims science makes. They should explain the success or failure of a scientific theory in the same terms. The outcome of all scientific controversies—successful or not—should be explained by social factors. As the climate change controversy unfolded in the US, journalists seemed to follow a similar principle in their reporting, thus giving a boost to the small number of climate change sceptics (Boykoff/Boykoff 2004). Could it not be that this political climate prevented sociologists from analysing the science and politics of climate change to any great extent? Might it not be that the political and social forces have become so strong that the very principles of the field of science studies have taken a back seat? In other words, we argue that there has been little engagement with climate change on the part of sociologists (and especially science studies scholars) because they are aware of the political implications and anxious of not wanting to play into the hands of climate change sceptics.

An example of framing is provided unwittingly by Lever-Tracy when she approvingly quotes Leggett saying that "The world had [*sic*] less than 10 more years, on its current global emissions trajectory, before a point of no return would be passed, and the possibility of stabilizing concentrations would be lost"

[4] A search for the key words 'climate' or 'warming' in the titles of articles appearing in *Science, Technology and Human Values* and *Social Studies of Science* yields just 14 articles. These were published in the period from 1992 to 2008.

(Lever-Tracy 2008: 450). She takes such statements at face value without analysing the rationale behind it. We should ask instead: Why has the time frame of 10 years been given? Doesn't this sound like an estimate based on political expediency? Who has come up with this formula of "we just have 10 more years to act before it will be too late"?[5] In spring 2009 several rumours were made public that IPCC members preparing for the Copenhagen summit were admitting that it was already too late to prevent 2 °C warming (see Monbiot 2009).[6]

22.5 Climate Change at the Heart of Sociology?

It is telling that in his recent book *The Politics of Climate Change*, Giddens (2009) makes very little use of sociological frameworks, including his own (e.g. structuration theory, or the notion of risk society). Instead, he applies theoretical tools from political science (such as the notion of policy entrepreneurs) and public policy (institutional decision rules and incentives). While the latter are clearly useful, it is regrettable to see a sophisticated notion of risk being reduced to the crude ritual of risk assessment and management. He advocates the use of risk-ranking exercises where the highest ranked risks get the most attention and resources. Giddens also dismisses two holy grails of the environmental discourse: the precautionary principle and sustainable development. One might argue that this book reflects more Lord Giddens's concerns with day-today politics than Tony Giddens, the sociological theorist. Be that as it may, it is still an open question if sociology will acknowledge the relevance of environmental issues (including climate change) for its core business.

It may become more frequent that core sociology journals will publish articles related to climate change. But what would editors be looking for? Probably for research results that are based and interpreted along the lines of the sociological canon or in critical discussion with it. In institutional terms this would mean to give more prominence to subfields like environmental sociology, studies in social science of knowledge or science and technology studies and to open the theoretical

[5] Lever-Tracy quotes Leggett who is citing an article in *The Independent* newspaper describing a report of a task force including the UK Institute for Public Policy Research, the US Center for American Progress, the Australia Institute and the chair of the IPCC. Again, sociologists should investigate primary sources and the construction of claims rather than taking statements at their face value.

[6] "Quietly in public, loudly in private, climate scientists everywhere are saying the same thing: it's over. The years in which more than 2 °C of global warming could have been prevented have passed, the opportunities squandered by denial and delay. […] This, at any rate, was the repeated whisper at the climate change conference in Copenhagen last week. It's more or less what Bob Watson, the environment department's chief scientific adviser, has been telling the British government" (Monbiot 2009).

debate about the core conceptual framework and identity of the discipline (see, for example, the sympathetic reception of Bruno Latour as a keynote speaker at the 2007 BSA conference). In a way, some of the "marginal fields" could move towards the center of the discipline. No doubt, there will be limits to such a move that ironically has to do with a certain reluctance of these fields to carry out their own programme. By this we mean the concern that anything that could be seen to cast doubt on the 'integrity' of the climate scientists has to be avoided in order to protect the political impetus behind it. There seems to be the curious conviction that lest you want to be accused of helping the fossil fuel lobbies and the climate sceptics, you better keep quiet.[7] In addition, these marginal fields would need to embrace the idea to move closer to the center of sociology, an aim which cannot be taken for granted.

Anthropogenic climate change is an example of a dialectical relationship between nature and society and therefore ought to move into the center of sociological concern. However, there lingers the question of what is an essential sociological methodology. Does it include the nature-society interaction or should it be confined to inner-social affairs *à la* Weber, Durkheim, Simmel, or Luhmann?

The inclusion of nature is a revisionist project and it is not clear that Lever-Tracy subscribes to it. She demands a respectful division of labor between natural and social scientists which indicates that the old separation of disciplines, C. P. Snow's two cultures, is to be maintained. Does she want to reaffirm Snow's political message of giving more weight to the natural sciences as "they have the future in their bones" (Snow 1964 [1959])? Efforts to protect the climate from society (mitigation) and measures to protect society from the climate (adaptation) both are efforts that have at their core the assumption that social conduct can alter natural processes and that natural processes can have a significant impact on social relations. From this it follows that sociologists should study the direct interchange between society and natural environments as mediated through technology (sociology of science and technology), the relation between knowledge and decision-making, the values that form our dealing with nature (sociology of culture), the social shifts that contribute or mitigate climate change, and the consequences a changed climate will have on social, economic and political affairs. Most importantly, if sociology wants to provide practical knowledge, then it needs to identify the social structures and dimensions of agency that will facilitate transitional pathways to a low carbon society.

Acknowledgements The authors would like to thank Jacqueline Luce, Jason Mast and two anonymous reviewers for their helpful comments and suggestions.

---

[7] Confronted with conspiracy theories after 9/11, Latour (2004) has become wary of too critical an attitude towards climate science: "Why does it burn my tongue to say that global warming is a fact whether you like it or not? Why can't I simply say that the argument is closed for good?"

References

Bloor, David. 1976. *Knowledge and Social Imagery*. London: Routledge.

Boykoff, Maxwell T. and Jules M. Boykoff. 2004. "Bias as Balance: Global Warming and the US Prestige Press", *Global Environmental Change*, 14: 125–136.

Demeritt, David. 2001. "The Construction of Global Warming and the Politics of Science", *Annals of the Association of American Geographers*, 91(2): 307–337.

Eastin, Josh, Reiner Grundmann, and Aseem Prakash. 2010 [2011]. "The Two 'Limits' Debates: Limits to Growth and Climate Change", *Futures*, 43: 16–26. https://doi.org/10.1016/j.futures.2010.03.001.

Elias, Norbert. 1956. "Problems on Involvement and Detachment", *British Journal of Sociology*, 7: 226–252.

Entman, Robert M. 1993. "Framing: Toward Clarification of a Fractured Paradigm", *Journal of Communication*, 43(4): 51–58.

Gamson, William A. and Andre Modigliani. 1989. "Media Discourse and Public Opinion on Nuclear Power: A Constructionist Approach", *American Journal of Sociology*, 95(1): 1–37.

Giddens, Anthony. 2009. *The Politics of Climate Change*. Cambridge: Polity.

Gitlin, Todd. 1980. *The Whole World is Watching: Mass Media in the Making and Unmaking of the New Left*. Berkeley and London: University of California Press.

Godin, Benoit. 2006. "The Linear Model of Innovation", *Science, Technology and Human Values*, 31(6): 639–667.

Grundmann, Reiner. 1991. *Marxism and Ecology*. Oxford: Oxford University Press.

Grundmann, Reiner and Ramesh Krishnamurthy. 2010. "The Discourse of Climate Change: A Corpus Based Approach", *Critical Approaches to Discourse Analysis Across Disciplines (CADAAD)*, 4(2).

Klein, Julie T. 1990. *Interdisciplinarity: History, Theory, and Practice*. Detroit, MI: Wayne State University Press.

Lahsen, Myanna. 2005. "Seductive Simulations? Uncertainty Distribution around Climate Models", *Social Studies of Science*, 35(6): 895–922.

Landes, David S. 1998. *The Wealth and Poverty of Nations. Why Some Are So Rich and Some So Poor*. New York: W. W. Norton.

Latour, Bruno. 2004. "Why Has Critique Run Out of Steam?", *Critical Inquiry*, 30(Winter): 238–239.

Lever-Tracy, Constance. 2008. "Global Warming and Sociology", *Current Sociology*, 56(3): 445 466.

Matthews, H. Damon and Ken Caldeira. 2008. "Stabilizing Climate Requires Near-Zero Emissions", *Geophysical Research Letters*, 35: L04705. https://doi.org/10.1029/2007GL032388.

Monbiot, George. 2009. "If We Behave as if It's Too Late, Then Our Prophecy is Bound to Come True", *The Guardian*, 17 March.

Prins, Gwyn and Steve Rayner. 2007. *The Wrong Trousers: Radically Rethinking Climate Policy*. Oxford: James Martin Institute for Science and Civilisation, Oxford University.

Prins, Gwyn et al. 2010. *The Hartwell Paper: A New Direction for Climate Policy After the Crash of 2009*. London, UK: Institute for Science, Innovation & Society, University of Oxford; LSE Mackinder Programme, London School of Economics and Political Science.

Sarewitz, Daniel. 2000. "Science and Environmental Policy: An Excess of Objectivity", in: Robert Frodeman (ed.), *Earth Matters: The Earth Sciences, Philosophy, and the Claims of Community*. Upper Saddle River, NJ: Prentice Hall: 79–98.

Sheehan, Peter. 2008. "The New Global Growth Path: Implications for Climate Change Analysis and Policy", *Climatic Change*, 91(3–4): 211–231.

Skodvin, Tora. 2000. *Structure and Agent in the Scientific Diplomacy of Climate Change: An Empirical Case Study of Science-Policy Interaction in the Intergovernmental Panel on Climate Change*. Amsterdam: Kluwer Academic.

Snow, Charles Percy. 1964 [1959]. *The Two Cultures, A Second Look: An Expanded Version of the Two Cultures and the Scientific Revolution*. Cambridge: Cambridge University Press.

Stern, Nicholas. 2006. *The Economics of Climate Change, The Stern Review*. London: HM Treasury.

Van der Sluijs, Jeroen, Josée van Eijndhoven, Simon Shackley, and Brian Wynne. 1998. "Anchoring Devices in Science for Policy: The Case of Consensus around Climate Sensitivity", *Social Studies of Science,* 28(2): 291–323.

Weingart, Peter and Nico Stehr. 2000. *Practising Interdisciplinarity.* Toronto: University of Toronto Press.

Wynne, Brian. 1996. "SSK's Identity Parade: Signing-Up, Off-and-On", *Social Studies of Science,* 26(2): 357–391.

Chapter 23
Exceptional Circumstances: Does Climate Change Trump Democracy?

Nico Stehr

The threats to democracy in the modern era are many. Not least is the risk posed by the widespread feeling among different segments of the public in contemporary democracies that no one from the political class is listening. Such discontent reaches from the Tea Party in the United States and the UK Independence Party (UKIP) in the United Kingdom to the Alternative for Germany (AfD) Party in Germany and the National Front in France. But worryingly, similar sentiments can be found in the scientific community.

The robustness of the consensus in the science community about human-caused climate change has in recent years not only increased in strength, but a number of current studies point to far more dramatic and long-lasting consequences of global warming than previously thought. Moreover, it is highly likely that the sophistication and depth of our knowledge about global and regional climates will substantially increase in the next couple of decades. Under such circumstances, how is it possible, many scientists ask, that such evidence does not motivate political action and behavior change in all societies around the world? Why are we waiting?

The well-known climate researcher James Hansen, who has been publicly sounding the alarm on global warming since his influential 1988 testimony before the U.S. Congress, summarized the general frustration when he asserted in 2007 that "the democratic process does not work." In his 2009 book, *The Vanishing Face of Gaia,* James Lovelock, another long-time scientific voice of warning, compares climate change to war, emphasizing that we need to abandon democracy to meet the challenges of climate change head on. To pull the world out of its state of lethargy, "nothing but blood, toil, tears, and sweat" is urgently needed.

This text was first published as: Stehr, Nico. 2016. "Exceptional Circumstances. Does Climate Change Trump Democracy?", *Issues in Science and Technology*, 32(2): 37–44. The permission to republish this text was granted on 24 July 2107 by Sonja Gold on behalf of the University of Texas in Dallas.

M. T. Adolf (ed.), *Nico Stehr: Pioneer in the Theory of Society and Knowledge*,
Pioneers in Arts, Humanities, Science, Engineering, Practice 16,
https://doi.org/10.1007/978-3-319-76995-0_23

Dale Jamieson, professor of environmental studies, philosophy, and law at New York University and author of *Reason in a Dark Time* (2014), exemplifies such a skeptical view about the obstacles faced by our present political order in coping with the consequences of global warming. He warns that climate change presents us "with the largest collective action problem that humanity has ever faced, [but] evolution did not design us to deal with such problems, and we have not designed political institutions that are conducive to solving them" He adds: "Sadly, it is not entirely clear that democracy is up to the challenge of climate change."

I do not disagree with Jamieson about the enormous challenge global warming likely offers. But I do disagree strongly about the implicit medicine, the rationale for which is beginning to come from scholars in diverse fields. The historian Eric Hobsbawm's long-time skepticism toward democracy extends in his 2008 book, *Globalisation, Democracy, and Terrorism,* to strong doubts about the effectiveness of democratic states in solving complex global problems such as global warming. And Nobel Laureate Daniel Kahneman says: "the bottom line is that I'm extremely skeptical that we can cope with climate change. To mobilize people, this has to become an emotional issue. It has to have immediacy and salience. A distant, abstract, and disputed threat just doesn't have the necessary characteristics for seriously mobilizing public opinion."

Al Gore's famous "inconvenient truth" begins now to imply the complements of an "inconvenient democracy" due to an "inconvenient mind," that is, the cognitive inability of ordinary citizens to incorporate longer-term transformation demanded by climate change into their thinking and decision making, and "inconvenient social institutions" that systematically ignore the future.

Climate scientists, social scientists concerned with climate change, and the media refer to a future of "exceptional circumstances." However, the same groups also assert that no one is listening to their diagnosis of potential incomparable dangers. An elite of climate scientists believes they are reading the evidence that others fail to acknowledge and know truths that others lack the courage to fully confront. In light of the extraordinary dangers to human civilization posed by climate change, democracy quickly becomes in their eyes an inconvenient form of governing.

Although the context is new, there is nothing new, as Cambridge University political scientist David Runciman has documented, "about this outburst of disgust with the workings of democracy. Nor is it distinctly American. Europeans (with the possible exception of [contemporary] Germans) are just as disenchanted with their elected politicians. Lamenting the failings of democracy is a permanent feature of democratic life, one that persists through governmental crises and successes alike." Surprisingly, however, scientists' disenchantment with democracy and the implication that political liberties might need to be suppressed in light of profound future environmental changes has not received much systematic attention in social science, politics, or the media.

I will bring this disenchantment into the spotlight. I will insist that there is no contradiction between democratic governance and scientific knowledge. Rather than lamenting the inconvenience of democratic governance, the need is to enhance democracy, not despite, but especially in light of, the massive challenges of a changing climate. We need to recognize our changing climate as an issue of political governance and not as an environmental or economic issue.

23.1 The Rise of Exceptional Circumstances

In the past, warlike conditions and major disasters typically were seen to justify the abolition of democratic liberties, if only temporarily. The term "exceptional circumstances" refers to conditions often invoked to grant governments additional powers to avert or tackle unforeseen but threatening political, economic, or environmental problems. The present appeal to exceptional circumstances echoes this sentiment, demanding the elevation of a single socio-political purpose—carbon emissions reductions to—ultimate political supremacy.

We are confronted with an entirely novel situation: Anthropogenic climate change is locked in. Most of the scientific discourse has been devoted to establishing the phenomenon. That issue has been settled. What is not settled in science is a range of important questions such as the speed of warming or the nature of the consequences of climate change on various significant attributes of human existence or, perhaps most importantly, what actually must be done, and how it can be done.

Except for reference to singular historical events such as war, there are no large-scale human experiences, or models of success, to which the claims of the climate science community can appeal as a precedent for the course it is considering. Governing the consequences of climate change refers to a time scale and to societal transformations that are clearly beyond the ability of human imagination and current political institutions to cope with.

Thus do Lovelock, Hobsbawm, and others appeal to extraordinary circumstances or a warlike footing that necessitates the suspension of freedoms and the political ascent of climate scientists. A growing chorus of critical voices within the scientific community, advocacy world, and the media seems certain that democratic societies are unable to effectively and speedily attack global environmental problems. The American political theorist and historian Timothy Mitchell, who has written some exceptional empirically based reflections on what he calls "carbon democracy" is equally pessimistic when he states, "faced with the threats of oil depletion and catastrophic climate change, the democratic machineries that emerged to govern the age of carbon energy seem to be unable to address the processes that may end it."

And the alternative? We are left with the central state, guided by scientists, as the source of security in the face of radical risk. Only an appeal to extraordinary circumstances—that is, to a threat to the very existence of humankind—"might be able to give capacity and [...] energy back to a failing or hampered [political] will" the French political scientist Pierre Rosanvallon argues.

But how does one govern well under exceptional circumstances? Doing so is allied to two assertions: that an "inconvenient mind" justifies imposing one's (superior) ideas on citizens and that "inconvenient social institutions" justify a strong state in the form of a command society.

Here one may note how history repeats itself. In the 1970s, a similar kind of skepticism arose around the question of limits to growth and the survival of humankind. Scientists warned about the essential slowness and inflexibility of democratic institutions and expressed their preference for authoritarian solutions. Dennis Meadows, the co-author of the original *Limits to Growth,* reiterated some 40 years later his strong suspicion about the barrier to needed action and solutions in the face of growing environmental threats to our civilization by virtue of the "slowness" and "short-sightedness of governance."

The implication of the position is that democratic governance of society must be subordinated to the defeat of the exceptional circumstances. The single purpose of defeating the exceptional circumstances legitimizes the suspension of political rights and liberties. But for how long can one defer liberties? At least in the case of war, in democratic societies the answer is that, in economist Friedrich Hayek's words, "it is sensible temporarily to sacrifice freedom in order to make it more secure in the future." However, is any massive absorption of powers in the hand of the state and its representatives easily reversible? And, are the potential consequences of climate change the equivalent of (abrupt) warlike conditions? How can one pinpoint the onset of exceptional circumstances? Or, perhaps even more troubling, their endpoint?

The deficiencies of, and the short-term as well as long-term challenges faced by, democratic governments are many and go far beyond the problem of climate change and its societal consequences. What alternatives do these impatient scholars have in mind? After all, authoritarian and totalitarian governments do not have a record of environmental accomplishments; nations that have followed the path of "authoritarian modernization" such as China and Russia cannot claim to have a better record, despite the high status of scientists and engineers in their societies. Nonetheless, the disenchantment with democracies continues to be advanced, perhaps becoming even more vocal as entrenched climate policy regimes such as the United Nations Framework Convention fail to live up to their promise and as one after another international conference on climate change fails to achieve the goals of a substantial global reduction of greenhouse gas emissions.

23.2 The Erosion of Democracy

The argument about upcoming exceptional circumstances due to climate change and the concomitant implication of an inconvenient democracy derives its intellectual sustenance from a range of considerations: a deep-seated pessimism about the psychological make-up of human beings; the specific temporality—that is, short-term perspective—of human thought; the failure to mobilize populations to support the cause of effective climate policies; the inability of government, given constitutional constraints, to attend to long-term goals; the fragility of political order, which depends for survival on mobilizing consensus around incremental change; the influence of vested interests on the political agendas of the day; the widespread social addiction to fossil fuel; and last but not least, the climate science community's sense that its message of evidence and rationality is not stimulating action.

The mass of citizens, it seems, simply cannot be won over to endorse and follow the course of policy options that scientists support. The large majority of citizens are basically inclined to act irrationally; people just do not appear to be able to comprehend what is good for them. As the German climate scientist Hans Joachim Schellnhuber complains,

"[My] own experience and everyday knowledge illustrate that comfort and ignorance are the biggest flaws of human character. This is a potentially deadly mix." Here Schellnhuber rehabilitates a sentiment from Immanuel Kant's 1784 essay "What is Enlightenment?" in which Kant notes: "Laziness and cowardice are the reasons why such a large proportion of men, even when nature has long freed them from external guidance, gladly remain immature all their lives."

Not just citizens, but their democratically elected politicians, seem unlikely to implement satisfactory policy: activist climate scientists, journalists, and many other observers agree that the recent climate summits in Copenhagen, Cancun, Durban, and Warsaw (and perhaps now Paris) were failures. The summits did not result in a new global agreement to cope with the emissions of greenhouse gases. Existing agreements seem to have no impact.

The typical short-term temporal perspectives in democracy apparently justify doubts about the effectiveness of democratic governance in the face of future risks and dangers of climate change. Problems of timeliness are central to the conditions of democratic governance. On the one hand, democratic governance is captivated by the immediacy of frequently changing 'events', such as economic crises or terrorist attacks, which attract the attention of voters and the media. On the other hand, it is constrained by constitutional rules of representation, like election and budget cycles, that prescribe relatively short time horizons for action. David Runciman in *The Confidence Trap,* his 2013 study of the history of democracy in crisis since World War I, sums up these often-noted deficiencies of democratic governance by observing that democracies "prioritize immediate over future experiences, simplicity over complexity, gut instinct over science."

The discussion in the climate science and policy community about the shortcomings of democratic governance resonates, at least superficially, with assessments coming from the social sciences of the present and future state of democracy, which have reached similar discouraging conclusions about the efficacy of democratic governance in many nations.

So, for example, political scientist and former UK Member of Parliament David Marquand sees "a hollowing out of citizenship; the marketization of the public sector; the soul-destroying targets and audits that go with it; the denigration of professionalism and the professional ethic; and the erosion of public trust." Many social science observers see contemporary democracy—whether by design of self-interested actors such as large corporations, or as an unintended outcome of structural economic, political, and moral changes—as tending toward increasingly autocratic forms of governance.

But social scientists and climate scientists diverge profoundly in their analyses of the necessary remedy. Social scientists such as political historian Rosanvallon and sociologist Colin Crouch see the need to restore the vitality of the core function of democracy through more active participation of large numbers of citizens in shaping the agenda of public life. Climate scientists and others whose chief concern is climate change seem instead to believe democratic governance to be inherently incapable of coping effectively with large-scale environmental problems. From this perspective, the very abolition of democracy would be a virtue and the establishment, for example, of "benevolent despotism" would be desirable. In some of the images of 'post-democracy', a return to aristocratic society has already been achieved: self-appointed elites claim to carry out the wishes of the masses. A direct, unequivocal commitment to authoritarian rule can rarely be found in scholarly communication. But the implicit message is clear when Dale Jamieson, for example, suggests that it is difficult to see "how to get populist democracies to accept constraints on the popular will that might help make climate stability possible" Jamieson (2014: 101).

To those who see climate change as a uniquely overwhelming threat to human well-being, democracy itself seems inappropriate, its slow procedures for implementation and management of specific, policy-relevant scientific knowledge leading to massive risks and dangers. The democratic system designed to balance divergent interests has failed in the face of these threats.

23.3 Enlightened Leadership?

Until recently, open and explicit expression of doubt about the virtues of democracy has been rare among intellectuals and politicians, with the obvious exception of certain leaders of decidedly undemocratic nations. In particular, scientists rarely have raised serious misgivings in public about democracy as a political system, most surely not in post-war Germany.

But times are changing. The irritation with democracy and the shifting understanding of the role of scientists goes hand-in-hand with a change in the function of the Intergovernmental Panel on Climate Change (IPCC). Increasingly, the IPCC behaves not like a scientific organization with the mandate to offer alternative policy options for political discussion and decision, but as a body prescribing actions that are essentially political in nature, such as limiting warming to no more than two degrees Celsius, as if such actions follow directly and unambiguously from the science.

Leading climate scientists insist that humanity is at a crossroads. A continuation of present economic and political trends could lead to disaster, if not a collapse of human civilization. To create a globally sustainable way of life, we immediately need, in the words of Schellnhuber, a "great transformation". What that statement exactly means is vague. Part, if not the core, of the required great transformation is a new political regime. Hansen, Mitchell, and Lovelock seem, in quotations I cited earlier, to be voicing similar sentiments.

23.4 Science, Knowledge, and Democracy

What should be the role of climate science knowledge and climate scientists in political deliberations about climate policy? Can science, and thus should scientists, tell us what to do? For the Massachusetts Institute of Technology historian and philosopher of science Evelyn Fox Keller, the answer is clear: "where the results of scientific research have a direct impact on the society in which they live, it becomes effectively impossible for scientists to separate their scientific analysis from the likely consequences of that analysis." To Keller, this seems to then add up to a compelling case for an immediately effective, practical political role of climate science, given the seriousness of the problem of global warming:

> There is no escaping our dependence on experts; we have no choice but to call on those (in this case, our climate scientists) who have the necessary expertise [...]. Furthermore, for the particular task of getting beyond our current impasse, I also suggest that climate scientists may be the only ones in a position to take the lead [...]. [G]iven the tacit contract between scientists and the state which supports them. I will also argue that climate scientists are not only in a position to take the lead, but also that they are obliged to do so. (Keller 2011)

Complementing the expectation that scientists must lead is the conviction that citizens are unprepared to act. We have already seen how some leading academics believe that the public is not cognitively capable of coming to the right conclusions about climate change's urgency. Robert Stavins, director of Harvard's Environmental Economics Program and an IPCC lead author, notes that a "bottom-up demand, which normally we always want to have and rely on in a representative democracy, is in my view unlikely to work in the case of climate change policy as it has for other environmental problems [...]. It's going to take enlightened leadership, leaders that take the lead."

But the idea that science and scientific leadership offer some sort of alternative to democracy has, to put it mildly, major weaknesses. To begin with, scientific knowledge does not and cannot dictate what to do. One of the fundamental flaws in the portrait of an inconvenient democracy is the failure to recognize that knowledge of nature must always enter society through politics (whether democratic or authoritarian)—through decisions about, as Harold Laswell famously put it, "who gets what, when, how." Knowledge about how such decisions are best made is not particularly available to scientists. Indeed, such knowledge is inherently and necessarily contestable.

The vision of a scientifically rational and beneficent authoritarian regime is thus incoherent because it treats a simple technical goal—the reduction of greenhouse gas emissions—as if the very fact of its articulation should automatically illuminate an optimal pathway for transforming the complex global energy system on which modern societies depend for their survival. But as stressed by Mike Hulme, a climate scientist who has come down clearly on the side of democracy, such notions may be favored by those "who are more likely to conceive of the planet as a machine amenable to control engineering" (Hulme 2014).

The pessimistic assessment of the ability of democratic governance to cope with and control exceptional circumstances seems to bring with it an optimistic assessment of the potential of large-scale social planning. Yet all evidence suggests that the capacity not only of governments, but societies, to plan their future is rather limited, perhaps non-existent. The problem is not one of democracy, but of the complexity of social change. From this perspective, the claims that the key uncertainties about the behavior of the natural climate processes have been eliminated does nothing whatsoever to address the uncertainties associated with the social and political processes for taking effective action. Consensus on the evidence of natural science, it is argued, should motivate a consensus on political action. The uncertainties of social, political, and economic events, the difficulty of anticipating the future, are treated as minor obstacles that can be managed by the experts. But contemporary societies show no evidence that these uncertainties are even comprehensible, let alone manageable.

Indeed, this is precisely why democracy, inconvenient as it may be, is not only necessary but, for a challenge of the magnitude and complexity of climate change, essential. To a far greater extent than authoritarian governance, democratic governance is flexible and capable of learning from policy mistakes, which are inevitable when trying to deal with something as complex as climate change. Democratic governments' ability to learn allows them, as Runciman explains in *The Confidence Trap*, "to keep experimenting and adapting to the challenge they encounter, so that no danger becomes overwhelming". Democracies "have the experimental adaptability and they have the collective resilience under duress". But Runciman offers a cautionary note, because "the knowledge that democracies have of their long-term strengths does not tell them how to access those strengths at the right moment. That is why climate change is so dangerous for democracies". Dangerous because the impatience of the climate science community leads it to imagine that other, less open forms of governance might do better than democracy.

23.5 Enhancing Democracy

What is good governance? And what is good governance under exceptional circumstances? Is good governance democratic governance?

The impatient scientists' disenchantment with democracy risks placing excessive power in the hands of states, transnational organizations, and multinational corporations. Participatory strategies are only rarely in evidence. Likewise, global mitigation has precedence over local adaptation. Global knowledge triumphs over local knowledge. However, societal trends appear to be moving in the opposite direction. The ability of large institutions, powerful as they may be, to impose their will on citizens is declining. People are mobilizing around local concerns and efforts, including responses to climate change. Opportunities for enhanced democratic governance are even now being created.

The reigning discussion of options for future climate policies seems grounded on the idea that after 25 years of failed international climate policy effort, the same failed climate policy regime must remain in place and is the only correct approach.

The problem is not the proposed and still imaginary governance regime, but the actual politics that prevent its realization. It follows that international negotiations must lead to a new agreement for concrete, but much deeper emission-reduction targets. Only a super-Kyoto can help us. But how the noble goals of comprehensive emission reduction can be practically and politically enforced at the international level remains in the fog of general declarations of intent. The more the international negotiations fail to lead to meaningful global emissions reductions, the more skeptical scientists seem to become of democratic governance at the national level.

A strong repudiation and falsification of the current line of attack for climate policy occurred during the recent global recession, which contributed to an unintended and unprecedented reduction in the growth rate of carbon dioxide emissions. The worldwide reaction to the economic crisis, most recently during the November 2014 G20 meeting in Australia, however, shows very clearly that no government conceives of a reduction in the growth of the wealth of its population as a useful mechanism to achieve a reduction in emissions.

On the contrary, all efforts worldwide aim at a resumption of economic growth. Apparently not a single nation believes that the political and social consequences of economic decline are worth trading for emissions reductions.

Proponents of the inconvenient democracy perspective draw the wrong conclusion from these developments, namely that only authoritarian political states, preferably guided by scientists, would be able to make effective and correct decisions on the climate issue. No evidence supports this view. Certainly today's China cannot serve as a model. On the contrary, the authoritarian Chinese government that has delivered enormous growth of both wealth and carbon emissions is also delivering an ever more educated and affluent population that, in turn, will demand participation in political decision making, including environmental policies.

An alternative model is therefore needed, and I submit that it will be found only through revitalized democratic interaction in which alternative perspectives can be presented and tested. A warlike footing has exactly the opposite effect, reducing the complexity of social and political life to focus on one outcome. As Rosanvallon explains, war "nationalizes people's life. Private activities [are] largely shaped by collective constraints" (Rosanvallon 2013: 183).

Climate policy needs to do the opposite. It must be compatible with democracy; otherwise the threat to civilization will be much more than just changes to our physical environment. The alternative to the abolition of democratic governance is more democracy—making not only democracy and solutions more complex, but also enhancing the worldwide empowerment and knowledgeability of individuals, groups, and movements who work on environmental issues. As the world gradually transitions toward further denationalization of governance, democracies will produce new, multiple forms of social solidarity and obligations, strengthening local and regional capacities to respond to climate change, and enhancing the awareness of social interdependence. Examples include the widespread community and regional support of renewable energy in Germany—and the success of wind energy in Texas.

Efforts to simplify the global approach to climate change through a single international governance regime have failed. Now is the time to commit to democratic complexification that fosters creativity and experimentation in the pursuit of multiple desired goals. For those who think that there can be only one global pathway to addressing climate change, the erosion of democracy might seem to be 'convenient'. History, of both recent decades and centuries, tells us that suppression of social complexity undermines the capacity of societies to solve problems. Friedrich Hayek points out a paradoxical development: As science advances, it tends to strengthen the observation shared by many scientists that we should "aim at more deliberate and comprehensive control of all human activities". Hayek (1960: 25) pessimistically adds, "It is for this reason that those intoxicated by the advance of knowledge so often become the enemies of freedom."

Recommended Reading

Fox Keller, Evelyn. 2011. "What Are Climate Scientists to Do?", *Spontaneous Generations: A Journal for the History and Philosophy of Science*, 5(1): 19–26.

Hansen, James. 2009. *Storms of My Grandchildren: The Truth About the Coming Climate Catastrophe and Our Last Chance to Save Humanity*. London: Bloomsbury.

Hayek, Friedrich A. 1960. *The Constitution of Liberty*. Chicago: University of Chicago Press.

Hobsbawm, Eric. 2007. *Globalisation, Democracy, and Terrorism*. London: Abacus.

Hulme, Mike. 2014. *Can Science Fix Climate Change? A Case Against Climate Engineering*. Oxford: Polity Press.

Jamieson, Dale. 2014. *Reason in a Dark Time: Why the Struggle Against Climate Change Failed —and What It Means for Our Future*. New York, NY: Oxford University Press.

Marquand, David. 2004. *The Decline of the Public: The Hollowing Out of Citizenship*. Cambridge: Polity Press.

Marshall, George. 2014. *Don't Even Think About It: Why Our Brains Are Wired to Ignore Climate Change*. New York, NY: Bloomsbury.

Mitchell, Timothy. 2014. "Ecomentality: How the Future Entered Government", *Critical Inquiry*, 40: 479–507.

Rosanvallon, Pierre. 2013. *The Society of Equals*. Cambridge, MA: Harvard University Press.

About the Author

Nico Stehr, Ph.D., FRSC, holds the Karl Mannheim-Chair of Cultural Studies at Zeppelin University, Friedrichshafen, Germany. He has previously held professorships at the University of Alberta, Edmonton, the University of British Columbia, Ludwig Maximilians University in Munich and was Paul F. Lazarsfeld Guest Professor at the University of Vienna. Nico Stehr is member of the Royal Society of Canada, and the European Academy of the Sciences and Arts and has published on a broad variety of topics, most notably on the sociology of knowledge, science and technology studies and sociological theory. A pioneer in the field of researching the social consequences of climate change, he was the founding director of the European Center for Sustainability Research in Friedrichshafen.

Key publications include: *Knowledge Societies* (Sage 1994), *The Fragility of Modern Societies* (Sage 2001), *Moral Markets* (Routledge 2008) and *Information, Power, and Democracy: Liberty is a Daughter of Knowledge* (Cambridge University Press 2015).

Address: Prof. Nico Stehr, Ph.D., FRSC, Zeppelin University, Am Seemooser Horn 20, D-88045, Friedrichshafen, Germany.
Email: nico.stehr@zu.de.
Website: https://www.zu.de/stehr.

M. T. Adolf (ed.), *Nico Stehr: Pioneer in the Theory of Society and Knowledge*,
Pioneers in Arts, Humanities, Science, Engineering, Practice 16,
https://doi.org/10.1007/978-3-319-76995-0

About the Editor

Marian Adolf is Professor of Media Culture at Zeppelin University in Friedrichshafen, Germany. His research is concerned with the interface of media and social change, especially the cultural consequences of the mediatization of modern life and emerging patterns of societal communication. Recent work focused on the future of the public sphere, the mediatization of society and the social mainsprings of knowledge and innovation. He has taught at numerous universities, among others in Vienna and Innsbruck and has held visiting professorships at Alpen-Adria-University Klagenfurt and Quest University in British Columbia.

Key publications include: *Die unverstandene Kultur* (Transcript 2006), Ist Wissen Macht? (with Nico Stehr, Velbrück Wissenschaft 2015), *Knowledge* (with Nico Stehr, Routledge 2017), *Zwischen Integration und Diversifikation* (with Olaf Jandura, Manuel Wendelin and Jeffrey Wimmer, Springer 2017).

Address: Prof. Dr. Marian Adolf, Zeppelin University, Department of Culture & Communication, Am Seemooser Horn 20, D-88045, Friedrichshafen, Germany.
Email: marian.adolf@zu.de.
Website: www.zu.de/adolf.

M. T. Adolf (ed.), *Nico Stehr: Pioneer in the Theory of Society and Knowledge*,
Pioneers in Arts, Humanities, Science, Engineering, Practice 16,
https://doi.org/10.1007/978-3-319-76995-0

About the Contributors

Gernot Böhme

Gernot Böhme is Director of the Institute for Practical Philosophy in Darmstadt, Germany. He studied Mathematics, Physics and Philosophy at the University of Hamburg, Heidelberg University and the University of Göttingen. His academic work comprises the philosophy of science, the theory of time, the philosophy of nature, ethics, aesthetics and the philosophy of technology. Among other positions as a guest professor, Gernot Böhme has taught at Harvard University, the University of Rotterdam and TU Wien and was a guest researcher at the University of Cambridge, England. From 1977 until 2002, Gernot Böhme was Professor for Philosophy at TU Darmstadt. During 1997 and 2001 he was speaker of the post-graduate college "Technology and Society".

Susan A. McDaniel

Susan A. McDaniel is Professor of Sociology at the University of Lethbridge, Canada. She studied Sociology at the University of Massachusetts and Cornell University and received her Ph.D. at the University of Alberta, Canada. Her research concerns the fields of demographic aging, generational relations, life course and the social impacts of technology. Susan McDaniel is frequently advising governments in Canada, the UK and the EU on social statistics, social policies, science and technology as well as innovation policies. Prior to her appointment as Director of the Prentice Institute and Prentice Research Chair in Global Population and Economy at the University of Lethbridge in 2009, she held professorships at the University of Alberta and Waterloo.

Reiner Grundmann

Reiner Grundmann is Professor in Science and Technology Studies at the University of Nottingham, United Kingdom. He studied Sociology at the Free University in Berlin, Germany and received his Ph.D. in Political and Social Sciences from the European University Institute, Florence. His research is concerned with the role of experts and knowledge in political decision-making

M. T. Adolf (ed.), *Nico Stehr: Pioneer in the Theory of Society and Knowledge*, Pioneers in Arts, Humanities, Science, Engineering, Practice 16,
https://doi.org/10.1007/978-3-319-76995-0

processes, climate change and the interaction between climate and society. Earlier, Reiner Grundman held post-doctoral positions at the Wissenschaftszentrum Berlin, at the Max Planck Institute for the Study of Societies, Cologne and was Deputy Director of the Aston Centre for Critical Infrastructures and Services at the Aston University, United Kingdom.

Alexander Ruser

Alexander Ruser is currently the scientific head of the Centre for Political Communication at Zeppelin University Friedrichshafen and a research fellow at the OIS Institute, Vienna. He holds a Ph.D. in Sociology from the Max-Weber-Institute of Sociology at Heidelberg University and was a Dahrendorf Postdoctoral Fellow at the Hertie School of Governance in Berlin, the London School of Economics and Political Sciences and most recently a visiting fellow at Punjab University, Chandigarh. His research focuses on the role of "science in society" and the impact of scientific expertise on climate and economic politics. Alexander Ruser has published in peer-reviewed journals such as *Global Policy*, *Innovation*, *Current Sociology* and the *Journal of Civil Society*.

Hans von Storch

Hans von Storch is Professor at the University of Hamburg and Director Emeritus of the Institute for Coastal Research of the Helmholtz-Zentrum Geesthacht Centre for Materials and Coastal Research. He studied Mathematics, Physics and Danish at the University of Hamburg, where he also obtained his Doctorate in Meteorology. His work is concerned with climate change and statistics, the interaction of climate and society as well as the role of science in climate debate. Prior to his professorship at the University of Hamburg, Hans von Storch held a research position at the Max Planck Institute for Meteorology. In recognition of his achievements he received an Honorary Doctorate from the University of Gothenburg and was made a member ("Zweitmitglied") of the Faculty of Social Sciences.

Hermann Strasser

Hermann Strasser is Professor Emeritus for Sociology at the University of Duisburg-Essen. He studied Economics at the University of Innsbruck and at the Free University of Berlin, received his Ph.D. in Sociology at Fordham University, New York and obtained his Habilitation in Sociology at the University of Klagenfurt. His work concerns sociological theory, social inequalities and the analysis of changing social structures. Before becoming Professor for Sociology at the University of Duisburg-Essen, Hermann Strasser held teaching and research positions at the University of Innsbruck, at Fordham University, New York, The University of Oklahoma, Norman and at the Institute for Advanced Studies in Vienna.

Index

M. T. Adolf (ed.), *Nico Stehr: Pioneer in the Theory of Society and Knowledge*, Pioneers in Arts, Humanities, Science, Engineering, Practice 16,
https://doi.org/10.1007/978-3-319-76995-0

The manufacturer's authorised representative in the EU is Springer Nature Customer Service Centre GmbH, Europaplatz 3, 69115 Heidelberg, Germany. If you have any concerns regarding our products, please contact ProductSafety@springernature.com

Printed and bound by CPI Group (UK) Ltd, Croydon, CR0 4YY
05/10/2026
02228122-0014